U0211773

Springer Handbook of Nanotechnology

Bharat Bhushan (Ed.)

3rd revised and extended edition

Springer Handbook of Nanotechnology

Since 2004 and with the 2nd edition in 2006, the Springer Handbook of Nanotechnology has established itself as the definitive reference in the nanoscience and nanotechnology area. It integrates the knowledge from nanofabrication, nanodevices, nanomechanics, nanotribology, materials science, and reliability engineering in just one volume. Beside the presentation of nanostructures, micro/nanofabrication, and micro/nanodevices, special emphasis is on scanning probe microscopy, nanotribology and nanomechanics, molecularly thick films, industrial applications and microdevice reliability, and on social aspects. In its 3rd edition, the book grew from 8 to 9 parts now including a part with chapters on biomimetics. More information is added to such fields as bionanotechnology, nanorobotics, and (bio) MEMS/NEMS, bio/nanotribology and bio/nanomechanics. The book is organized by an experienced editor with a universal knowledge and written by an international team of over 145 distinguished experts. It addresses mechanical and electrical engineers, materials scientists, physicists and chemists who work either in the nano area or in a field that is or will be influenced by this new key technology.

"The strong point is its focus on many of the practical aspects of nanotechnology... Anyone working in or learning about the field of nanotechnology would find this an excellent working handbook."

IEEE Electrical Insulation Magazine

"Outstandingly succeeds in its aim... It really is a magnificent volume and every scientific library and nanotechnology group should have a copy."

Materials World

"The integrity and authoritativeness... is guaranteed by an experienced editor and an international team of authors which have well summarized in their chapters information on fundamentals and applications."

Polymer News

使用说明

1.《纳米技术手册》原版为一册，分为A～I部分。考虑到使用方便以及内容一致，影印版分为7册：第1册—Part A，第2册—Part B，第3册—Part C，第4册—Part D，第5册—Part E，第6册—Part F, 第7册—Part G、H、I。

2.各册在页脚重新编排页码，该页码对应中文目录。保留了原书页眉及页码，其页码对应原书目录及主题索引。

3.各册均给出完整7册书的章目录。

4.作者及其联系方式、缩略语表各册均完整呈现。

5.主题索引安排在第7册。

6.目录等采用中英文对照形式给出，方便读者快速浏览。

材料科学与工程图书工作室

联系电话　0451-86412421

　　　　　0451-86414559

邮　　箱　yh_bj@yahoo.com.cn

　　　　　xuyaying81823@gmail.com

　　　　　zhxh6414559@yahoo.com.cn

Springer 手册精选系列

纳米技术手册

仿 生 学

【第6册】

Springer Handbook of Nanotechnology

〔美〕Bharat Bhushan 主编

（第三版影印版）

黑版贸审字 08-2013-001号

图书在版编目（CIP）数据

纳米技术手册：第3版. 6, 仿生学 = Handbook of Nanotechnology. 6, Biomimetics：英文 / (美) 布尚(Bhushan,B.) 主编. —影印本. —哈尔滨：哈尔滨工业大学出版社, 2013.1
（Springer手册精选系列）
ISBN 978-7-5603-3952-8

Ⅰ. ①纳… Ⅱ. ①布… Ⅲ. ①纳米技术－手册－英文②纳米材料－仿生－手册－英文 Ⅳ. ①TB303-62②TB383-62

中国版本图书馆CIP数据核字(2013)第004228号

材料科学与工程
图书工作室

责任编辑 杨 桦 许雅莹 张秀华
出版发行 哈尔滨工业大学出版社
社　　址 哈尔滨市南岗区复华四道街10号 邮编 150006
传　　真 0451-86414749
网　　址 http://hitpress.hit.edu.cn
印　　刷 哈尔滨市石桥印务有限公司
开　　本 787mm×960mm 1/16 印张 15.25
版　　次 2013年1月第1版 2013年1月第1次印刷
书　　号 ISBN 978-7-5603-3952-8
定　　价 48.00元

Foreword by Neal Lane

In a January 2000 speech at the California Institute of Technology, former President W.J. Clinton talked about the exciting promise of *nanotechnology* and the importance of expanding research in nanoscale science and engineering and, more broadly, in the physical sciences. Later that month, he announced in his State of the Union Address an ambitious US$ 497 million federal, multiagency national nanotechnology initiative (NNI) in the fiscal year 2001 budget; and he made the NNI a top science and technology priority within a budget that emphasized increased investment in US scientific research. With strong bipartisan support in Congress, most of this request was appropriated, and the NNI was born. Often, federal budget initiatives only last a year or so. It is most encouraging that the NNI has remained a high priority of the G.W. Bush Administration and Congress, reflecting enormous progress in the field and continued strong interest and support by industry.

Nanotechnology is the ability to manipulate individual atoms and molecules to produce nanostructured materials and submicron objects that have applications in the real world. Nanotechnology involves the production and application of physical, chemical and biological systems at scales ranging from individual atoms or molecules to about 100 nm, as well as the integration of the resulting nanostructures into larger systems. Nanotechnology is likely to have a profound impact on our economy and society in the early 21st century, perhaps comparable to that of information technology or cellular and molecular biology. Science and engineering research in nanotechnology promises breakthroughs in areas such as materials and manufacturing, electronics, medicine and healthcare, energy and the environment, biotechnology, information technology and national security. Clinical trials are already underway for nanomaterials that offer the promise of cures for certain cancers. It is widely felt that nanotechnology will be the next industrial revolution.

Nanometer-scale features are built up from their elemental constituents. Micro- and nanosystems components are fabricated using batch-processing techniques that are compatible with integrated circuits and range in size from micro- to nanometers. Micro- and nanosystems include micro/nanoelectro-mechanical systems (MEMS/NEMS), micromechatronics, optoelectronics, microfluidics and systems integration. These systems can sense, control, and activate on the micro/nanoscale and can function individually or in arrays to generate effects on the macroscale. Due to the enabling nature of these systems and the significant impact they can have on both the commercial and defense applications, industry as well as the federal government have taken special interest in seeing growth nurtured in this field. Micro- and nanosystems are the next logical step in the *silicon revolution*.

Prof. Neal Lane
Malcolm Gillis University Professor,
Department of Physics and Astronomy,
Senior Fellow,
James A. Baker III Institute for Public Policy
Rice University
Houston, Texas

Served in the Clinton Administration as Assistant to the President for Science and Technology and Director of the White House Office of Science and Technology Policy (1998–2001) and, prior to that, as Director of the National Science Foundation (1993–1998). While at the White House, he was a key figure in the creation of the NNI.

The discovery of novel materials, processes, and phenomena at the nanoscale and the development of new experimental and theoretical techniques for research provide fresh opportunities for the development of innovative nanosystems and nanostructured materials. There is an increasing need for a multidisciplinary, systems-oriented approach to manufacturing micro/nanodevices which function reliably. This can only be achieved through the cross-fertilization of ideas from different disciplines and the systematic flow of information and people among research groups.

Nanotechnology is a broad, highly interdisciplinary, and still evolving field. Covering even the most important aspects of nanotechnology in a single book that reaches readers ranging from students to active researchers in academia and industry is an enormous challenge. To prepare such a wide-ranging book on nanotechnology, Prof. Bhushan has harnessed his own knowledge and experience, gained in several industries and universities, and has assembled internationally recognized authorities from four continents to write chapters covering a wide array of nanotechnology topics, including the latest advances. The authors come from both academia and industry. The topics include major advances in many fields where nanoscale science and engineering is being pursued and illustrate how the field of nanotechnology has continued to emerge and blossom. Given the accelerating pace of discovery and applications in nanotechnology, it is a challenge to capture it all in one volume. As in earlier editions, professor Bhushan does an admirable job.

Professor Bharat Bhushan's comprehensive book is intended to serve both as a textbook for university courses as well as a reference for researchers. The first and second editions were timely additions to the literature on nanotechnology and stimulated further interest in this important new field, while serving as invaluable resources to members of the international scientific and industrial community. The increasing demand for up-to-date information on this fast moving field led to this third edition. It is increasingly important that scientists and engineers, whatever their specialty, have a solid grounding in the fundamentals and potential applications of nanotechnology. This third edition addresses that need by giving particular attention to the widening audience of readers. It also includes a discussion of the social, ethical and political issues that tend to surround any emerging technology.

The editor and his team are to be warmly congratulated for bringing together this exclusive, timely, and useful nanotechnology handbook.

Foreword by James R. Heath

Nanotechnology has become an increasingly popular buzzword over the past five years or so, a trend that has been fueled by a global set of publicly funded nanotechnology initiatives. Even as researchers have been struggling to demonstrate some of the most fundamental and simple aspects of this field, the term nanotechnology has entered into the public consciousness through articles in the popular press and popular fiction. As a consequence, the expectations of the public are high for nanotechnology, even while the actual public definition of nanotechnology remains a bit fuzzy.

Why shouldn't those expectations be high? The late 1990s witnessed a major information technology (IT) revolution and a minor biotechnology revolution. The IT revolution impacted virtually every aspect of life in the western world. I am sitting on an airplane at 30 000 feet at the moment, working on my laptop, as are about half of the other passengers on this plane. The plane itself is riddled with computational and communications equipment. As soon as we land, many of us will pull out cell phones, others will check e-mail via wireless modem, some will do both. This picture would be the same if I was landing in Los Angeles, Beijing, or Capetown. I will probably never actually print this text, but will instead submit it electronically. All of this was unthinkable a dozen years ago. It is therefore no wonder that the public expects marvelous things to happen quickly. However, the science that laid the groundwork for the IT revolution dates back 60 years or more, with its origins in fundamental solid-state physics.

By contrast, the biotech revolution was relatively minor and, at least to date, not particularly effective. The major diseases that plagued mankind a quarter century ago are still here. In some third-world countries, the average lifespan of individuals has actually decreased from where it was a full century ago. While the costs of electronics technologies have plummeted, health care costs have continued to rise. The biotech revolution may have a profound impact, but the task at hand is substantially more difficult than what was required for the IT revolution. In effect, the IT revolution was based on the advanced engineering of two-dimensional digital circuits constructed from relatively simple components – extended solids. The biotech revolution is really dependent upon the ability to reverse engineer three-dimensional analog systems constructed from quite complex components – proteins. Given that the basic science behind biotech is substantially younger than the science that has supported IT, it is perhaps not surprising that the biotech revolution has not really been a proper revolution yet, and it likely needs at least another decade or so to come into fruition.

Prof. James R. Heath

Department of Chemistry
California Institute of Technology
Pasadena, California

Worked in the group of Nobel Laureate Richard E. Smalley at Rice University (1984–88) and co-invented Fullerene molecules which led to a revolution in Chemistry including the realization of nanotubes. The work on Fullerene molecules was cited for the 1996 Nobel Prize in Chemistry. Later he joined the University of California at Los Angeles (1994–2002), and co-founded and served as a Scientific Director of The California Nanosystems Institute.

Where does nanotechnology fit into this picture? In many ways, nanotechnology depends upon the ability to engineer two- and three-dimensional systems constructed from complex components such as macromolecules, biomolecules, nanostructured solids, etc. Furthermore, in terms of patents, publications, and other metrics that can be used to gauge the birth and evolution of a field, nanotech lags some 15–20 years behind biotech. Thus, now is the time that the fundamental science behind nanotechnology is being explored and developed. Nevertheless, progress with that science is moving forward at a dramatic pace. If the scientific community can keep up this pace and if the public sector will continue to support this science, then it is possible, and even perhaps likely, that in 20 years we may be speaking of the nanotech revolution.

The first edition of Springer Handbook of Nanotechnology was timely to assemble chapters in the broad field of nanotechnology. Given the fact that the second edition was in press one year after the publication of the first edition in April 2004, it is clear that the handbook has shown to be a valuable reference for experienced researchers as well as for a novice in the field. The third edition has one Part added and an expanded scope should have a wider appeal.

Preface to the 3rd Edition

On December 29, 1959 at the California Institute of Technology, Nobel Laureate Richard P. Feynman gave at talk at the Annual meeting of the American Physical Society that has become one of the 20th century classic science lectures, titled *There's Plenty of Room at the Bottom*. He presented a technological vision of extreme miniaturization in 1959, several years before the word *chip* became part of the lexicon. He talked about the problem of manipulating and controlling things on a small scale. Extrapolating from known physical laws, Feynman envisioned a technology using the ultimate toolbox of nature, building nanoobjects atom by atom or molecule by molecule. Since the 1980s, many inventions and discoveries in fabrication of nanoobjects have been testament to his vision. In recognition of this reality, National Science and Technology Council (NSTC) of the White House created the Interagency Working Group on Nanoscience, Engineering and Technology (IWGN) in 1998. In a January 2000 speech at the same institute, former President W.J. Clinton talked about the exciting promise of *nanotechnology* and the importance of expanding research in nanoscale science and technology, more broadly. Later that month, he announced in his State of the Union Address an ambitious US$ 497 million federal, multi-agency national nanotechnology initiative (NNI) in the fiscal year 2001 budget, and made the NNI a top science and technology priority. The objective of this initiative was to form a broad-based coalition in which the academe, the private sector, and local, state, and federal governments work together to push the envelop of nanoscience and nanoengineering to reap nanotechnology's potential social and economic benefits.

The funding in the US has continued to increase. In January 2003, the US senate introduced a bill to establish a National Nanotechnology Program. On December 3, 2003, President George W. Bush signed into law the 21st Century Nanotechnology Research and Development Act. The legislation put into law programs and activities supported by the National Nanotechnology Initiative. The bill gave nanotechnology a permanent home in the federal government and authorized US$ 3.7 billion to be spent in the four year period beginning in October 2005, for nanotechnology initiatives at five federal agencies. The funds would provide grants to researchers, coordinate R&D across five federal agencies (National Science Foundation (NSF), Department of Energy (DOE), NASA, National Institute of Standards and Technology (NIST), and Environmental Protection Agency (EPA)), establish interdisciplinary research centers, and accelerate technology transfer into the private sector. In addition, Department of Defense (DOD), Homeland Security, Agriculture and Justice as well as the National Institutes of Health (NIH) also fund large R&D activities. They currently account for more than one-third of the federal budget for nanotechnology.

European Union (EU) made nanosciences and nanotechnologies a priority in Sixth Framework Program (FP6) in 2002 for a period of 2003–2006. They had dedicated small funds in FP4 and FP5 before. FP6 was tailored to help better structure European research and to cope with the strategic objectives set out in Lisbon in 2000. Japan identified nanotechnology as one of its main research priorities in 2001. The funding levels increases sharply from US$ 400 million in 2001 to around US$ 950 million in 2004. In 2003, South Korea embarked upon a ten-year program with around US$ 2 billion of public funding, and Taiwan has committed around US$ 600 million of public funding over six years. Singapore and China are also investing on a large scale. Russia is well funded as well.

Nanotechnology literally means any technology done on a nanoscale that has applications in the real world. Nanotechnology encompasses production and application of physical, chemical and biological systems at scales, ranging from individual atoms or molecules to submicron dimensions, as well as the integration of the resulting nanostructures into larger systems. Nanotechnology is likely to have a profound impact on our economy and society in the early 21st century, comparable to that of semiconductor technology, information technology, or cellular and molecular biology. Science and technology research in nanotechnology promises breakthroughs in areas such as materials and manufacturing, nanoelectronics, medicine and healthcare, energy, biotechnology, information technology and national security. It is widely felt that nanotechnology will be the next industrial revolution.

There is an increasing need for a multidisciplinary, system-oriented approach to design and manufacturing of micro/nanodevices which function reliably. This can only be achieved through the cross-fertilization of ideas from different disciplines and the systematic flow of information and people among research groups. Reliability is a critical technology for many micro- and nanosystems and nanostructured materials. A broad based handbook was needed, and the first edition of Springer Handbook of Nanotechnology was published in April 2004. It presented an overview of nanomaterial synthesis, micro/nanofabrication, micro- and nanocomponents and systems, scanning probe microscopy, reliability issues (including nanotribology and nanomechanics) for nanotechnology, and industrial applications. When the handbook went for sale in Europe, it was sold out in ten days. Reviews on the handbook were very flattering.

Given the explosive growth in nanoscience and nanotechnology, the publisher and the editor decided to develop a second edition after merely six months of publication of the first edition. The second edition (2007) came out in December 2006. The publisher and the editor again decided to develop a third edition after six month of publication of the second edition. This edition of the handbook integrates the knowledge from nanostructures, fabrication, materials science, devices, and reliability point of view. It covers various industrial applications. It also addresses social, ethical, and political issues. Given the significant interest in biomedical applications, and biomimetics a number of additional chapters in this arena have been added. The third edition consists of 53 chapters (new 10, revised 28, and as is 15). The chapters have been written by 139 internationally recognized experts in the field, from academia, national research labs, and industry, and from all over the world.

This handbook is intended for three types of readers: graduate students of nanotechnology, researchers in academia and industry who are active or intend to become active in this field, and practicing engineers and scientists who have encountered a problem and hope to solve it as expeditiously as possible. The handbook should serve as an excellent text for one or two semester graduate courses in nanotechnology in mechanical engineering, materials science, applied physics, or applied chemistry.

We embarked on the development of third edition in June 2007, and we worked very hard to get all the chapters to the publisher in a record time of about 12 months. I wish to sincerely thank the authors for offering to write comprehensive chapters on a tight schedule. This is generally an added responsibility in the hectic work schedules of researchers today. I depended on a large number of reviewers who provided critical reviews. I would like to thank Dr. Phillip J. Bond, Chief of Staff and Under Secretary for Technology, US Department of Commerce, Washington, D.C. for suggestions for chapters as well as authors in the handbook. Last but not the least, I would like to thank my secretary Caterina Runyon-Spears for various administrative duties and her tireless efforts are highly appreciated.

I hope that this handbook will stimulate further interest in this important new field, and the readers of this handbook will find it useful.

February 2010

Bharat Bhushan
Editor

Preface to the 2nd Edition

On 29 December 1959 at the California Institute of Technology, Nobel Laureate Richard P. Feynman gave at talk at the Annual meeting of the American Physical Society that has become one of the 20th century classic science lectures, titled "There's Plenty of Room at the Bottom." He presented a technological vision of extreme miniaturization in 1959, several years before the word "chip" became part of the lexicon. He talked about the problem of manipulating and controlling things on a small scale. Extrapolating from known physical laws, Feynman envisioned a technology using the ultimate toolbox of nature, building nanoobjects atom by atom or molecule by molecule. Since the 1980s, many inventions and discoveries in the fabrication of nanoobjects have been a testament to his vision. In recognition of this reality, the National Science and Technology Council (NSTC) of the White House created the Interagency Working Group on Nanoscience, Engineering and Technology (IWGN) in 1998. In a January 2000 speech at the same institute, former President W. J. Clinton talked about the exciting promise of "nanotechnology" and the importance of expanding research in nanoscale science and, more broadly, technology. Later that month, he announced in his State of the Union Address an ambitious $497 million federal, multiagency national nanotechnology initiative (NNI) in the fiscal year 2001 budget, and made the NNI a top science and technology priority. The objective of this initiative was to form a broad-based coalition in which the academe, the private sector, and local, state, and federal governments work together to push the envelope of nanoscience and nanoengineering to reap nanotechnology's potential social and economic benefits.

The funding in the U.S. has continued to increase. In January 2003, the U. S. senate introduced a bill to establish a National Nanotechnology Program. On 3 December 2003, President George W. Bush signed into law the 21st Century Nanotechnology Research and Development Act. The legislation put into law programs and activities supported by the National Nanotechnology Initiative. The bill gave nanotechnology a permanent home in the federal government and authorized $3.7 billion to be spent in the four year period beginning in October 2005, for nanotechnology initiatives at five federal agencies. The funds would provide grants to researchers, coordinate R&D across five federal agencies (National Science Foundation (NSF), Department of Energy (DOE), NASA, National Institute of Standards and Technology (NIST), and Environmental Protection Agency (EPA)), establish interdisciplinary research centers, and accelerate technology transfer into the private sector. In addition, Department of Defense (DOD), Homeland Security, Agriculture and Justice as well as the National Institutes of Health (NIH) would also fund large R&D activities. They currently account for more than one-third of the federal budget for nanotechnology.

The European Union made nanosciences and nanotechnologies a priority in the Sixth Framework Program (FP6) in 2002 for the period of 2003-2006. They had dedicated small funds in FP4 and FP5 before. FP6 was tailored to help better structure European research and to cope with the strategic objectives set out in Lisbon in 2000. Japan identified nanotechnology as one of its main research priorities in 2001. The funding levels increased sharply from $400 million in 2001 to around $950 million in 2004. In 2003, South Korea embarked upon a ten-year program with around $2 billion of public funding, and Taiwan has committed around $600 million of public funding over six years. Singapore and China are also investing on a large scale. Russia is well funded as well.

Nanotechnology literally means any technology done on a nanoscale that has applications in the real world. Nanotechnology encompasses production and application of physical, chemical and biological systems at scales, ranging from individual atoms or molecules to submicron dimensions, as well as the integration of the resulting nanostructures into larger systems. Nanotechnology is likely to have a profound impact on our economy and society in the early 21st century, comparable to that of semiconductor technology, information technology, or cellular and molecular biology. Science and technology research in nanotechnology promises breakthroughs in areas such as materials and manufacturing, nanoelectronics, medicine and healthcare, energy, biotechnology, information technology and national security. It is widely felt that nanotechnology will be the next industrial revolution.

There is an increasing need for a multidisciplinary, system-oriented approach to design and manufacturing of micro/nanodevices that function reliably. This can only be achieved through the cross-fertilization of ideas from different disciplines and the systematic flow of information and people among research groups. Reliability is a critical technology for many micro- and nanosystems and nanostructured materials. A broad-based handbook was needed, and thus the first edition of Springer Handbook of Nanotechnology was published in April 2004. It presented an overview of nanomaterial synthesis, micro/nanofabrication, micro- and nanocomponents and systems, scanning probe microscopy, reliability issues (including nanotribology and nanomechanics) for nanotechnology, and industrial applications. When the handbook went for sale in Europe, it sold out in ten days. Reviews on the handbook were very flattering.

Given the explosive growth in nanoscience and nanotechnology, the publisher and the editor decided to develop a second edition merely six months after publication of the first edition. This edition of the handbook integrates the knowledge from the nanostructure, fabrication, materials science, devices, and reliability point of view. It covers various industrial applications. It also addresses social, ethical, and political issues. Given the significant interest in biomedical applications, a number of chapters in this arena have been added. The second edition consists of 59 chapters (new: 23; revised: 27; unchanged: 9). The chapters have been written by 154 internationally recognized experts in the field, from academia, national research labs, and industry.

This book is intended for three types of readers: graduate students of nanotechnology, researchers in academia and industry who are active or intend to become active in this field, and practicing engineers and scientists who have encountered a problem and hope to solve it as expeditiously as possible. The handbook should serve as an excellent text for one or two semester graduate courses in nanotechnology in mechanical engineering, materials science, applied physics, or applied chemistry.

We embarked on the development of the second edition in October 2004, and we worked very hard to get all the chapters to the publisher in a record time of about 7 months. I wish to sincerely thank the authors for offering to write comprehensive chapters on a tight schedule. This is generally an added responsibility to the hectic work schedules of researchers today. I depended on a large number of reviewers who provided critical reviews. I would like to thank Dr. Phillip J. Bond, Chief of Staff and Under Secretary for Technology, US Department of Commerce, Washington, D.C. for chapter suggestions as well as authors in the handbook. I would also like to thank my colleague, Dr. Zhenhua Tao, whose efforts during the preparation of this handbook were very useful. Last but not the least, I would like to thank my secretary Caterina Runyon-Spears for various administrative duties; her tireless efforts are highly appreciated.

I hope that this handbook will stimulate further interest in this important new field, and the readers of this handbook will find it useful.

May 2005

Bharat Bhushan
Editor

Preface to the 1st Edition

On December 29, 1959 at the California Institute of Technology, Nobel Laureate Richard P. Feynman gave a talk at the Annual meeting of the American Physical Society that has become one classic science lecture of the 20th century, titled "There's Plenty of Room at the Bottom." He presented a technological vision of extreme miniaturization in 1959, several years before the word "chip" became part of the lexicon. He talked about the problem of manipulating and controlling things on a small scale. Extrapolating from known physical laws, Feynman envisioned a technology using the ultimate toolbox of nature, building nanoobjects atom by atom or molecule by molecule. Since the 1980s, many inventions and discoveries in fabrication of nanoobjects have been a testament to his vision. In recognition of this reality, in a January 2000 speech at the same institute, former President W. J. Clinton talked about the exciting promise of "nanotechnology" and the importance of expanding research in nanoscale science and engineering. Later that month, he announced in his State of the Union Address an ambitious $497 million federal, multi-agency national nanotechnology initiative (NNI) in the fiscal year 2001 budget, and made the NNI a top science and technology priority. Nanotechnology literally means any technology done on a nanoscale that has applications in the real world. Nanotechnology encompasses production and application of physical, chemical and biological systems at size scales, ranging from individual atoms or molecules to submicron dimensions as well as the integration of the resulting nanostructures into larger systems. Nanofabrication methods include the manipulation or self-assembly of individual atoms, molecules, or molecular structures to produce nanostructured materials and sub-micron devices. Micro- and nanosystems components are fabricated using top-down lithographic and nonlithographic fabrication techniques. Nanotechnology will have a profound impact on our economy and society in the early 21st century, comparable to that of semiconductor technology, information technology, or advances in cellular and molecular biology. The research and development in nanotechnology will lead to potential breakthroughs in areas such as materials and manufacturing, nanoelectronics, medicine and healthcare, energy, biotechnology, information technology and national security. It is widely felt that nanotechnology will lead to the next industrial revolution.

Reliability is a critical technology for many micro- and nanosystems and nanostructured materials. No book exists on this emerging field. A broad based handbook is needed. The purpose of this handbook is to present an overview of nanomaterial synthesis, micro/nanofabrication, micro- and nanocomponents and systems, reliability issues (including nanotribology and nanomechanics) for nanotechnology, and industrial applications. The chapters have been written by internationally recognized experts in the field, from academia, national research labs and industry from all over the world.

The handbook integrates knowledge from the fabrication, mechanics, materials science and reliability points of view. This book is intended for three types of readers: graduate students of nanotechnology, researchers in academia and industry who are active or intend to become active in this field, and practicing engineers and scientists who have encountered a problem and hope to solve it as expeditiously as possible. The handbook should serve as an excellent text for one or two semester graduate courses in nanotechnology in mechanical engineering, materials science, applied physics, or applied chemistry.

We embarked on this project in February 2002, and we worked very hard to get all the chapters to the publisher in a record time of about 1 year. I wish to sincerely thank the authors for offering to write comprehensive chapters on a tight schedule. This is generally an added responsibility in the hectic work schedules of researchers today. I depended on a large number of reviewers who provided critical reviews. I would like to thank Dr. Phillip J. Bond, Chief of Staff and Under Secretary for Technology, US Department of Commerce, Washington, D.C. for suggestions for chapters as well as authors in the handbook. I would also like to thank my colleague, Dr. Huiwen Liu, whose efforts during the preparation of this handbook were very useful.

I hope that this handbook will stimulate further interest in this important new field, and the readers of this handbook will find it useful.

September 2003

Bharat Bhushan
Editor

Editors Vita

Dr. Bharat Bhushan received an M.S. in mechanical engineering from the Massachusetts Institute of Technology in 1971, an M.S. in mechanics and a Ph.D. in mechanical engineering from the University of Colorado at Boulder in 1973 and 1976, respectively, an MBA from Rensselaer Polytechnic Institute at Troy, NY in 1980, Doctor Technicae from the University of Trondheim at Trondheim, Norway in 1990, a Doctor of Technical Sciences from the Warsaw University of Technology at Warsaw, Poland in 1996, and Doctor Honouris Causa from the National Academy of Sciences at Gomel, Belarus in 2000. He is a registered professional engineer. He is presently an Ohio Eminent Scholar and The Howard D. Winbigler Professor in the College of Engineering, and the Director of the Nanoprobe Laboratory for Bio- and Nanotechnology and Biomimetics (NLB^2) at the Ohio State University, Columbus, Ohio. His research interests include fundamental studies with a focus on scanning probe techniques in the interdisciplinary areas of bio/nanotribology, bio/nanomechanics and bio/nanomaterials characterization, and applications to bio/nanotechnology and biomimetics. He is an internationally recognized expert of bio/nanotribology and bio/nanomechanics using scanning probe microscopy, and is one of the most prolific authors. He is considered by some a pioneer of the tribology and mechanics of magnetic storage devices. He has authored 6 scientific books, more than 90 handbook chapters, more than 700 scientific papers (*h* factor – 45+; ISI Highly Cited in Materials Science, since 2007), and more than 60 technical reports, edited more than 45 books, and holds 17 US and foreign patents. He is co-editor of Springer NanoScience and Technology Series and co-editor of Microsystem Technologies. He has given more than 400 invited presentations on six continents and more than 140 keynote/plenary addresses at major international conferences.

Dr. Bhushan is an accomplished organizer. He organized the first symposium on Tribology and Mechanics of Magnetic Storage Systems in 1984 and the first international symposium on Advances in Information Storage Systems in 1990, both of which are now held annually. He is the founder of an ASME Information Storage and Processing Systems Division founded in 1993 and served as the founding chair during 1993–1998. His biography has been listed in over two dozen Who's Who books including Who's Who in the World and has received more than two dozen awards for his contributions to science and technology from professional societies, industry, and US government agencies. He is also the recipient of various international fellowships including the Alexander von Humboldt Research Prize for Senior Scientists, Max Planck Foundation Research Award for Outstanding Foreign Scientists, and the Fulbright Senior Scholar Award. He is a foreign member of the International Academy of Engineering (Russia), Byelorussian Academy of Engineering and Technology and the Academy of Triboengineering of Ukraine, an honorary member of the Society of Tribologists of Belarus, a fellow of ASME, IEEE, STLE, and the New York Academy of Sciences, and a member of ASEE, Sigma Xi and Tau Beta Pi.

Dr. Bhushan has previously worked for the R&D Division of Mechanical Technology Inc., Latham, NY; the Technology Services Division of SKF Industries Inc., King of Prussia, PA; the General Products Division Laboratory of IBM Corporation, Tucson, AZ; and the Almaden Research Center of IBM Corporation, San Jose, CA. He has held visiting professor appointments at University of California at Berkeley, University of Cambridge, UK, Technical University Vienna, Austria, University of Paris, Orsay, ETH Zurich and EPFL Lausanne.

List of Authors

Chong H. Ahn
University of Cincinnati
Department of Electrical
and Computer Engineering
Cincinnati, OH 45221, USA
e-mail: *chong.ahn@uc.edu*

Boris Anczykowski
nanoAnalytics GmbH
Münster, Germany
e-mail: *anczykowski@nanoanalytics.com*

W. Robert Ashurst
Auburn University
Department of Chemical Engineering
Auburn, AL 36849, USA
e-mail: *ashurst@auburn.edu*

Massood Z. Atashbar
Western Michigan University
Department of Electrical
and Computer Engineering
Kalamazoo, MI 49008-5329, USA
e-mail: *massood.atashbar@wmich.edu*

Wolfgang Bacsa
University of Toulouse III (Paul Sabatier)
Laboratoire de Physique des Solides (LPST),
UMR 5477 CNRS
Toulouse, France
e-mail: *bacsa@ramansco.ups-tlse.fr; bacsa@lpst.ups-tlse.fr*

Kelly Bailey
University of Adelaide
CSIRO Human Nutrition
Adelaide SA 5005, Australia
e-mail: *kelly.bailey@csiro.au*

William Sims Bainbridge
National Science Foundation
Division of Information, Science and Engineering
Arlington, VA, USA
e-mail: *wsbainbridge@yahoo.com*

Antonio Baldi
Institut de Microelectronica de Barcelona (IMB)
Centro National Microelectrónica (CNM-CSIC)
Barcelona, Spain
e-mail: *antoni.baldi@cnm.es*

Wilhelm Barthlott
University of Bonn
Nees Institute for Biodiversity of Plants
Meckenheimer Allee 170
53115 Bonn, Germany
e-mail: *barthlott@uni-bonn.de*

Roland Bennewitz
INM – Leibniz Institute for New Materials
66123 Saarbrücken, Germany
e-mail: *roland.bennewitz@inm-gmbh.de*

Bharat Bhushan
Ohio State University
Nanoprobe Laboratory for Bio- and
Nanotechnology and Biomimetics (NLB^2)
201 W. 19th Avenue
Columbus, OH 43210-1142, USA
e-mail: *bhushan.2@osu.edu*

Gerd K. Binnig
Definiens AG
Trappentreustr. 1
80339 Munich, Germany
e-mail: *gbinnig@definiens.com*

Marcie R. Black
Bandgap Engineering Inc.
1344 Main St.
Waltham, MA 02451, USA
e-mail: *marcie@alum.mit.edu; marcie@bandgap.com*

Donald W. Brenner
Department of Materials Science and Engineering
Raleigh, NC, USA
e-mail: *brenner@ncsu.edu*

Jean-Marc Broto
Institut National des Sciences Appliquées of Toulouse
Laboratoire National des Champs Magnétiques Pulsés (LNCMP)
Toulouse, France
e-mail: *broto@lncmp.fr*

Guozhong Cao
University of Washington
Dept. of Materials Science and Engineering
302M Roberts Hall
Seattle, WA 98195-2120, USA
e-mail: *gzcao@u.washington.edu*

Edin (I-Chen) Chen
National Central University
Institute of Materials Science and Engineering
Department of Mechanical Engineering
Chung-Li, 320, Taiwan
e-mail: *ichen@ncu.edu.tw*

Yu-Ting Cheng
National Chiao Tung University
Department of Electronics Engineering & Institute of Electronics
1001, Ta-Hsueh Rd.
Hsinchu, 300, Taiwan, R.O.C.
e-mail: *ytcheng@mail.nctu.edu.tw*

Giovanni Cherubini
IBM Zurich Research Laboratory
Tape Technologies
8803 Rüschlikon, Switzerland
e-mail: *cbi@zurich.ibm.com*

Mu Chiao
Department of Mechanical Engineering
6250 Applied Science Lane
Vancouver, BC V6T 1Z4, Canada
e-mail: *muchiao@mech.ubc.ca*

Jin-Woo Choi
Louisiana State University
Department of Electrical and Computer Engineering
Baton Rouge, LA 70803, USA
e-mail: *choi@ece.lsu.edu*

Tamara H. Cooper
University of Adelaide
CSIRO Human Nutrition
Adelaide SA 5005, Australia
e-mail: *tamara.cooper@csiro.au*

Alex D. Corwin
GE Global Research
1 Research Circle
Niskayuna, NY 12309, USA
e-mail: *corwin@ge.com*

Maarten P. de Boer
Carnegie Mellon University
Department of Mechanical Engineering
5000 Forbes Avenue
Pittsburgh, PA 15213, USA
e-mail: *mpdebo@andrew.cmu.edu*

Dietrich Dehlinger
Lawrence Livermore National Laboratory
Engineering
Livermore, CA 94551, USA
e-mail: *dehlinger1@llnl.gov*

Frank W. DelRio
National Institute of Standards and Technology
100 Bureau Drive, Stop 8520
Gaithersburg, MD 20899-8520, USA
e-mail: *frank.delrio@nist.gov*

Michel Despont
IBM Zurich Research Laboratory
Micro- and Nanofabrication
8803 Rüschlikon, Switzerland
e-mail: *dpt@zurich.ibm.com*

Lixin Dong
Michigan State University
Electrical and Computer Engineering
2120 Engineering Building
East Lansing, MI 48824-1226, USA
e-mail: *ldong@egr.msu.edu*

Gene Dresselhaus
Massachusetts Institute of Technology
Francis Bitter Magnet Laboratory
Cambridge, MA 02139, USA
e-mail: *gene@mgm.mit.edu*

Mildred S. Dresselhaus
Massachusetts Institute of Technology
Department of Electrical Engineering
and Computer Science
Department of Physics
Cambridge, MA, USA
e-mail: *millie@mgm.mit.edu*

Urs T. Dürig
IBM Zurich Research Laboratory
Micro-/Nanofabrication
8803 Rüschlikon, Switzerland
e-mail: *drg@zurich.ibm.com*

Andreas Ebner
Johannes Kepler University Linz
Institute for Biophysics
Altenberger Str. 69
4040 Linz, Austria
e-mail: *andreas.ebner@jku.at*

Evangelos Eleftheriou
IBM Zurich Research Laboratory
8803 Rüschlikon, Switzerland
e-mail: *ele@zurich.ibm.com*

Emmanuel Flahaut
Université Paul Sabatier
CIRIMAT, Centre Interuniversitaire de Recherche
et d'Ingénierie des Matériaux, UMR 5085 CNRS
118 Route de Narbonne
31062 Toulouse, France
e-mail: *flahaut@chimie.ups-tlse.fr*

Anatol Fritsch
University of Leipzig
Institute of Experimental Physics I
Division of Soft Matter Physics
Linnéstr. 5
04103 Leipzig, Germany
e-mail: *anatol.fritsch@uni-leipzig.de*

Harald Fuchs
Universität Münster
Physikalisches Institut
Münster, Germany
e-mail: *fuchsh@uni-muenster.de*

Christoph Gerber
University of Basel
Institute of Physics
National Competence Center for Research
in Nanoscale Science (NCCR) Basel
Klingelbergstr. 82
4056 Basel, Switzerland
e-mail: *christoph.gerber@unibas.ch*

Franz J. Giessibl
Universität Regensburg
Institute of Experimental and Applied Physics
Universitätsstr. 31
93053 Regensburg, Germany
e-mail: *franz.giessibl@physik.uni-regensburg.de*

Enrico Gnecco
University of Basel
National Center of Competence in Research
Department of Physics
Klingelbergstr. 82
4056 Basel, Switzerland
e-mail: *enrico.gnecco@unibas.ch*

Stanislav N. Gorb
Max Planck Institut für Metallforschung
Evolutionary Biomaterials Group
Heisenbergstr. 3
70569 Stuttgart, Germany
e-mail: *s.gorb@mf.mpg.de*

Hermann Gruber
University of Linz
Institute of Biophysics
Altenberger Str. 69
4040 Linz, Austria
e-mail: *hermann.gruber@jku.at*

Jason Hafner
Rice University
Department of Physics and Astronomy
Houston, TX 77251, USA
e-mail: *hafner@rice.edu*

Judith A. Harrison
U.S. Naval Academy
Chemistry Department
572 Holloway Road
Annapolis, MD 21402-5026, USA
e-mail: *jah@usna.edu*

Martin Hegner
CRANN – The Naughton Institute
Trinity College, University of Dublin
School of Physics
Dublin, 2, Ireland
e-mail: *martin.hegner@tcd.ie*

Thomas Helbling
ETH Zurich
Micro and Nanosystems
Department of Mechanical
and Process Engineering
8092 Zurich, Switzerland
e-mail: *thomas.helbling@micro.mavt.ethz.ch*

Michael J. Heller
University of California San Diego
Department of Bioengineering
Dept. of Electrical and Computer Engineering
La Jolla, CA, USA
e-mail: *mjheller@ucsd.edu*

Seong-Jun Heo
Lam Research Corp.
4650 Cushing Parkway
Fremont, CA 94538, USA
e-mail: *seongjun.heo@lamrc.com*

Christofer Hierold
ETH Zurich
Micro and Nanosystems
Department of Mechanical
and Process Engineering
8092 Zurich, Switzerland
e-mail: *christofer.hierold@micro.mavt.ethz.ch*

Peter Hinterdorfer
University of Linz
Institute for Biophysics
Altenberger Str. 69
4040 Linz, Austria
e-mail: *peter.hinterdorfer@jku.at*

Dalibor Hodko
Nanogen, Inc.
10498 Pacific Center Court
San Diego, CA 92121, USA
e-mail: *dhodko@nanogen.com*

Hendrik Hölscher
Forschungszentrum Karlsruhe
Institute of Microstructure Technology
Linnéstr. 5
76021 Karlsruhe, Germany
e-mail: *hendrik.hoelscher@imt.fzk.de*

Hirotaka Hosoi
Hokkaido University
Creative Research Initiative Sousei
Kita 21, Nishi 10, Kita-ku
Sapporo, Japan
e-mail: *hosoi@cris.hokudai.ac.jp*

Katrin Hübner
Staatliche Fachoberschule Neu-Ulm
89231 Neu-Ulm, Germany
e-mail: *katrin.huebner1@web.de*

Douglas L. Irving
North Carolina State University
Materials Science and Engineering
Raleigh, NC 27695-7907, USA
e-mail: *doug_irving@ncsu.edu*

Jacob N. Israelachvili
University of California
Department of Chemical Engineering
and Materials Department
Santa Barbara, CA 93106-5080, USA
e-mail: *jacob@engineering.ucsb.edu*

Guangyao Jia
University of California, Irvine
Department of Mechanical
and Aerospace Engineering
Irvine, CA, USA
e-mail: *gjia@uci.edu*

Sungho Jin
University of California, San Diego
Department of Mechanical
and Aerospace Engineering
9500 Gilman Drive
La Jolla, CA 92093-0411, USA
e-mail: *jin@ucsd.edu*

Anne Jourdain
Interuniversity Microelectronics Center (IMEC)
Leuven, Belgium
e-mail: *jourdain@imec.be*

Yong Chae Jung
Samsung Electronics C., Ltd.
Senior Engineer Process Development Team
San #16 Banwol-Dong, Hwasung-City
Gyeonggi-Do 445-701, Korea
e-mail: *yc423.jung@samsung.com*

Harold Kahn
Case Western Reserve University
Department of Materials Science and Engineering
Cleveland, OH , USA
e-mail: *kahn@cwru.edu*

Roger Kamm
Massachusetts Institute of Technology
Department of Biological Engineering
77 Massachusetts Avenue
Cambridge, MA 02139, USA
e-mail: *rdkamm@mit.edu*

Ruti Kapon
Weizmann Institute of Science
Department of Biological Chemistry
Rehovot 76100, Israel
e-mail: *ruti.kapon@weizmann.ac.il*

Josef Käs
University of Leipzig
Institute of Experimental Physics I
Division of Soft Matter Physics
Linnéstr. 5
04103 Leipzig, Germany
e-mail: *jkaes@physik.uni-leipzig.de*

Horacio Kido
University of California at Irvine
Mechanical and Aerospace Engineering
Irvine, CA, USA
e-mail: *hkido@uci.edu*

Tobias Kießling
University of Leipzig
Institute of Experimental Physics I
Division of Soft Matter Physics
Linnéstr. 5
04103 Leipzig, Germany
e-mail: *Tobias.Kiessling@uni-leipzig.de*

Jitae Kim
University of California at Irvine
Department of Mechanical
and Aerospace Engineering
Irvine, CA, USA
e-mail: *jitaekim@uci.edu*

Jongbaeg Kim
Yonsei University
School of Mechanical Engineering
1st Engineering Bldg.
Seoul, 120-749, South Korea
e-mail: *kimjb@yonsei.ac.kr*

Nahui Kim
Samsung Advanced Institute of Technology
Research and Development
Seoul, South Korea
e-mail: *nahui.kim@samsung.com*

Kerstin Koch
Rhine-Waal University of Applied Science
Department of Life Science, Biology
and Nanobiotechnology
Landwehr 4
47533 Kleve, Germany
e-mail: *kerstin.koch@hochschule.rhein-waal.de*

Jing Kong
Massachusetts Institute of Technology
Department of Electrical Engineering
and Computer Science
Cambridge, MA, USA
e-mail: *jingkong@mit.edu*

Tobias Kraus
Leibniz-Institut für Neue Materialien gGmbH
Campus D2 2
66123 Saarbrücken, Germany
e-mail: *tobias.kraus@inm-gmbh.de*

Anders Kristensen
Technical University of Denmark
DTU Nanotech
2800 Kongens Lyngby, Denmark
e-mail: *anders.kristensen@nanotech.dtu.dk*

Ratnesh Lal
University of Chicago
Center for Nanomedicine
5841 S Maryland Av
Chicago, IL 60637, USA
e-mail: *rlal@uchicago.edu*

Jan Lammerding
Harvard Medical School
Brigham and Women's Hospital
65 Landsdowne St
Cambridge, MA 02139, USA
e-mail: *jlammerding@rics.bwh.harvard.edu*

Hans Peter Lang
University of Basel
Institute of Physics, National Competence Center for Research in Nanoscale Science (NCCR) Basel
Klingelbergstr. 82
4056 Basel, Switzerland
e-mail: *hans-peter.lang@unibas.ch*

Carmen LaTorre
Owens Corning Science and Technology
Roofing and Asphalt
2790 Columbus Road
Granville, OH 43023, USA
e-mail: *carmen.latorre@owenscorning.com*

Christophe Laurent
Université Paul Sabatier
CIRIMAT UMR 5085 CNRS
118 Route de Narbonne
31062 Toulouse, France
e-mail: *laurent@chimie.ups-tlse.fr*

Abraham P. Lee
University of California Irvine
Department of Biomedical Engineering
Department of Mechanical
and Aerospace Engineering
Irvine, CA 92697, USA
e-mail: *aplee@uci.edu*

Stephen C. Lee
Ohio State University
Biomedical Engineering Center
Columbus, OH 43210, USA
e-mail: *lee@bme.ohio-state.edu*

Wayne R. Leifert
Adelaide Business Centre
CSIRO Human Nutrition
Adelaide SA 5000, Australia
e-mail: *wayne.leifert@csiro.au*

Liwei Lin
UC Berkeley
Mechanical Engineering Department
5126 Etcheverry
Berkeley, CA 94720-1740, USA
e-mail: *lwlin@me.berkeley.edu*

Yu-Ming Lin
IBM T.J. Watson Research Center
Nanometer Scale Science & Technology
1101 Kitchawan Road
Yorktown Heigths, NY 10598, USA
e-mail: *yming@us.ibm.com*

Marc J. Madou
University of California Irvine
Department of Mechanical and Aerospace
and Biomedical Engineering
Irvine, CA, USA
e-mail: *mmadou@uci.edu*

Othmar Marti
Ulm University
Institute of Experimental Physics
Albert-Einstein-Allee 11
89069 Ulm, Germany
e-mail: *othmar.marti@uni-ulm.de*

Jack Martin
66 Summer Street
Foxborough, MA 02035, USA
e-mail: *jack.martin@alumni.tufts.edu*

Shinji Matsui
University of Hyogo
Laboratory of Advanced Science
and Technology for Industry
Hyogo, Japan
e-mail: *matsui@lasti.u-hyogo.ac.jp*

Mehran Mehregany
Case Western Reserve University
Department of Electrical Engineering
and Computer Science
Cleveland, OH 44106, USA
e-mail: *mxm31@cwru.edu*

Etienne Menard
Semprius, Inc.
4915 Prospectus Dr.
Durham, NC 27713, USA
e-mail: *etienne.menard@semprius.com*

Ernst Meyer
University of Basel
Institute of Physics
Basel, Switzerland
e-mail: *ernst.meyer@unibas.ch*

Robert Modliñski
Baolab Microsystems
Terrassa 08220, Spain
e-mail: *rmodlinski@gmx.com*

Mohammad Mofrad
University of California, Berkeley
Department of Bioengineering
Berkeley, CA 94720, USA
e-mail: *mofrad@berkeley.edu*

Marc Monthioux
CEMES - UPR A-8011 CNRS
Carbones et Matériaux Carbonés,
Carbons and Carbon-Containing Materials
29 Rue Jeanne Marvig
31055 Toulouse 4, France
e-mail: *monthiou@cemes.fr*

Markus Morgenstern
RWTH Aachen University
II. Institute of Physics B and JARA-FIT
52056 Aachen, Germany
e-mail: *mmorgens@physik.rwth-aachen.de*

Seizo Morita
Osaka University
Department of Electronic Engineering
Suita-City
Osaka, Japan
e-mail: *smorita@ele.eng.osaka-u.ac.jp*

Koichi Mukasa
Hokkaido University
Nanoelectronics Laboratory
Sapporo, Japan
e-mail: *mukasa@nano.eng.hokudai.ac.jp*

Bradley J. Nelson
Swiss Federal Institute of Technology (ETH)
Institute of Robotics and Intelligent Systems
8092 Zurich, Switzerland
e-mail: *bnelson@ethz.ch*

Michael Nosonovsky
University of Wisconsin-Milwaukee
Department of Mechanical Engineering
3200 N. Cramer St.
Milwaukee, WI 53211, USA
e-mail: *nosonovs@uwm.edu*

Hiroshi Onishi
Kanagawa Academy of Science and Technology
Surface Chemistry Laboratory
Kanagawa, Japan
e-mail: *oni@net.ksp.or.jp*

Alain Peigney
Centre Inter-universitaire de Recherche
sur l'Industrialisation des Matériaux (CIRIMAT)
Toulouse 4, France
e-mail: *peigney@chimie.ups-tlse.fr*

Oliver Pfeiffer
Individual Computing GmbH
Ingelsteinweg 2d
4143 Dornach, Switzerland
e-mail: *oliver.pfeiffer@gmail.com*

Haralampos Pozidis
IBM Zurich Research Laboratory
Storage Technologies
Rüschlikon, Switzerland
e-mail: *hap@zurich.ibm.com*

Robert Puers
Katholieke Universiteit Leuven
ESAT/MICAS
Leuven, Belgium
e-mail: *bob.puers@esat.kuleuven.ac.be*

Calvin F. Quate
Stanford University
Edward L. Ginzton Laboratory
450 Via Palou
Stanford, CA 94305-4088, USA
e-mail: *quate@stanford.edu*

Oded Rabin
University of Maryland
Department of Materials Science and Engineering
College Park, MD, USA
e-mail: *oded@umd.edu*

Françisco M. Raymo
University of Miami
Department of Chemistry
1301 Memorial Drive
Coral Gables, FL 33146-0431, USA
e-mail: *fraymo@miami.edu*

Manitra Razafinimanana
University of Toulouse III (Paul Sabatier)
Centre de Physique des Plasmas
et leurs Applications (CPPAT)
Toulouse, France
e-mail: *razafinimanana@cpat.ups-tlse.fr*

Ziv Reich
Weizmann Institute of Science Ha'Nesi Ha'Rishon
Department of Biological Chemistry
Rehovot 76100, Israel
e-mail: *ziv.reich@weizmann.ac.il*

John A. Rogers
University of Illinois
Department of Materials Science and Engineering
Urbana, IL, USA
e-mail: *jrogers@uiuc.edu*

Cosmin Roman
ETH Zurich
Micro and Nanosystems Department of Mechanical
and Process Engineering
8092 Zurich, Switzerland
e-mail: *cosmin.roman@micro.mavt.ethz.ch*

Marina Ruths
University of Massachusetts Lowell
Department of Chemistry
1 University Avenue
Lowell, MA 01854, USA
e-mail: *marina_ruths@uml.edu*

Ozgur Sahin
The Rowland Institute at Harvard
100 Edwin H. Land Blvd
Cambridge, MA 02142, USA
e-mail: *sahin@rowland.harvard.edu*

Akira Sasahara
Japan Advanced Institute
of Science and Technology
School of Materials Science
1-1 Asahidai
923-1292 Nomi, Japan
e-mail: *sasahara@jaist.ac.jp*

Helmut Schift
Paul Scherrer Institute
Laboratory for Micro- and Nanotechnology
5232 Villigen PSI, Switzerland
e-mail: *helmut.schift@psi.ch*

André Schirmeisen
University of Münster
Institute of Physics
Wilhelm-Klemm-Str. 10
48149 Münster, Germany
e-mail: *schirmeisen@uni-muenster.de*

Christian Schulze
Beiersdorf AG
Research & Development
Unnastr. 48
20245 Hamburg, Germany
e-mail: *christian.schulze@beiersdorf.com;*
christian.schulze@uni-leipzig.de

Alexander Schwarz
University of Hamburg
Institute of Applied Physics
Jungiusstr. 11
20355 Hamburg, Germany
e-mail: *aschwarz@physnet.uni-hamburg.de*

Udo D. Schwarz
Yale University
Department of Mechanical Engineering
15 Prospect Street
New Haven, CT 06520-8284, USA
e-mail: *udo.schwarz@yale.edu*

Philippe Serp
Ecole Nationale Supérieure d'Ingénieurs
en Arts Chimiques et Technologiques
Laboratoire de Chimie de Coordination (LCC)
118 Route de Narbonne
31077 Toulouse, France
e-mail: *philippe.serp@ensiacet.fr*

Huamei (Mary) Shang
GE Healthcare
4855 W. Electric Ave.
Milwaukee, WI 53219, USA
e-mail: *huamei.shang@ge.com*

Susan B. Sinnott
University of Florida
Department of Materials Science and Engineering
154 Rhines Hall
Gainesville, FL 32611-6400, USA
e-mail: *ssinn@mse.ufl.edu*

Anisoara Socoliuc
SPECS Zurich GmbH
Technoparkstr. 1
8005 Zurich, Switzerland
e-mail: *socoliuc@nanonis.com*

Olav Solgaard
Stanford University
E.L. Ginzton Laboratory
450 Via Palou
Stanford, CA 94305-4088, USA
e-mail: *solgaard@stanford.edu*

Dan Strehle
University of Leipzig
Institute of Experimental Physics I
Division of Soft Matter Physics
Linnéstr. 5
04103 Leipzig, Germany
e-mail: *dan.strehle@uni-leipzig.de*

Carsten Stüber
University of Leipzig
Institute of Experimental Physics I
Division of Soft Matter Physics
Linnéstr. 5
04103 Leipzig, Germany
e-mail: *stueber@rz.uni-leipzig.de*

Yu-Chuan Su
ESS 210
Department of Engineering and System Science 101
Kuang-Fu Road
Hsinchu, 30013, Taiwan
e-mail: *ycsu@ess.nthu.edu.tw*

Kazuhisa Sueoka
Graduate School of Information Science
and Technology
Hokkaido University
Nanoelectronics Laboratory
Kita-14, Nishi-9, Kita-ku
060-0814 Sapporo, Japan
e-mail: *sueoka@nano.isthokudai.ac.jp*

Yasuhiro Sugawara
Osaka University
Department of Applied Physics
Yamada-Oka 2-1, Suita
565-0871 Osaka, Japan
e-mail: *sugawara@ap.eng.osaka-u.ac.jp*

Benjamin Sullivan
TearLab Corp.
11025 Roselle Street
San Diego, CA 92121, USA
e-mail: *bdsulliv@TearLab.com*

Paul Swanson
Nexogen, Inc.
Engineering
8360 C Camino Santa Fe
San Diego, CA 92121, USA
e-mail: *pswanson@nexogentech.com*

Yung-Chieh Tan
Washington University School of Medicine
Department of Medicine
Division of Dermatology
660 S. Euclid Ave.
St. Louis, MO 63110, USA
e-mail: *ytanster@gmail.com*

Shia-Yen Teh
University of California at Irvine
Biomedical Engineering Department
3120 Natural Sciences II
Irvine, CA 92697-2715, USA
e-mail: *steh@uci.edu*

W. Merlijn van Spengen
Leiden University
Kamerlingh Onnes Laboratory
Niels Bohrweg 2
Leiden, CA 2333, The Netherlands
e-mail: *spengen@physics.leidenuniv.nl*

Peter Vettiger
University of Neuchâtel
SAMLAB
Jaquet-Droz 1
2002 Neuchâtel, Switzerland
e-mail: *peter.vettiger@unine.ch*

Franziska Wetzel
University of Leipzig
Institute of Experimental Physics I
Division of Soft Matter Physics
Linnéstr. 5
04103 Leipzig, Germany
e-mail: *franziska.wetzel@uni-leipzig.de*

Heiko Wolf
IBM Research GmbH
Zurich Research Laboratory
Säumerstr. 4
8803 Rüschlikon, Switzerland
e-mail: *hwo@zurich.ibm.com*

Darrin J. Young
Case Western Reserve University
Department of EECS, Glennan 510
10900 Euclid Avenue
Cleveland, OH 44106, USA
e-mail: *djy@po.cwru.edu*

Babak Ziaie
Purdue University
Birck Nanotechnology Center
1205 W. State St.
West Lafayette, IN 47907-2035, USA
e-mail: *bziaie@purdue.edu*

Christian A. Zorman
Case Western Reserve University
Department of Electrical Engineering
and Computer Science
10900 Euclid Avenue
Cleveland, OH 44106, USA
e-mail: *caz@case.edu*

Jim V. Zoval
Saddleback College
Department of Math and Science
28000 Marguerite Parkway
Mission Viejo, CA 92692, USA
e-mail: *jzoval@saddleback.edu*

Acknowledgements

F.43 Biological and Biologically Inspired Attachment Systems

by Stanislav N. Gorb

This work, as part of the European Science Foundation EUROCORES programme FANAS, was supported by funds from the German Science Foundation (DFG, contract no. GO 995/4-1) and the EC Sixth Framework Programme (contract no. ERAS-CT-2003-980409).

目 录

Contents

List of Abbreviations

μCP	microcontact printing
1-D	one-dimensional
18-MEA	18-methyl eicosanoic acid
2-D	two-dimensional
2-DEG	two-dimensional electron gas
3-APTES	3-aminopropyltriethoxysilane
3-D	three-dimensional

A

a-BSA	anti-bovine serum albumin
a-C	amorphous carbon
A/D	analog-to-digital
AA	amino acid
AAM	anodized alumina membrane
ABP	actin binding protein
AC	alternating-current
AC	amorphous carbon
ACF	autocorrelation function
ADC	analog-to-digital converter
ADXL	analog devices accelerometer
AFAM	atomic force acoustic microscopy
AFM	atomic force microscope
AFM	atomic force microscopy
AKD	alkylketene dimer
ALD	atomic layer deposition
AM	amplitude modulation
AMU	atomic mass unit
AOD	acoustooptical deflector
AOM	acoustooptical modulator
AP	alkaline phosphatase
APB	actin binding protein
APCVD	atmospheric-pressure chemical vapor deposition
APDMES	aminopropyldimethylethoxysilane
APTES	aminopropyltriethoxysilane
ASIC	application-specific integrated circuit
ASR	analyte-specific reagent
ATP	adenosine triphosphate

B

BAP	barometric absolute pressure
BAPDMA	behenyl amidopropyl dimethylamine glutamate
bcc	body-centered cubic
BCH	brucite-type cobalt hydroxide
BCS	Bardeen–Cooper–Schrieffer
BD	blu-ray disc
BDCS	biphenyldimethylchlorosilane
BE	boundary element
BFP	biomembrane force probe
BGA	ball grid array
BHF	buffered HF
BHPET	1,1'-(3,6,9,12,15-pentaoxapentadecane-1,15-diyl)bis(3-hydroxyethyl-1H-imidazolium-1-yl) di[bis(trifluoromethanesulfonyl)imide]
BHPT	1,1'-(pentane-1,5-diyl)bis(3-hydroxyethyl-1H-imidazolium-1-yl) di[bis(trifluoromethanesulfonyl)imide]
BiCMOS	bipolar CMOS
bioMEMS	biomedical microelectromechanical system
bioNEMS	biomedical nanoelectromechanical system
BMIM	1-butyl-3-methylimidazolium
BP	bit pitch
BPAG1	bullous pemphigoid antigen 1
BPT	biphenyl-4-thiol
BPTC	cross-linked BPT
BSA	bovine serum albumin
BST	barium strontium titanate
BTMAC	behentrimonium chloride

C

CA	constant amplitude
CA	contact angle
CAD	computer-aided design
CAH	contact angle hysteresis
cAMP	cyclic adenosine monophosphate
CAS	Crk-associated substrate
CBA	cantilever beam array
CBD	chemical bath deposition
CCD	charge-coupled device
CCVD	catalytic chemical vapor deposition
CD	compact disc
CD	critical dimension
CDR	complementarity determining region
CDW	charge density wave
CE	capillary electrophoresis
CE	constant excitation
CEW	continuous electrowetting
CG	controlled geometry
CHO	Chinese hamster ovary
CIC	cantilever in cantilever
CMC	cell membrane complex
CMC	critical micelle concentration
CMOS	complementary metal–oxide–semiconductor
CMP	chemical mechanical polishing

CNF	carbon nanofiber
CNFET	carbon nanotube field-effect transistor
CNT	carbon nanotube
COC	cyclic olefin copolymer
COF	chip-on-flex
COF	coefficient of friction
COG	cost of goods
CoO	cost of ownership
COS	CV-1 in origin with SV40
CP	circularly permuted
CPU	central processing unit
CRP	C-reactive protein
CSK	cytoskeleton
CSM	continuous stiffness measurement
CTE	coefficient of thermal expansion
Cu-TBBP	Cu-tetra-3,5 di-tertiary-butyl-phenyl porphyrin
CVD	chemical vapor deposition

D

DBR	distributed Bragg reflector
DC-PECVD	direct-current plasma-enhanced CVD
DC	direct-current
DDT	dichlorodiphenyltrichloroethane
DEP	dielectrophoresis
DFB	distributed feedback
DFM	dynamic force microscopy
DFS	dynamic force spectroscopy
DGU	density gradient ultracentrifugation
DI	FESPdigital instrument force modulation etched Si probe
DI	TESPdigital instrument tapping mode etched Si probe
DI	digital instrument
DI	deionized
DIMP	diisopropylmethylphosphonate
DIP	dual inline packaging
DIPS	industrial postpackaging
DLC	diamondlike carbon
DLP	digital light processing
DLVO	Derjaguin–Landau–Verwey–Overbeek
DMD	deformable mirror display
DMD	digital mirror device
DMDM	1,3-dimethylol-5,5-dimethyl
DMMP	dimethylmethylphosphonate
DMSO	dimethyl sulfoxide
DMT	Derjaguin–Muller–Toporov
DNA	deoxyribonucleic acid
DNT	2,4-dinitrotoluene
DOD	Department of Defense
DOE	Department of Energy
DOE	diffractive optical element
DOF	degree of freedom
DOPC	1,2-dioleoyl-sn-glycero-3-phosphocholine
DOS	density of states
DP	decylphosphonate
DPN	dip-pen nanolithography
DRAM	dynamic random-access memory
DRIE	deep reactive ion etching
ds	double-stranded
DSC	differential scanning calorimetry
DSP	digital signal processor
DTR	discrete track recording
DTSSP	3,3'-dithio-bis(sulfosuccinimidylproprionate)
DUV	deep-ultraviolet
DVD	digital versatile disc
DWNT	double-walled CNT

E

EAM	embedded atom method
EB	electron beam
EBD	electron beam deposition
EBID	electron-beam-induced deposition
EBL	electron-beam lithography
ECM	extracellular matrix
ECR-CVD	electron cyclotron resonance chemical vapor deposition
ED	electron diffraction
EDC	1-ethyl-3-(3-diamethylaminopropyl) carbodiimide
EDL	electrostatic double layer
EDP	ethylene diamine pyrochatechol
EDTA	ethylenediamine tetraacetic acid
EDX	energy-dispersive x-ray
EELS	electron energy loss spectra
EFM	electric field gradient microscopy
EFM	electrostatic force microscopy
EHD	elastohydrodynamic
EO	electroosmosis
EOF	electroosmotic flow
EOS	electrical overstress
EPA	Environmental Protection Agency
EPB	electrical parking brake
ESD	electrostatic discharge
ESEM	environmental scanning electron microscope
EU	European Union
EUV	extreme ultraviolet
EW	electrowetting
EWOD	electrowetting on dielectric

F

F-actin	filamentous actin
FA	focal adhesion
FAA	formaldehyde–acetic acid–ethanol
FACS	fluorescence-activated cell sorting

FAK focal adhesion kinase
FBS fetal bovine serum
FC flip-chip
FCA filtered cathodic arc
fcc face-centered cubic
FCP force calibration plot
FCS fluorescence correlation spectroscopy
FD finite difference
FDA Food and Drug Administration
FE finite element
FEM finite element method
FEM finite element modeling
FESEM field emission SEM
FESP force modulation etched Si probe
FET field-effect transistor
FFM friction force microscope
FFM friction force microscopy
FIB-CVD focused ion beam chemical vapor deposition
FIB focused ion beam
FIM field ion microscope
FIP feline coronavirus
FKT Frenkel–Kontorova–Tomlinson
FM frequency modulation
FMEA failure-mode effect analysis
FP6 Sixth Framework Program
FP fluorescence polarization
FPR *N*-formyl peptide receptor
FS force spectroscopy
FTIR Fourier-transform infrared
FV force–volume

G

GABA γ-aminobutyric acid
GDP guanosine diphosphate
GF gauge factor
GFP green fluorescent protein
GMR giant magnetoresistive
GOD glucose oxidase
GPCR G-protein coupled receptor
GPS global positioning system
GSED gaseous secondary-electron detector
GTP guanosine triphosphate
GW Greenwood and Williamson

H

HAR high aspect ratio
HARMEMS high-aspect-ratio MEMS
HARPSS high-aspect-ratio combined poly- and single-crystal silicon
HBM human body model
hcp hexagonal close-packed
HDD hard-disk drive
HDT hexadecanethiol
HDTV high-definition television
HEK human embryonic kidney 293
HEL hot embossing lithography
HEXSIL hexagonal honeycomb polysilicon
HF hydrofluoric
HMDS hexamethyldisilazane
HNA hydrofluoric-nitric-acetic
HOMO highest occupied molecular orbital
HOP highly oriented pyrolytic
HOPG highly oriented pyrolytic graphite
HOT holographic optical tweezer
HP hot-pressing
HPI hexagonally packed intermediate
HRTEM high-resolution transmission electron microscope
HSA human serum albumin
HtBDC hexa-*tert*-butyl-decacyclene
HTCS high-temperature superconductivity
HTS high throughput screening
HUVEC human umbilical venous endothelial cell

I

IBD ion beam deposition
IC integrated circuit
ICA independent component analysis
ICAM-1 intercellular adhesion molecules 1
ICAM-2 intercellular adhesion molecules 2
ICT information and communication technology
IDA interdigitated array
IF intermediate filament
IF intermediate-frequency
IFN interferon
IgG immunoglobulin G
IKVAV isoleucine–lysine–valine–alanine–valine
IL ionic liquid
IMAC immobilized metal ion affinity chromatography
IMEC Interuniversity MicroElectronics Center
IR infrared
ISE indentation size effect
ITO indium tin oxide
ITRS International Technology Roadmap for Semiconductors
IWGN Interagency Working Group on Nanoscience, Engineering, and Technology

J

JC jump-to-contact
JFIL jet-and-flash imprint lithography
JKR Johnson–Kendall–Roberts

K

KASH	Klarsicht, ANC-1, Syne Homology
KPFM	Kelvin probe force microscopy

L

LA	lauric acid
LAR	low aspect ratio
LB	Langmuir–Blodgett
LBL	layer-by-layer
LCC	leadless chip carrier
LCD	liquid-crystal display
LCoS	liquid crystal on silicon
LCP	liquid-crystal polymer
LDL	low-density lipoprotein
LDOS	local density of states
LED	light-emitting diode
LFA-1	leukocyte function-associated antigen-1
LFM	lateral force microscope
LFM	lateral force microscopy
LIGA	Lithographie Galvanoformung Abformung
LJ	Lennard-Jones
LMD	laser microdissection
LMPC	laser microdissection and pressure catapulting
LN	liquid-nitrogen
LoD	limit-of-detection
LOR	lift-off resist
LPC	laser pressure catapulting
LPCVD	low-pressure chemical vapor deposition
LSC	laser scanning cytometry
LSN	low-stress silicon nitride
LT-SFM	low-temperature scanning force microscope
LT-SPM	low-temperature scanning probe microscopy
LT-STM	low-temperature scanning tunneling microscope
LT	low-temperature
LTM	laser tracking microrheology
LTO	low-temperature oxide
LTRS	laser tweezers Raman spectroscopy
LUMO	lowest unoccupied molecular orbital
LVDT	linear variable differential transformer

M

MALDI	matrix assisted laser desorption ionization
MAP	manifold absolute pressure
MAPK	mitogen-activated protein kinase
MAPL	molecular assembly patterning by lift-off
MBE	molecular-beam epitaxy
MC	microcantilever
MC	microcapillary
MCM	multi-chip module
MD	molecular dynamics
ME	metal-evaporated
MEMS	microelectromechanical system
MExFM	magnetic exchange force microscopy
MFM	magnetic field microscopy
MFM	magnetic force microscope
MFM	magnetic force microscopy
MHD	magnetohydrodynamic
MIM	metal–insulator–metal
MIMIC	micromolding in capillaries
MLE	maximum likelihood estimator
MOCVD	metalorganic chemical vapor deposition
MOEMS	microoptoelectromechanical system
MOS	metal–oxide–semiconductor
MOSFET	metal–oxide–semiconductor field-effect transistor
MP	metal particle
MPTMS	mercaptopropyltrimethoxysilane
MRFM	magnetic resonance force microscopy
MRFM	molecular recognition force microscopy
MRI	magnetic resonance imaging
MRP	molecular recognition phase
MscL	mechanosensitive channel of large conductance
MST	microsystem technology
MT	microtubule
mTAS	micro total analysis system
MTTF	mean time to failure
MUMP	multiuser MEMS process
MVD	molecular vapor deposition
MWCNT	multiwall carbon nanotube
MWNT	multiwall nanotube
MYD/BHW	Muller–Yushchenko–Derjaguin/Burgess–Hughes–White

N

NA	numerical aperture
NADIS	nanoscale dispensing
NASA	National Aeronautics and Space Administration
NC-AFM	noncontact atomic force microscopy
NEMS	nanoelectromechanical system
NGL	next-generation lithography
NHS	*N*-hydroxysuccinimidyl
NIH	National Institute of Health
NIL	nanoimprint lithography
NIST	National Institute of Standards and Technology
NMP	no-moving-part
NMR	nuclear magnetic resonance
NMR	nuclear mass resonance
NNI	National Nanotechnology Initiative

NOEMS	nanooptoelectromechanical system
NP	nanoparticle
NP	nanoprobe
NSF	National Science Foundation
NSOM	near-field scanning optical microscopy
NSTC	National Science and Technology Council
NTA	nitrilotriacetate
nTP	nanotransfer printing

O

ODA	octadecylamine
ODDMS	*n*-octadecyldimethyl(dimethylamino)silane
ODMS	*n*-octyldimethyl(dimethylamino)silane
ODP	octadecylphosphonate
ODTS	octadecyltrichlorosilane
OLED	organic light-emitting device
OM	optical microscope
OMVPE	organometallic vapor-phase epitaxy
OS	optical stretcher
OT	optical tweezers
OTRS	optical tweezers Raman spectroscopy
OTS	octadecyltrichlorosilane
oxLDL	oxidized low-density lipoprotein

P

P–V	peak-to-valley
PAA	poly(acrylic acid)
PAA	porous anodic alumina
PAH	poly(allylamine hydrochloride)
PAPP	*p*-aminophenyl phosphate
Pax	paxillin
PBC	periodic boundary condition
PBS	phosphate-buffered saline
PC	polycarbonate
PCB	printed circuit board
PCL	polycaprolactone
PCR	polymerase chain reaction
PDA	personal digital assistant
PDMS	polydimethylsiloxane
PDP	2-pyridyldithiopropionyl
PDP	pyridyldithiopropionate
PE	polyethylene
PECVD	plasma-enhanced chemical vapor deposition
PEEK	polyetheretherketone
PEG	polyethylene glycol
PEI	polyethyleneimine
PEN	polyethylene naphthalate
PES	photoemission spectroscopy
PES	position error signal
PET	poly(ethyleneterephthalate)
PETN	pentaerythritol tetranitrate
PFDA	perfluorodecanoic acid
PFDP	perfluorodecylphosphonate
PFDTES	perfluorodecyltriethoxysilane
PFM	photonic force microscope
PFOS	perfluorooctanesulfonate
PFPE	perfluoropolyether
PFTS	perfluorodecyltricholorosilane
PhC	photonic crystal
PI3K	phosphatidylinositol-3-kinase
PI	polyisoprene
PID	proportional–integral–differential
PKA	protein kinase
PKC	protein kinase C
PKI	protein kinase inhibitor
PL	photolithography
PLC	phospholipase C
PLD	pulsed laser deposition
PMAA	poly(methacrylic acid)
PML	promyelocytic leukemia
PMMA	poly(methyl methacrylate)
POCT	point-of-care testing
POM	polyoxy-methylene
PP	polypropylene
PPD	*p*-phenylenediamine
PPMA	poly(propyl methacrylate)
PPy	polypyrrole
PS-PDMS	poly(styrene-b-dimethylsiloxane)
PS/clay	polystyrene/nanoclay composite
PS	polystyrene
PSA	prostate-specific antigen
PSD	position-sensitive detector
PSD	position-sensitive diode
PSD	power-spectral density
PSG	phosphosilicate glass
PSGL-1	P-selectin glycoprotein ligand-1
PTFE	polytetrafluoroethylene
PUA	polyurethane acrylate
PUR	polyurethane
PVA	polyvinyl alcohol
PVD	physical vapor deposition
PVDC	polyvinylidene chloride
PVDF	polyvinyledene fluoride
PVS	polyvinylsiloxane
PWR	plasmon-waveguide resonance
PZT	lead zirconate titanate

Q

QB	quantum box
QCM	quartz crystal microbalance
QFN	quad flat no-lead
QPD	quadrant photodiode
QWR	quantum wire

R

RBC	red blood cell
RCA	Radio Corporation of America
RF	radiofrequency
RFID	radiofrequency identification
RGD	arginine–glycine–aspartic
RH	relative humidity
RHEED	reflection high-energy electron diffraction
RICM	reflection interference contrast microscopy
RIE	reactive-ion etching
RKKY	Ruderman–Kittel–Kasuya–Yoshida
RMS	root mean square
RNA	ribonucleic acid
ROS	reactive oxygen species
RPC	reverse phase column
RPM	revolutions per minute
RSA	random sequential adsorption
RT	room temperature
RTP	rapid thermal processing

S

SAE	specific adhesion energy
SAM	scanning acoustic microscopy
SAM	self-assembled monolayer
SARS-CoV	syndrome associated coronavirus
SATI	self-assembly, transfer, and integration
SATP	(*S*-acetylthio)propionate
SAW	surface acoustic wave
SB	Schottky barrier
SCFv	single-chain fragment variable
SCM	scanning capacitance microscopy
SCPM	scanning chemical potential microscopy
SCREAM	single-crystal reactive etching and metallization
SDA	scratch drive actuator
SEcM	scanning electrochemical microscopy
SEFM	scanning electrostatic force microscopy
SEM	scanning electron microscope
SEM	scanning electron microscopy
SFA	surface forces apparatus
SFAM	scanning force acoustic microscopy
SFD	shear flow detachment
SFIL	step and flash imprint lithography
SFM	scanning force microscope
SFM	scanning force microscopy
SGS	small-gap semiconducting
SICM	scanning ion conductance microscopy
SIM	scanning ion microscope
SIP	single inline package
SKPM	scanning Kelvin probe microscopy
SL	soft lithography
SLIGA	sacrificial LIGA
SLL	sacrificial layer lithography
SLM	spatial light modulator
SMA	shape memory alloy
SMM	scanning magnetic microscopy
SNOM	scanning near field optical microscopy
SNP	single nucleotide polymorphisms
SNR	signal-to-noise ratio
SOG	spin-on-glass
SOI	silicon-on-insulator
SOIC	small outline integrated circuit
SoS	silicon-on-sapphire
SP-STM	spin-polarized STM
SPM	scanning probe microscope
SPM	scanning probe microscopy
SPR	surface plasmon resonance
sPROM	structurally programmable microfluidic system
SPS	spark plasma sintering
SRAM	static random access memory
SRC	sampling rate converter
SSIL	step-and-stamp imprint lithography
SSRM	scanning spreading resistance microscopy
STED	stimulated emission depletion
SThM	scanning thermal microscope
STM	scanning tunneling microscope
STM	scanning tunneling microscopy
STORM	statistical optical reconstruction microscopy
STP	standard temperature and pressure
STS	scanning tunneling spectroscopy
SUN	Sad1p/UNC-84
SWCNT	single-wall carbon nanotube
SWCNT	single-walled carbon nanotube
SWNT	single wall nanotube
SWNT	single-wall nanotube

T

TA	tilt angle
TASA	template-assisted self-assembly
TCM	tetracysteine motif
TCNQ	tetracyanoquinodimethane
TCP	tricresyl phosphate
TEM	transmission electron microscope
TEM	transmission electron microscopy
TESP	tapping mode etched silicon probe
TGA	thermogravimetric analysis
TI	Texas Instruments
TIRF	total internal reflection fluorescence
TIRM	total internal reflection microscopy
TLP	transmission-line pulse
TM	tapping mode
TMAH	tetramethyl ammonium hydroxide
TMR	tetramethylrhodamine
TMS	tetramethylsilane

TMS	trimethylsilyl
TNT	trinitrotoluene
TP	track pitch
TPE-FCCS	two-photon excitation fluorescence cross-correlation spectroscopy
TPI	threads per inch
TPMS	tire pressure monitoring system
TR	torsional resonance
TREC	topography and recognition
TRIM	transport of ions in matter
TSDC	thermally stimulated depolarization current
TTF	tetrathiafulvalene
TV	television

U

UAA	unnatural AA
UHV	ultrahigh vacuum
ULSI	ultralarge-scale integration
UML	unified modeling language
UNCD	ultrananocrystalline diamond
UV	ultraviolet
UVA	ultraviolet A

V

VBS	vinculin binding site
VCO	voltage-controlled oscillator
VCSEL	vertical-cavity surface-emitting laser
vdW	van der Waals
VHH	variable heavy–heavy
VLSI	very large-scale integration
VOC	volatile organic compound
VPE	vapor-phase epitaxy
VSC	vehicle stability control

X

XPS	x-ray photon spectroscopy
XRD	x-ray powder diffraction

Y

YFP	yellow fluorescent protein

Z

Z-DOL	perfluoropolyether

Biomime

Part F

Part F Biomimetics

41. Multifunctional Plant Surfaces and Smart Materials

Kerstin Koch, Bharat Bhushan, Wilhelm Barthlott

The surfaces of plants represent multifunctional interfaces between the organisms and their biotic (living) and the nonbiotic solid, liquid, and gaseous environment. The diversity of plant surface structures has evolved over several hundred million years of evolution. Evolutionary processes have led to a large variety of functional plant surfaces which exhibit, for example, superhydrophobicity, self-cleaning, superhydrophilicity, and reduction of adhesion and light reflection. The primary surface of nearly all parts of land plants is the epidermis. The outer part of epidermal cells is an extracellular membrane called the cuticle. The cuticle, with its associated waxes, is a stabilization element, has a barrier function, and is responsible for various kinds of surface structuring by cuticular folding or deposition of three-dimensional wax crystals on the cuticle. Surface properties, such as superhydrophobicity, self-cleaning, reduction of adhesion and light reflection, and absorption of harmful ultraviolet (UV) radiation, are based on the existence of three-dimensional waxes. Waxes form different morphologies, such as tubules, platelets or rodlets, by self-assembly. The ability of plant waxes to self-assemble into three-dimensional nanostructures can be used to create hierarchical roughness of various kinds of surfaces. The structures and principles which nature uses to develop functional surfaces are of special interest in biomimetics. Hierarchical structures play a key role in surface wetting and are discussed in the context of superhydrophobic and self-cleaning plants and for the development of biomimetic surfaces. Superhydrophobic biomimetic surfaces are introduced and their use for self-cleaning or development of air-retaining surfaces, for, e.g., drag reduction at surfaces moving in water, are discussed. This chapter presents an overview of plant structures, combines the structural basis of plant surfaces with their functions, and introduces existing biomimetic superhydrophobic surfaces and their fabrication.

Plant surfaces are multifunctional interfaces between the plant and the physical and biological environment. They have developed over several million years of evolution by a long-lasting game of mutation and selection, or trial and error, for the development of adaptations to their environment. These evolutionary processes led to huge structural variety and the development of multifunctional, protective interfaces.

Even in a cursory look at plants it is obvious that plant surfaces appear different. Surface structures on micro- and nanoscales are responsible for these different optical appearances. A selection of common plant surfaces and their surface structures, imaged by scanning electron microscopy (SEM), is shown in Fig. 41.1. Based on a microscopically smooth surface, the leaves of *Magnolia grandiflora* appear glossy (Fig. 41.1a,b). These and many other glossy leaves are covered by a thin layer of hydrophobic waxes, not or only rarely detectable in SEM images. Other leaves appear velvety, and under SEM they show a much rougher surface structure, built up by convex cells, with a nanostructure superimposed on them. Velvety surfaces are common on flower petals of roses, daisies, and *Dahlia*, as shown in Fig. 41.1c,d. A dense layer of air-filled hairs make the leaves of *Leucadendron argenteum* appear silvery (Fig. 41.1e,f). Hairy surfaces are common on all known life forms, as trees, shrubs, and herbaceous plants, but are absent in plants growing underwater. The leaves of *Eucalyptus macrocarpa* (Fig. 41.1g) and the surfaces of some fruits such as plums appear bluish because of a dense coverage of three-dimensional waxes, such as those shown in Fig. 41.1h. The waxes, hairs, and convex cells shown here are very common in plant surfaces but occur in different morphologies, which are discussed in detail in this chapter.

Fig. 41.1a–h Macroscopic optical appearance of plant surfaces, and their surface microstructures shown in SEM micrographs. In (a) leaves of *Magnolia grandiflora* appear glossy because of a flat surface structure, shown in figure (b). In (c) the flower petals of *Dahlia* appear velvety, because of the convex microstructure of the epidermis cells, shown in (d). In (e) the silvery appearance of *Leucadendron argenteum* leaves is caused by a dense layer of light-reflecting hairs, shown in (f). In (g) the leaf and flower bud surfaces of *Eucalyptus macrocarpa* appear white or bluish, caused by a dense covering with three-dimensional waxes, shown in (h)

The diversity of plant surface structures arises from the variability of cell shapes, micro- and nanostructures of the cell surfaces, and by the formation of multicellular structures. Based on these cellular and subcellular units a nearly unlimited combination of structures leads to a large structural biodiversity of plant surfaces [41.1–4]. The epidermis, as the outermost cell layer of the primary tissues of all leaves and several other organs of plants, plays an important role in environmental interactions and surface structuring. The simplified model presented in Fig. 41.2 shows a layered stratification of the outermost part of epidermis cells. Starting with the outside, one finds a highly functional thin outermost layer, called the cuticle. This outermost layer covers nearly all aerial tissues of land-living plants as a continuous extracellular membrane, but is absent in roots. One of the most important attributes of the cuticle is its function as a transpiration barrier, which enables plants to overcome the physical and physiological problems connected to an ambient environment, such as desiccation.

The cuticle is basically a biopolymer made of a polyester called cutin, impregnated with integrated (intracuticular) waxes. Waxes on the cuticle surface (epicuticular waxes) play an important role in surface structuring at the subcellular scale. They occur in different morphologies, show a large variability in their chemistry, and are able to self-assemble into three-dimensional crystals. Intracuticular waxes function as the main transport barrier to reduce the loss of water and small molecules such as ions from inside the cell, and also reduce the uptake of liquids and molecules from the outside [41.5]. The next layer shown in Fig. 41.2 is the pectin layer. It connects the cuticle to the much thicker underlying cellulose wall, which is built by single cellulose fibrils. Pectin is not always formed as a layer, but in some species, especially during the early ontogeny of the cuticle, a layered structure has been shown by transmission electron microscopy. Additionally polysaccharides, not shown in this schematic, are integrated into the cellulose wall. The last layer shown is the plasma membrane, which separates the living compartment of the water-containing cell from the outer, nonliving part of the epidermis.

Plant surfaces provide multifunctional interfaces with their environments. Their properties include the reflection and adsorption of solar radiation. These properties have been optimized in plants which are exposed to intense radiation, e.g., in deserts, to reduce the uptake of harmful radiation and heat energy. The most important structural units for these functions are plant waxes and hairs [41.2]. In other plants the surfaces are optimized for the reduction of adhesion, e.g., insect adhesion. Optimized surfaces with anti-adhesive properties for insects can be found in carnivorous plants, which developed slippery surfaces to catch insects. The structural basis for slippery surfaces is given by three-dimensional waxes, which reduce adhesion by developing a sufficient surface roughness. Other surfaces reduce the adhesion of insects by providing a water film on their surfaces. In plants, we find all kinds of surface wettability. Superhydrophilicity lets water completely spread on the surface. Such surfaces have been developed in plants which use their complete surface for the uptake of water and nutrients from the environment. Therefore plants developed porous or specialized hairy surfaces. On superhydrophobic surfaces wetting is hindered since water forms spherical droplets and the contact to the surface is minimized. These properties have been developed by different groups of plants to prevent water films on the surface from reducing gas exchange and to reduce the growth of pathogenic microorganisms by maintaining a dry surface. Some plants with leaves that float on water, such as *Azolla* and *Salvinia* species, use their superhydrophobicity to stay dry, even when they are submerged in water. The structural basics of the leaf surfaces of *Azolla* and *Salvinia* are large multicellular hairs, covered with three-dimensional waxes. Some plants, such as the lotus (*Nelumbo nucifera*) leaves, have superhydrophobic surfaces which are also self-cleaning. On such surfaces, the attachment and growth of pathogens is reduced, and a water droplet rolling over the surface collects pathogens and other dirt particles and thereby cleans the surface. The superhydrophobic and self-cleaning properties of plant surfaces are always correlated with the presence of three-dimensional waxes, but are common and optimized in surfaces with convex epidermal cells [41.6].

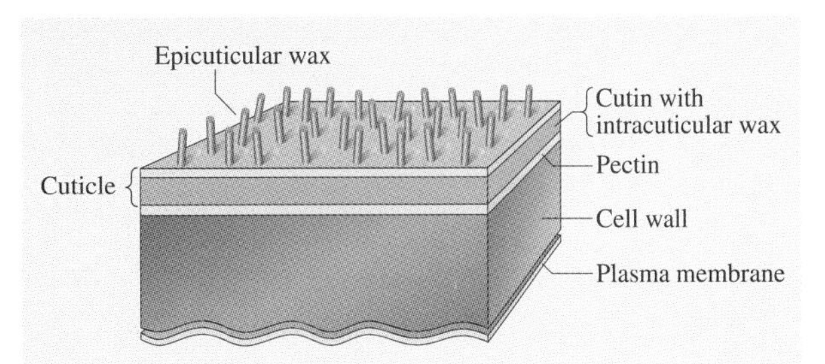

Fig. 41.2 A simplified model of the stratification of the outermost layers of plant epidermis cells. In this scheme, the outermost wax layer is shown in its commonest form, as a composite of three-dimensional waxes with an underlying wax film. Below this layer is the cuticula, made of a cutin network and integrated waxes. The cuticula is connected with the underlying cellulose wall by pectin, here simply visualized as a layer. Below the cell wall, the plasma membrane is shown. This membrane separates the water-containing part of the cells from the outermost structure-forming components of the epidermis above

Plant surfaces provide a large diversity of structures and functions which can be useful for the development of biomimetic surfaces and materials. In biomimetics (synonym *bionics*) a wide spectra of research fields combines biological innovations with technical approaches. Some successful biomimetic surfaces, such as water-repellent and self-cleaning surfaces, have already been developed.

The cuticle and its waxes are the most important materials used by plants to provide multifunctional properties, thus both are introduced in this chapter. Special attention is given to the epicuticular wax crystals, their diverse chemistry and morphology, and their ability to self-assemble into micro- and nanostructures. The origin and diversity of plant surface structures is based on the epidermis cells, and their shapes and surface structures. The morphological variations leading to various surface structures are introduced, and several examples are given. The functional properties of plant surfaces are presented and correlated with their specific surface structures. In this chapter the structural basics for superhydrophobicity and its existing and potential technical uses are discussed.

SEM micrographs presented in the chapter were taken from a database of some 250 000 SEM micrographs at the University of Bonn, which has been built up as a result of 30 years of SEM research on biological surfaces by the senior author and his collaborators.

41.1 The Architecture of Plant Surfaces

41.1.1 The Plant Cuticle: a Multifunctional Interface

Approximately 460 million years ago, when the first plants moved from their aqueous environment to the drier atmosphere on land [41.8], plants needed a protective skin to survive in the drier gaseous environment. This, for most plant surfaces, is a continuous membrane, called the cuticle. The cuticle serves as a multifunctional interface of the primary tissue of all land-living plants, which have vessels for the transport of water, e.g., ferns, conifers, and flowering plants. Roots and some secondary plant tissues, such as wood and bark, are not covered by a cuticle. The cuticle is a hydrophobic composite material, made basically of a polymer called cutin, and integrated and superimposed lipids called *waxes*, which enable plants to overcome the physical and physiological problems connected with an ambient environment, such as desiccation [41.5]. The cuticle network is formed by cutin, a polyester-like biopolymer composed of hydroxyl and hydroxyepoxy fatty acids, and sometimes also by cutan, which is built by polymethylene chains. Nonlipidic compounds of the cuticle are cellulose, pectin, phenolics, and proteins. Large differences in the chemical composition and microstructure of the waxes and the cutin network have been found by comparing different species and different developmental stages of organ ontogeny. Chemical composition, microstructure, and biosynthesis of the cuticle have been reviewed by several authors [41.9–11], and a few books summarize the intensive research on plant cuticles [41.12–15].

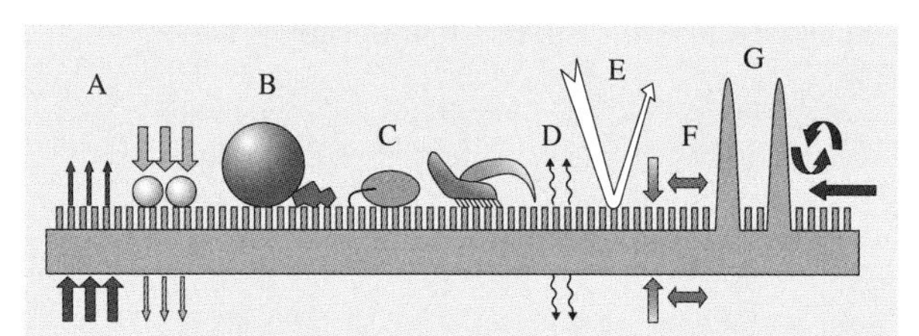

Fig. 41.3 Schematic of the most prominent functions of the plant boundary layer on a hydrophobic microstructured surface: (A) transport barrier: limitation of uncontrolled water loss/leaching from the interior and foliar uptake; (B) surface wettability; (C) antiadhesive, self-cleaning properties: reduction of contamination, pathogen attack, and attachment/locomotion of insects; (D) signaling: cues for host pathogens/insect recognition and epidermal cell development; (E) optical properties: protection against harmful radiation; (F) mechanical properties: resistance against mechanical stress and maintenance of physiological integrity, and (G) reduction of surface temperature by increasing turbulent air flow over the air boundary layer (after *Bargel* et al. [41.7])

Based on their hydrophobic chemistry and diversity of micro- and nanostructures, the cuticle provides several protective properties, which are summarized in Fig. 41.3. One of the most important properties is the function as a transpiration barrier (A in Fig. 41.3), which is based on the hydrophobic integrated and superimposed waxes. In this, the intracuticular waxes, which are located in the cutin network, form the main transport barrier to reduce the loss of water by transpiration and reduce leaching of molecules from inside the cells [41.16]. In addition to the reduction of water loss, the cuticle prevents leaching of ions from the inside of the cells to the environment and the uptake of molecules from the outside. Epi- and intracuticular waxes reduce the uptake of molecules from the environment, which might become a crucial factor in agriculture when the uptake of, e.g., nutrients or fungicides is desired. The cuticle structure, as well as chemistry, has a strong influence on the wettability (B in Fig. 41.3) and adherence of particles, pathogens, and insects (C in Fig. 41.3). The epicuticular waxes are responsible for the maintenance of wettability and self-cleaning properties [41.17, 18]. The three-dimensional waxes play an important role in surface structuring, e.g., formation of hierarchical surface, and they can reduce the adhesion of particles [41.6, 19]. These aspects are of great interest for the development of water-repellent and self-cleaning technical surfaces and will be discussed later in detail. Three-dimensional waxes not only reduce the adhesion of particles; they also reduce the adhesion of insects. This property has been developed very efficiently by some carnivorous plants to trap insects. The structural basics of waxy and other slippery plant surfaces are introduced in Sect. 41.2. The plant cuticle also plays an important role for insect and microorganism interaction, whereby some surface structures have a signaling function (D in Fig. 41.3). In some species epicuticular waxes are responsible for reflection of visible light; others are effective in adsorption of harmful UV radiation (E in Fig. 41.3). However, there are various strategies in plants to increase light reflection. Thus, this part is also introduced in detail in Sect. 41.2. Cuticles are thin membranes, with thicknesses between a few nanometers and some micrometers, but they are an important structural stabilization component for the primary epidermis tissue (F in Fig. 41.3) [41.7], with elasticity modulus in the same range as that of technical polypropylene membranes of comparable thickness [41.20, 21]. Structural components at larger scales increase the heat transfer via induction of turbulent air flow and convection (G in Fig. 41.3). Examples of several functional surface structures are introduced in more detail later.

41.1.2 Waxes – Hydrophobic Structures of Plant Surfaces

Plant waxes are sometimes visible as a white or bluish coloration of leaves and fruits, such as grapes and plums. These colorations are induced by reflection of part of the visible light spectrum by a dense coverage of three-dimensional wax structures. The fan palm *Copernicia cerifera*, the natural source of carnauba wax, has massive crusts of epicuticular wax, weighing several mg/cm^2. Carnauba wax is commercially used, e.g., for car and furniture polishes, medical products, and candy. Even when there is not a bluish coloration visible, three-dimensional waxes can be present in lower amounts, or the waxes form thin films. In plants, three-dimensional waxes are responsible for several surface functions, such as reduction of wettability and decrease of energy uptake from the environment. Waxes are an integrative part of the plant cuticle, which covers the primary tissue of all aboveground parts of lower and higher land-living plants. Waxes occur as fillers of the basic cutin network (intracuticular), but are also on top of the cuticle (epicuticular). The epicuticular waxes occur in different morphologies, which all are crystalline, and it is well known that the wax morphologies originate by self-assembly. However, waxes of different plants, and also waxes of different parts of a plant, can vary in their morphology and chemical composition. In general, plant waxes are mixtures of long-chain hydrocarbons and their derivates, and in some species they also contain cyclic compounds. Because of the strong correlation between the wax crystal morphology and their chemical composition, some waxes, such as the nonacosanol tubules of the lotus leaves, have been named after their main wax constitution. The most common wax morphologies, their chemical composition, their self-assembly, and functions are introduced here.

Chemistry and Morphology of Plant Waxes

The term *wax* is used for a variety of natural or artificial commercial products that contain fatty materials of various kinds. Well-known examples are beeswax, paraffin, and carnauba wax from wax palms (*Copernicia cerifera*) (*Arecaceae*). Plant waxes are mixtures of aliphatic hydrocarbons and their derivatives with carbon chain lengths between 20 and 40, and in the case of esters (two connected chains) about 60 atoms. Several reviews have addressed the chemical composition of plant waxes [41.11, 22, 23]. The chemical composition of plant waxes is highly variable amongst plant species, the organs of one species (e.g., different leaves), and during organ ontogeny [41.24]. The main component classes are primary and secondary alcohols, ketones, fatty acids, and aldehydes. Alkanes are very common in plant waxes, but usually occur in low concentrations. Other compounds are more rarely found in plant waxes, but in those waxes where they occur, they may be the dominating compound. The most common wax compounds and their typical chain length are listed in Table 41.1. Examples of frequently existing waxes and their major compounds are presented in Table 41.2. For some of those waxes, it has been shown that their dominating compounds crystallize in the same morphology as the complete wax mixture. Examples are the primary alcohols and the β-diketone waxes found on different parts of wheat plants. However, an increasing number of publications report the discovery of new wax components, and a long list of rare and uncommon ingredients, such as methyl-branched aliphatics, have been reported [41.25]. Environmental factors, such as temperature or light intensity, may change the amount of waxes rather than their chemical composition [41.26–28].

Many plant waxes do not match the chemical definition of true waxes. For example, triterpenoids are cyclic hydrocarbons, which occur in high concentrations in the epicuticular coatings of grapes (*Vitis vinifera*) [41.26]. Other plant waxes contain polymeric components such

Table 41.1 The commonest chemical compounds in plant waxes and their spectrum of chain length. Further examples are given in Table 41.2

Aliphatic compounds Frequently existing in waxes, but mostly as minor compounds		Chain length
Alkanes	$CH_3-(CH_2)_n-CH_3$	Odd $C_{19}-C_{37}$
Primary alcohols*	$CH_3-(CH_2)_n-CH_2-OH$	Even $C_{12}-C_{36}$
Esters	$CH_3-(CH_2)_n-CO-O-(CH_2)_m-CH_3$	Even $C_{30}-C_{60}$
Fatty acids	$CH_3-(CH_2)_n-COOH$	Even $C_{12}-C_{36}$
Aldehydes	$CH_3-(CH_2)_n-CHO$	Even $C_{14}-C_{34}$
Rarely existing in waxes, but if present, than major wax compounds		
Ketones, e.g. palmitones	$CH_3-(CH_2)_n-CO-(CH_2)_m-CH_3$	Odd $C_{25}-C_{33}$
β-Diketones	$CH_3-(CH_2)_n-CO-CH_2-CO-(CH_2)_m-CH_3$	Odd $C_{27}-C_{35}$
Secondary alcohols, e.g. nonacosan-10-ol	$CH_3-(CH_2)_n-CH_2OH-(CH_2)_m-CH_3$	Odd $C_{21}-C_{33}$
Cyclic compounds		
Flavonoids	e.g. Quercetin	
Triterpene	e.g. β-Amyrin	

* Primary alcohols are common minor constitutions in waxes, but can occur as major compounds in the wax, e.g. of grasses, eucalypts, clover and other legumes, etc. [41.11]

Table 41.2 Common wax types in plant species and their major chemical compounds. With the exception of the fruit surface of *Benincasa hispida*, data represent the waxes on the leaves of the species. All references for the chemical data are listed in *Ensikat* et al. [41.33], and examples of the wax types listed here are shown in Fig. 41.4

Wax type	Species	Dominating chemical compound(s)
Films	*Hedera helix*	Primary alcohols, aldehydes
Films	*Magnolia grandiflora*	Fatty acids $C_{24}-C_{30}$, primary alcohols $C_{24}-C_{28}$
Films	*Prunus laurocerasus*	Alkanes C_{29}, C_{31}
Crust	*Crassula ovata*	Aldehydes C_{30}, C_{32}, alkanes C_{31}
Diketone tubules	*Eucalyptus globulus*	β-diketones C_{33}
Diketone tubules	*Leymus arenarius*	β-diketone C_{31}, hydroxy-β-diketone C_{31}
Nonacosanol tubules	*Ginkgo biloba*	Secondary alcohols C_{29}
Nonacosanol tubules	*Nelumbo nucifera*	Secondary alkanediols C_{29}
Nonacosanol tubules	*Thalictrum flavum glaucum*	Secondary alcohols C_{29}
Nonacosanol tubules	*Tropaeolum majus*	Secondary alcohols C_{29}
Nonacosanol tubules	*Tulipa gesneriana*	Secondary alcohols C_{29}
Platelets	*Convallaria majalis*	Primary alcohols C_{26}, C_{28}, aldehydes
Platelets	*Euphorbia myrsinites*	Primary alcohols C_{26}, aldehydes
Platelets	*Galanthus nivalis*	Primary alcohols C_{26}
Platelets	*Iris germanica*	Primary alcohols C_{26}
Platelets	*Triticum aestivum*	Primary alcohols C_{28}
Transversaly ridged rodlets	*Aristolochia tomentosa*	Ketones
Transversaly ridged rodlets	*Gypsophila acutifolia*	Alkanes C_{31}
Transversaly ridged rodlets	*Liriodendron chinense*	Ketones
Longitudinal ridged rodlets	*Benincasa hispida*	Triterpenol acetates

as polymerized aldehydes which are only slightly soluble in chloroform [41.29, 30]. It should be noted that nearly all the existing data on the chemical composition of plant waxes is based on solvent-extracted waxes. These are mixtures of epicuticular and intracuticular waxes, which may be chemically different, as shown

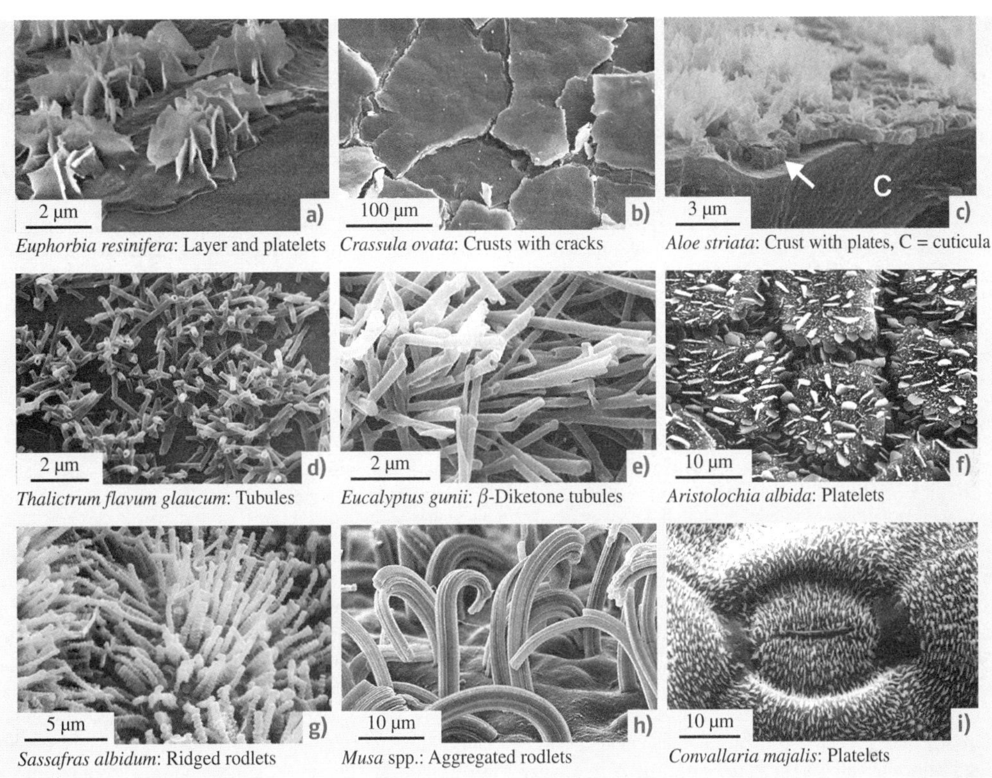

Fig. 41.4a–i SEM micrographs of epicuticular waxes. (**a**) Waxes on a leaf of *Euphorbia resinifera* have been partially removed to show the composite structure of the basal wax layer with three-dimensional wax platelets on it. (**b**) A wax crust with fissures on a leaf of *Crassula ovata*. A cross-section through the periclinal wall of *Aloe striata*. (**c**) The cuticle (indicated by C) and a wax layer (indicated by an *arrow*) with wax platelets on top. (**d**) Nonacosanol tubules on *Thalictrum flavum glaucum* leaves and (**e**) β-diketone wax tubules on *Eucalyptus gunnii* leaves. (**f**) Wax platelets on *Aristolochia albida* leaf and (**g**) transversely ridged rodlets on a leaf of *Sassafras albidum*. In (**h**) longitudinally aggregated wax threads form large aggregated rodlets on the lower side of the leaves of *Musa* spp. In *Convallaria majalis* leaves, shown in (**i**), wax platelets are arranged in a pattern similar to magnetic field lines around the stomata. Thin wax films are not visible under SEM, but are present below and between the three-dimensional waxes shown here

for the waxes of *Prunus laurocerasus* by *Jetter* and *Schäffer* [41.24] and by *Wen* et al. [41.31] for *Taxus baccata*. The development of more selective methods of wax sampling allows selective removal of the epicuticular waxes and their analysis separately from the intracuticular wax fractions [41.24, 32].

Epicuticular wax structures usually occur in the size range from 0.2 to 100 µm; thus, the appropriate microscopic techniques for investigation of their morphology are SEM and low-pressure or environmental SEM. Several SEM investigations showed that most of the epicuticular waxes form three-dimensional structures, with great variations of their morphologies. Comprehensive overviews of the terminology and micromorphology of epicuticular waxes are given by *Barthlott* et al. [41.34] and *Jeffree* [41.11]. The classification of *Barthlott* et al. [41.34] includes 23 different wax types. It is based on chemical and morphological features and also considers orientation of single crystals on the surface and the orientation of the waxes to each other (pattern formation). In this classification, the wax morphologies include thin films and several three-dimensional structures such as crusts, platelets, filaments, rods, and tubules which have a hollow center. Morphological subtypes are, for example, entire and nonentire wax platelets. A further subclassification is based on the arrangement of the crystals, e.g., whether they are randomly distributed, in clusters, in parallel orientation, or in specific arrangements around stomata, as in the *Convallaria* type. *Jeffree* [41.11] distinguishes six main morphological wax types, but suggests many more subtypes, based on chemical differences, found, e.g., for wax tubules. The most common wax morphologies are introduced in the following part and in Table 41.2.

That the three-dimensional wax crystals appear together with an underlying wax film shown in Fig. 41.4a, for the waxes of *Euphorbia resinifera* and has been reported for several species [41.34–38]. Wax films are often incorrectly referred to as an *amorphous* layer, more of a morphological description than a crystallographic one [41.33]. On several plant surfaces the wax film is limited to a few molecular layers which are hardly visible in the SEM. However, by mechanical isolation of the epicuticular waxes, e.g., freezing in glycerol, the waxes can be removed from the cuticle, and transferred onto a smooth artificial substrate for microscopic investigations [41.32]. By this method, the edges of the wax film can be detected, and the film thicknesses can be determined. Wax film formation has been investigated on a living plant surface by atomic force microscopy (AFM) [41.37]. Such investigations showed that wax films are composed of several monomolecular layers, with thicknesses up to several hundred nanometers. In the following, these relatively thin wax films ($< 0.5\,\mu m$) are called two-dimensional (2-D) waxes, and the thicker wax layers ($0.5–1\,\mu m$) and wax crusts ($> 1\,\mu m$) are called three-dimensional (3-D) waxes. Wax crusts are often found in succulent plants, as on the leaves of *Crassula ovata*, shown in Fig. 41.4b. In some species, three-dimensional waxes occur on a wax layer or wax film. Such a multilayered assembly of waxes is detectable by a cross-section through the epidermis, as shown in Fig. 41.4c for *Aloe striata*.

Three-dimensional waxes occur in different morphologies. The most common ones are tubules, platelets, rodlets, and longitudinally aggregated rodlets, as shown in Fig. 41.4d–i and introduced below. Wax tubules are hollow structures, which can be distinguished chemically as well as morphologically. The first type, called nonacosanol tubules, contains high amounts of asymmetrical secondary alcohols, predominantly nonacosan-10-ol and its homologs and to a certain degree also asymmetrical diols [41.39, 40]. These tubules can be found on most conifers, on lotus leaves, and in nearly all members of the *Papaveraceae* and *Ranunculaceae*, as shown on *Thalictrum flavum glaucum* in Fig. 41.4d [41.41]. Nonacosanol tubules are usually $0.3–1.1\,\mu m$ long and $0.1–0.2\,\mu m$ wide. The second type of tubules contains high amounts of β-diketones, such as hentriacontan-14,16-dione [41.42]. This particular kind of wax tubule is characteristic of several species of *Poaceae* but also occurs in various other groups [41.41]. Figure 41.4e shows that the β-diketone tubules are two to five times longer than the nonacosanol tubules shown in Fig. 41.4d. Their length ranges from 2 to $5\,\mu m$, and diameters vary between 0.2 and $0.3\,\mu m$. Platelets, as shown in Fig. 41.4f, are the most common wax structures, found in all major groups of plants. Following the terminology of *Barthlott* et al. [41.34], waxes are termed platelets when flat crystals are connected with their narrow side to the surface. Platelets can be further differentiated by their outline into, e.g., entire or undulated ones. In contrast, plates are polygonal crystalloids with distinct edges and are attached to the surface at varying angles. Platelets vary considerably in shape, chemical composition, and spatial pattern. For platelets, only limited information about the connection between morphology and chemical composition is available. In some species, wax platelets are dominated by high amounts of a single chemical compound, which can be primary alcohols, alkanes, aldehydes, esters, secondary alcohols or flavonoids [41.11]. The morphology of three-dimensional wax structures is not necessarily determined by the dominating chemical compound or compound class. One example of wax crystals determined by a minor component of a complex mixture are the transversely ridged rodlets, shown in Fig. 41.4g, which contain high amounts of hentriacontan-16-one (palmitone) [41.43]. Wax rodlets are massive sculptures which are irregular, polygonal, triangular or circular in cross-section. They have a distinct longitudinal axis, with a length-to-width ratio usually not exceeding 50 : 1. In addition, rodlets may have variable diameter along the length of their axis. More complex structures are the longitudinally ridged rodlets, such as those found on banana leaves (*Musa* species), shown in Fig. 41.4h. These waxes consist exclusively of aliphatic compounds, with high amounts of wax esters and less of hydrocarbons, aldehydes, primary alcohols, and fatty acids. The origin of these wax aggregates is still not clear, and so far all attempts to recrystallize these wax types have failed. As a consequence, it is assumed that their origin is connected to structural properties of the underlying plant cuticle. A very complex wax crystal morphology is known from *Brassica oleracea*, in which several cultivars form several different wax types, and where several different wax morphologies can occur on the same cell surface [41.26]. It is still an unanswered question, why do different three-dimensional wax morphologies coexist even on the surface of a single cell, and whether these different morphologies are built up by phase separation of different compounds or by the same compound.

The last example shown in Fig. 41.4i represents plant surfaces on which waxes are arranged in a specific pattern. Examples are parallel rows of longitudinally aligned platelets, which always exceed one cell (e.g., in *Convallaria majalis*), or rosettes, in which the arrangements of platelets is more or less in radially assembled clusters. In particular, the parallel orientation of platelets on the leaves of several plant species leads to the question of how the orientation is controlled by the plant. It is assumed that the cutin network functions as a template for the growth of the three-dimensional wax crystals, but there is still a lack of information about the molecular structure of the cuticle, so this question cannot be answered yet.

Certain surface wax morphologies and their orientation patterns are characteristic for certain groups of plants; thus, patterns and the morphology of plant waxes have been used in plant systematics. *Barthlott* et al. [41.41] provide an overview of the existence of the most important wax types in plants, based on SEM analysis of at least 13 000 species, representing all major groups of vascular plants.

Crystallinity of Plant Waxes

All aliphatic plant surface waxes have crystalline order. The classical definition of crystals implies a periodic structure in three dimensions, but with the increasing importance of liquid crystals and the detection of quasicrystals it has become necessary to extend the definition, so that certain less periodic and helical structures, as found for some waxes, are included [41.44].

The crystal structure of the epicuticular waxes can be examined by electron diffraction (ED), nuclear mass resonance (NMR) spectroscopy, and x-ray powder diffraction (XRD). Electron diffraction with the transmission electron microscope provides structural information for single wax crystals of less than 1 μm size, as shown in Fig. 41.5a,b for a single wax platelet. However, even with a low-intensity imaging system, the crystal structure is rapidly destroyed by the electron beam intensity. Therefore, XRD is useful to determine the crystal symmetry, and it provides information about different types of disorder. Very thin mono- or bimolecular layers of waxes, as shown in Fig. 41.5c,d, are of course not periodic in three dimensions, but form, in regard to the entire molecules, two-dimensional (2-D) crystals. Besides the planar wax structures such as films and platelets, many natural plant waxes develop irregular three-dimensional morphologies, or structures such as threads and tubules with large extension in one direction. These morphologically different waxes were found to occur in three different crystal structures. The majority of waxes exhibit an orthorhombic structure, which is the most common for pure aliphatic compounds. Tubules containing mainly secondary alcohols show diffraction reflections of a triclinic phase, with relatively large disorder, and β-diketone tubules show hexagonal structure [41.33].

Self-Assembly of Plant Waxes

That different wax morphologies on plant surfaces originate by self-assembly of the wax molecules has been shown by the recrystallization of waxes isolated from plant surfaces [41.36, 37, 40, 45–47]. In these studies, most waxes recrystallized in their original morphology, as found before on the plant surfaces.

Self-assembly processes resulting in nano- and microstructures are found in nature as well as in engineering. They are the basis for highly efficient ways

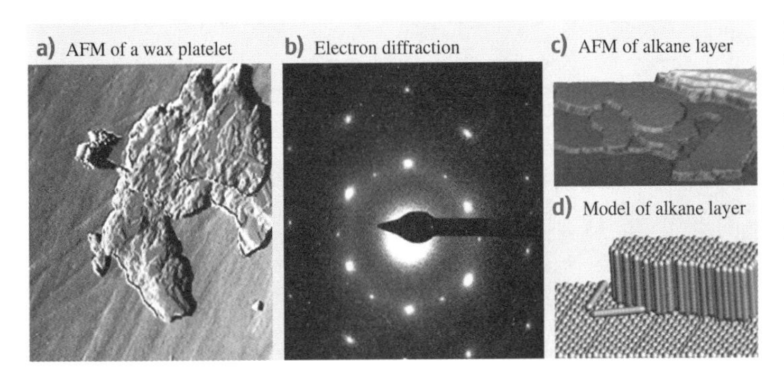

Fig. 41.5a–d The layered and crystalline structure of alkane waxes is demonstrated by an AFM map of a single wax platelet (**a**) and the corresponding electron diffraction pattern (**b**). In (**a**) the steps visible on the crystal surface are caused by perpendicular orientation of the molecules. Such steps can be monomolecular, e.g., for alkanes, or in some waxes bilayers are formed by polar molecules such as primary alcohols. The AFM map of recrystallized alkanes (**c**) and the model shown in (**d**) demonstrate the layered orthorhombic wax structure

of structuring surfaces reaching down to the molecular level. Self-assembly is a general process of structuring in which atoms, molecules, particles or other building units interact and self-organize to form well-defined structures [41.48]. The processes of self-assembly in molecular systems are determined by five characteristics: the components, interactions, reversibility, environment, and mass transport with agitation [41.49]. The most important driving forces are weak and noncovalent intermolecular interactions, such as van der Waals and Coulomb interactions, hydrophobic interactions, and hydrogen bonds. During self-assembly, their interactions start from a less-ordered state, e.g., dissolved waxes in a solution, to a final more-ordered state, a crystal [41.50, 51]. Environmental factors such as temperature, solvent, and substrate might influence the self-assembly process, and in the case of waxes, their morphology.

The most suitable microscopy technique for studying the self-assembly process of waxes under environmental conditions is AFM because it combines sufficient resolving power to image nanostructures with the ability to work at standard temperature and pressure (STP) with living plant material. Self-assembly of waxes has been studied directly on plant surfaces, but also by recrystallization of waxes and single wax compounds on artificial surfaces. However, AFM is not suitable for all plant surfaces. Within a leaf surface large structures such as hairs with dimensions of several tens of micrometers can emerge out of the epidermis and pose a barrier against the surface scanning probe. Additionally, high-aspect-ratio structures caused by cell surface structures might cause artifacts in the resulting images. Species with smooth or slightly convex cell surface sculptures are most appropriate for AFM investigations. The process of wax regeneration occurs over several hours; thus the loss of water from inside the plant has to be minimized to reduce specimen drift by material shrinkage during investigation. This precondition limits the range of specimens for AFMs with a small specimen chamber, because the sizes and shapes of the leaves must allow them to be mounted in the AFM without cutting them. An experimental setup where the complete plant is placed close to the AFM and a leaf is fixed on the AFM specimen holder is shown in Fig. 41.6. In this, the leaf was fixed at its lower side to the specimen holder with a drop of a two-compound glue, and waxes on the upper leaf side were removed by embedding them into a drop of water-based glue. After hardening, the glue and the embedded waxes can be removed from the leaf surface and the process of wax regeneration can be studied. Temperature increase in long-term investigations, caused by the laser beam on top of the cantilever, induces expansion of the water-containing leaf, and therewith a drift of the specimen. To minimize this, reflective cantilevers must be used, and the laser beam intensity should be reduced by integrating an attenuation filter above the cantilever [41.37]. However, the waxes themselves are fragile; thus appropriate scan conditions at scan sizes of 3–20 μm are tapping mode and scan rates of 0.7–2 Hz, encompassing 256 lines per image and a set point near the upper limit to minimize the interaction between tip and sample. Figure 41.7 shows the regeneration of a wax film on a leaf of snowdrop (*Galanthus nivalis*) by formation of a multilayered wax film and the growth of three-dimensional wax platelets. This and further investigation showed that the growth of the three-dimensional wax crystals occurs by apical accumulation of new wax molecules on only one side of the crystal. The regeneration of the wax film results in a multilayered crystalline coverage on the plant cuticle. The time needed to regenerate removed waxes shows large variations in different species and some species do not even regenerate removed waxes. In these plants, wax synthesis seems to be inactive when leaves are mature [41.52].

Self-assembly of plant waxes can alternatively be studied by recrystallization of the waxes on smooth

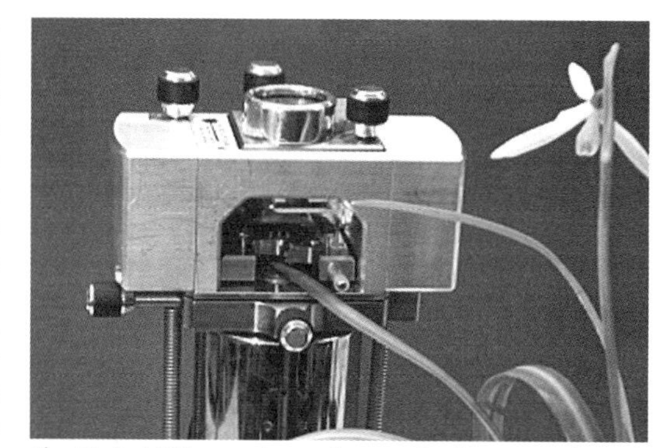

Fig. 41.6 AFM experimental setup for long-term investigations of wax crystallization on a living plant surface. The tip of the leaf of *Galanthus nivalis* has been fixed onto the specimen holder with a drop of two-compound glue. Existing waxes have been removed and the rebuilding (self-healing) of the wax can be studied over several hours. Appropriate scan conditions for living plant surfaces are given in the text and the method of wax removal is described in detail by *Koch* et al. [41.37]

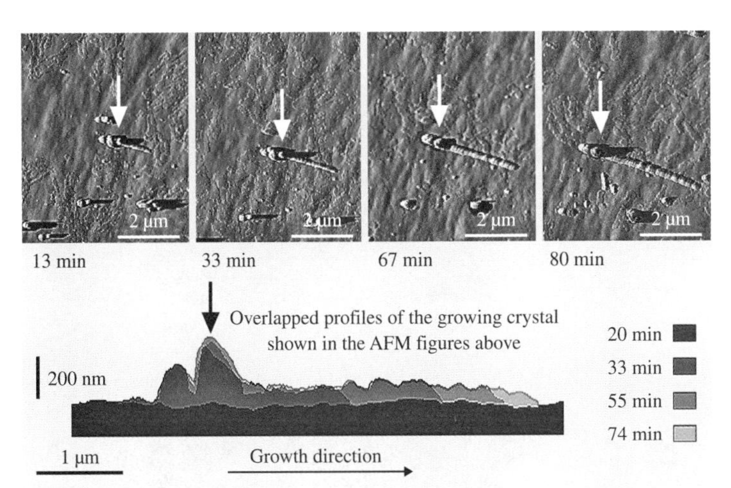

Fig. 41.7 AFM maps and a series of profile lines, taken from repeated scans during crystal growth on a leaf of snow belt (*Galanthus nivalis*). The first AFM map represents wax regeneration within 13 min; the last map was taken 80 min after wax removal. The *white arrows* mark the same position of the crystal as the *black arrow* marks in the profile figure. In the lower figure, the outlines of the growing crystal have been overlapped to demonstrate that the extension is occurring at the distal end of the growing crystal and that at this time the growth in height is limited to a few nanometers. Outlines were taken from four AFM scans, 20, 33, 55, and 74 min after the wax regeneration process started. The experimental setup is shown in Fig. 41.6

artificial substrates. Based on those studies, the formation of wax tubules and platelets has been described in detail. Wax platelets, characteristic of wheat leaves (*Triticum aestivum*), are constructed from the primary alcohol octacosan-1-ol [41.46]. Crystallization of the wax mixture isolated from the leaves and of pure octacosan-1-ol on different artificial substrates showed substrate-dependent growth. On a nonpolar, crystalline substrate (highly ordered pyrolytic graphite, HOPG) platelets grow with a vertical orientation to the substrates, whereas on a polar surface, such as mica, crystals grow horizontally to the substrate surface. On amorphous polar glass only amorphous wax layers grow. This substrate dependence demonstrates epitaxial control of crystal growth depending on the orientation and order of the first layers of molecules adhering on the substrate surface. Octacosan-1-ol forms ordered bilayer structures on the substrate. In these, the first layers of molecules lie flat on nonpolar substrates, but stand upright (perpendicular) on crystalline polar surfaces. The grown platelet morphology results from an anisotropic crystal growth, caused by faster parallel assembly of the molecules at the length side of already existing molecules than at the ends of the molecules [41.53]. AFM micrographs in Fig. 41.8 and schematics of the molecule orientation demonstrate the differences of growth on polar and nonpolar substrates for octacosan-1-ol molecules. In both cases, flat crystals with different orientation grow. Crystals grown horizontal to the substrate surface are called plates (Fig. 41.4a,b), whereas those grown perpendicular to the substrate surface are termed platelets (Fig. 41.4c,d). The substrates on which crystals grow influence the crystal morphology and their orientation. This fact can be used to create different kinds of nano- and micropatterns on technical surfaces [41.46, 47, 54, 55]. In summary, substrates can have a direct influence on the self-assembly processes of wax crystals, and can function as a template on the molecular level. In such a case, the substrate organizes the

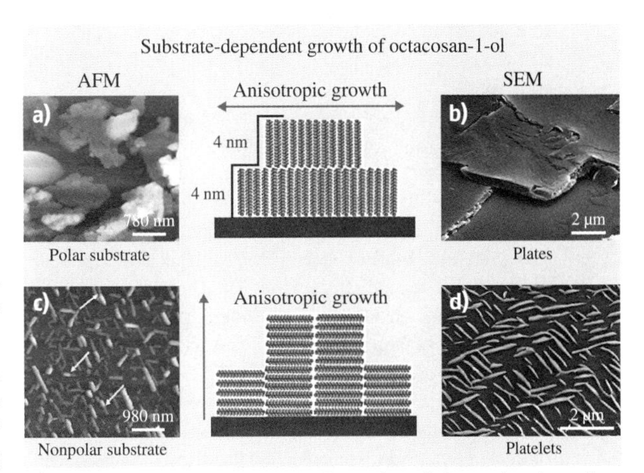

Fig. 41.8a–d AFM maps and schematics of the molecular orientation demonstrate the differences between growth on polar and nonpolar substrates for octacosan-1-ol molecules. AFM figures (a,c) show growing crystals, whereas the SEM figures (b,d) show the final crystal morphology. On both substrates flat crystals with different orientations were grown. Crystals grown horizontal to the substrate surface (a,b), whereas those grown perpendicular to the substrate surface (c,d) are termed platelets. The principle of anisotropic crystal growth is shown schematically for both preferred growth directions and is described in the text

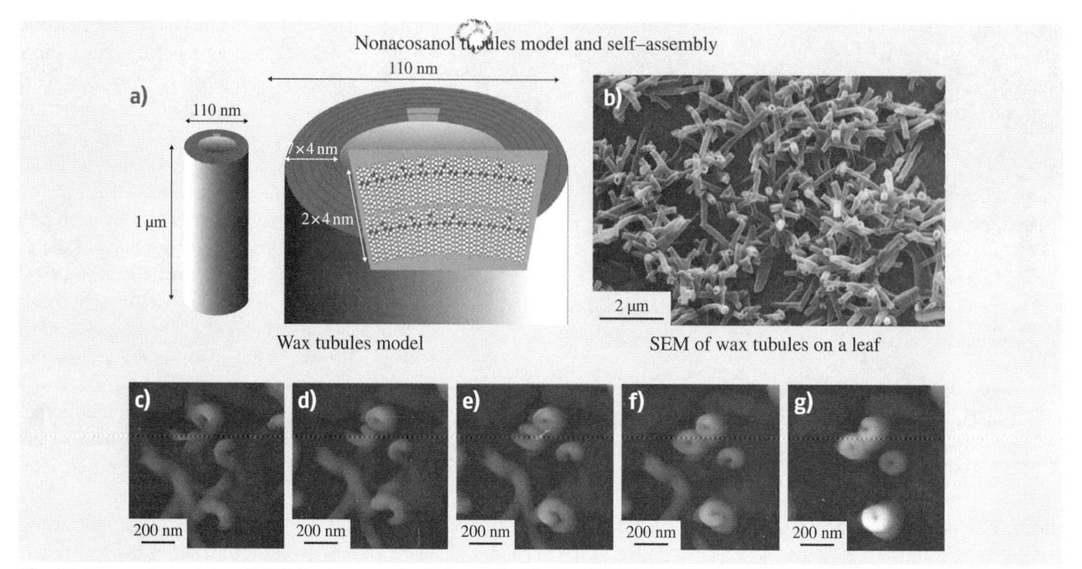

Fig. 41.9a–g A model and SEM micrograph of the molecular order of nonacosanol tubules, and AFM analysis of their self-assembly. Based on SEM characterization, chemical analysis, single-compound crystallization, and crystallographic data, a model of nonacosan-10-ol tubules has been developed (**a**). Original nonacosan-10-ol tubules are shown in the SEM micrograph (**b**) for *Thalictrum flavum glaucum* leaves. Consecutive AFM figures of tubule formation (nonacosan-10-ol wax from *Tropaeolum majus*) were made after applying a wax solution on HOPG. After 65 min (**c**) the waxes mainly formed curved rodlets, which were arranged horizontally to the substrate. The same area of the HOPG substrate shows that waxes start to form (**d–f**) circles and, after 223 min (**g**), the initially observed rodlets were dissolved and short tubules were formed

assembly of the molecules in a specific spatial arrangement [41.56,57]. Such a template effect was reported for wax platelets formed by primary alcohols [41.46]. On HOPG substrate, the spatial pattern of the reassembled wax platelets strictly followed the hexagonal symmetry of the crystalline substrate. However, the cutin matrix of the cuticle, which acts as a substrate in plant surfaces, is supposed to be amorphous, and epitaxial growth on an amorphous substrate seems unusual.

Wax crystals which are composed of more than one compound are, for example, the transversely ridged rodlets. These waxes can be recrystallized from the total wax mixture, but not from individual compounds such as alkanes or palmitones. For these waxes, it is assumed that their morphology is also formed by a self-assembly-based crystallization process, but the presence of minor amounts of other compounds is required as an additive for crystal growth [41.43].

The origin of wax tubules, shown schematically in Fig. 41.9a and in SEM micrographs in Fig. 41.9b, has been debated for a long time. Several observations, such as spiral lines on the surfaces of some nonacosanol tubules [41.45], led to the assumption that tubules arise from a twisting or folding of a platelet-like precursor form. Recrystallization experiments with nonacosanol waxes showed that these tubules grow perpendicular to the substrate surface when recrystallized on HOPG. This vertical orientation of the tubules allows detailed study of the growth process by AFM and showed that the building of nonacosan-10-ol tubules from lotus (*Nelumbo nucifera*) and nasturtium (*Tropaeolum majus*) leaves is based on a continuous growth of a small circular precursor structure by supplementation of the wax on top of it [41.47]. The micrographs in Fig. 41.9c–g show consecutive AFM images of growing tubules, made during the tubule formation process. The terminal ends of growing tubules are asymmetric in height. This asymmetry seems to be caused by an accumulation of new wax molecules at edges found at the terminal end of the tubules and indicates a helical growth mechanism for the tubules. The pure nonacosan-10-ol alcohol, the dominating compound of wax tubules, can crystallize in different forms [41.36, 40,45]. *Jetter* and *Riederer* [41.40] showed that a range of alkanediols, present in the waxes of many secondary alcohol tube-forming species, also have tube-forming capability.

Chemical analysis of the leaf waxes of lotus and nasturtium *(Tropaeolum majus)* showed that waxes of both species are composed of a mixture of aliphatic compounds, with nonacosan-10-ol (a secondary alcohol) and nonacosandiols (an C_{29} alkane with two alcohol groups) as their main components [41.47]. These compounds have been separated from the rest of the wax compounds and used for recrystallization experiments. It could be shown with mixtures of nonacosan-10-ol and nonacosandiols components that a minimum amount of 2% of nonacosandiols supports tubule formation [41.58].

Analysis of wax chemistry, crystalline order, and their self-assembly has led to better understanding of the molecular architecture of three-dimensional waxes. Based on these data, a model of nonacosan-10-ol tubule structure has been developed, as shown in Fig. 41.9a. Here it is assumed that the lateral oxygen atoms at the side of the straight molecules hinder the formation of the normal, densely packed, orthorhombic structure and require additional space, causing local disorder between the molecules and thus spiral growth, leading to the tubule form.

41.1.3 From Single Cell to Multicellular Surface Structuring

Millions of years of plant evolution have resulted in a variety of plant functional surface structures. Optimized structures can be found in different environments, for example, the water-repellent and self-cleaning leaves of lotus (*Nelumbo nucifera*), which are specialized surfaces for pathogen defense in humid environments. In plants, surface structuring arises at different hierarchical levels. The epidermal cells create sculpturing of the surfaces on the microscale, but on a smaller scale the surfaces of individual cells are structured as well. In the following, the origin of surface sculpturing in plants is introduced, and the basic terminology for their description is given. The first part gives an overview about plant surface structures which are formed by a single cell, including the shape and sculpture of the cell and the structures of the cell surface. The second part includes multicellular sculptures. For a broader understanding, we try to avoid the use of many specialist terms here and use more common names, and scientific designations are placed in brackets.

Cellular Surface Structures in Plants

The Outlines of Cells. The description of plant micro- and nanostructures requires the use of some basic uniform terms, for example, to describe the outline of a single epidermis cell. Several variations are known and introduced in detail by *Barthlott* and *Ehler* [41.59], *Barthlott* [41.60], and *Koch* et al. [41.1–4]. In the following, a brief introduction is given.

The boundary walls between two adjacent epidermal cells are called anticlinal walls, whereas the outer wall forming the cell surface is called the periclinal wall. The primary sculpture of a single cell encompasses the outline, including the shape and relief of the anticlines and curvature of the outer periclinal wall. There are two basic forms of cells, the tetragonal and polygonal form, both of which can have a uniform length of their sides or be elongated. Additionally, the course of the anticlinal walls can be straight or un-

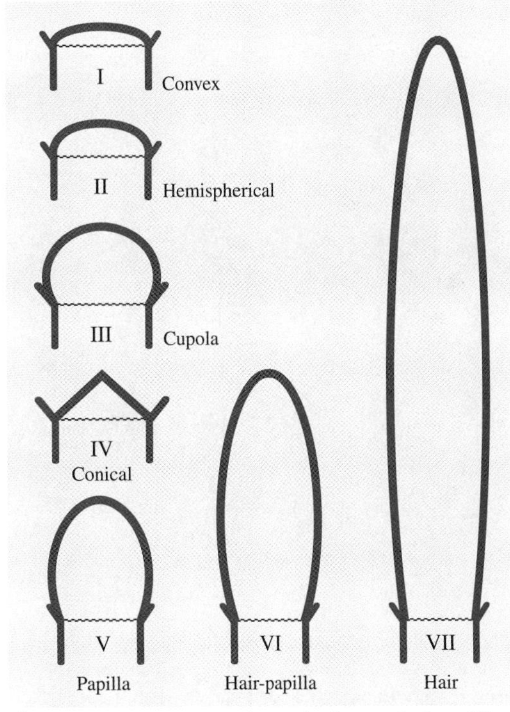

Fig. 41.10 Schematics and terminology of the different convex cell outlines and their aspect ratios ($\beta =$ width/height): (I) convex ($\beta \geq 3/1$), (II) hemispherical ($\beta \approx 2/1$), (III) cupola ($\beta < 3/2$), (IV) conical ($\beta > 3/2$), (V) papilla ($\beta < 3/2$ and $> 1/2$), (VI) hair-papilla ($\beta < 1/3$ and $> 1/6$), and (VII) hair ($\beta < 1/7$)

even. It is assumed that the outline of anticlines has an influence on the mechanical stability of the epidermis tissue, but experimental evidence for this hypothesis is not available. The cell sculptures or curvature of the outer epidermis wall (periclinal wall) can be tabular (flat), convex (arced to the outside) or concave (arced to the inside), and has a great influence on the surface roughness on the micrometer scale. Additionally, only the central area of a cell can form a convex outgrowth and form a papilla or hair-like structure. The convex cell type is the most common cell type of epidermal surfaces, often found on flower-leaves, stems, and leaves [41.61]. These cell morphologies originate by expansion of the outer side (periclinal wall) of the epidermis cells. They can be divided into several subtypes, depending on the outline of the epidermis cells and their aspect ratio (width to height), which determines their designation. In Fig. 41.10 a schematic of different convex cell outlines and their designations is given. The terminology is based on the cell outline and aspect ratios (β = width/height) of the cells and includes: convex ($\beta \geq 3/1$), hemispherical ($\beta \approx 2/1$), cupola ($\beta < 3/2$),

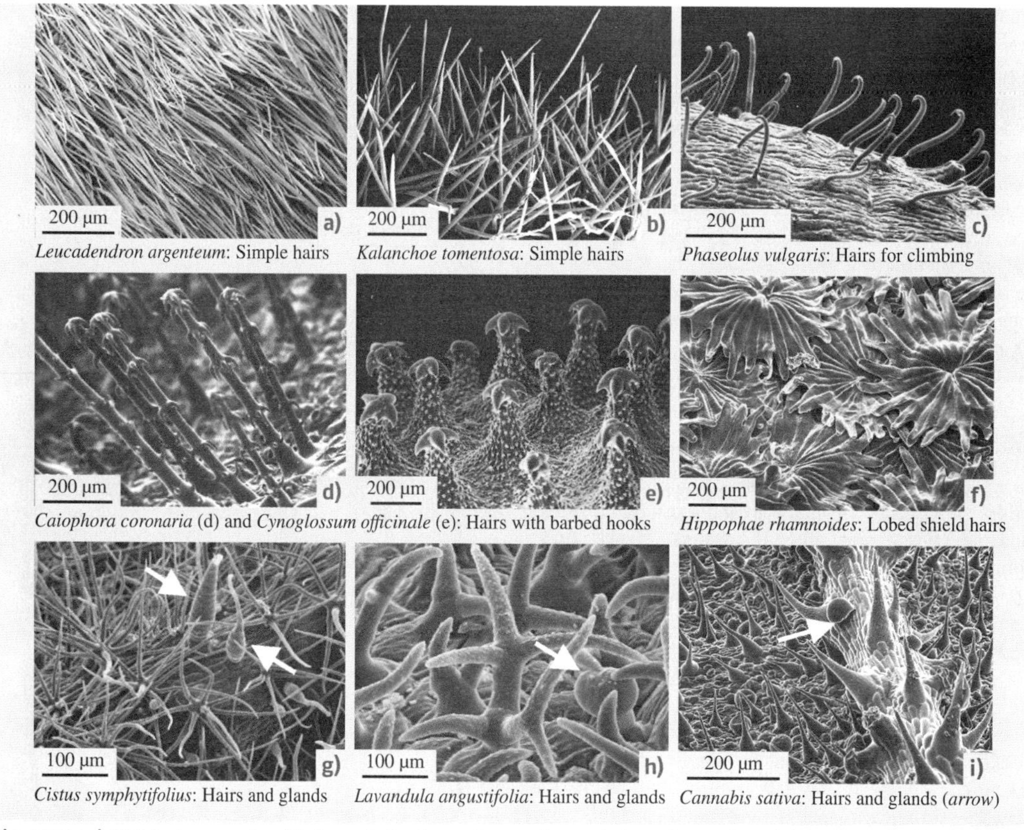

Fig. 41.11a–i SEM micrographs of hairs and glands on plant surfaces. A dense layer of straight, unbranched hairs, almost orientated parallel to the leaf surface on *Leucadendron argenteum* leaf (**a**). The unbranched hairs of *Kalanchoe tomentosa* shown in (**b**) are orientated upright. The shoot surface of the climbing bean plant *Phaseolus vulgaris* with terminal hooks is shown in (**c**). Single hairs on a leaf of *Caiophora coronaria* (**d**) and those on the seed surface of *Cynoglossum officinale* (**e**) are characterized by terminal and lateral barbed hooks. The peltate hairs of *Hippophae rhamnoides* (**f**) form lobed shields. Simple branched star-like hairs and two morphologically different glands (*arrows*) on the leaf of *Cistus symphytifolius* are shown in (**g**). Multiple ramified hairs and short-stalked glands (*arrow*) on a leaf of *Lavandula angustifolia* are shown in (**h**). In (**i**) an exposed leaf vessel with hair-papilla and unstalked glands (*arrow*) of *Cannabis sativa* is shown. (**a–h**) were obtained from upper (adaxial) leaf sides and (**i**) from the lower (abaxial) leaf side

conical ($\beta > 3/2$), papilla ($\beta < 3/2$ and $> 1/2$), hair-papilla ($\beta < 1/3$ and $> 1/6$), and hairs ($\beta < 1/7$). In these, hairs are built by the outgrowth of a single surface cell. Hairs are often named trichome (Greek: *trichoma*).

The leaf surfaces of *Leucadendron argenteum* and *Kalanchoe tomentosa*, shown in the SEM micrographs in Fig. 41.11, are two representative surfaces with hairs. Hairs can decrease, but also increase, the loss of water and influence the wettability of the surfaces [41.62]. The wide spectrum of functions of plant hairs has been reviewed by *Wagner* et al. [41.63], and more recently by *Martin* and *Glover* [41.61]. With respect to their functions, it is important to notice that hairs can be glandular or nonglandular (nonsecreting), dead or living, and hairs can also be built up by several cells (multicellular), which are introduced later. Unicellular trichomes can be found on the aerial surfaces of most flower-plants (angiosperms), some conifers (gymnosperms), and on some mosses (bryophytes) [41.63]. Many plants of dry habitats show a dense cover of dead, air-filled hairs to reflect visible light, which makes the surfaces appear white. Two examples are the South-African protea tree *Leucadendron argenteum* (Fig. 41.11a) and *Kalanchoe tomentosa* (Fig. 41.11b). The structures of hairs are often more complex; thus, the definition based on the aspect ratio fits well only for simple, undivided hairs. On kidney shoots (*Phaseolus vulgaris*), hairs form hooks to get better adhesion for climbing (Fig. 41.11c) and in *Caiophora coronaria* (Fig. 41.11d) and *Cynoglossum officinale* (Fig. 41.11e) the hairs have lateral barbed hooks. The hairs of *Hippophae rhamnoides* form lobed shields (Fig. 41.11f). Further trichomes are the simple or double-branched hairs and secretion glands on the leaves of *Cistus symphytifolius*, *Lavandula angustifolia*, and *Cannabis sativa* (Fig. 41.11g–i). These complex hair structures require a more differentiated description than the aspect ratio used for simple hairs [41.64, 65]. The sizes and morphologies of trichomes are often species specific, which makes some trichomes useful as morphological features in plant systematics [41.65].

A different surface structure is formed by concave cells. Cell shrinking induced by water loss from the cells might lead to concave cell morphology, thus concave cells are seldom found on water-containing epidermal cells, but occur often on dry seed surfaces (testa cells) [41.66]. They are characterized by complete or particular deflection of the outer epidermis wall. The concave cell type is characteristic for small, wind-dispersed seeds, as for *Aeginetia indica* and *Triphora trianthophora* (Fig. 41.12).

The Structures of Cell Surfaces. Surface structuring is also related to the fine structures of cell surfaces. In plants, four frequently found morphological modifications of the outermost epidermis cell layers are known. These modifications are shown schematically in Fig. 41.13. In the first case, shown in Fig. 41.13a, the surface structure is induced by concavities of the cell wall which lead to coves and folding of the surface. The second kind of structuring originates by subcuticular inserts of mineral crystals, such as silicon oxides (Fig. 41.13b). The third kind of surface structuring results from folding of the cuticle itself (Fig. 41.13c). Additionally, on many plants, waxes on top of the cuticle lead to surface structuring, as shown in Fig. 41.13d.

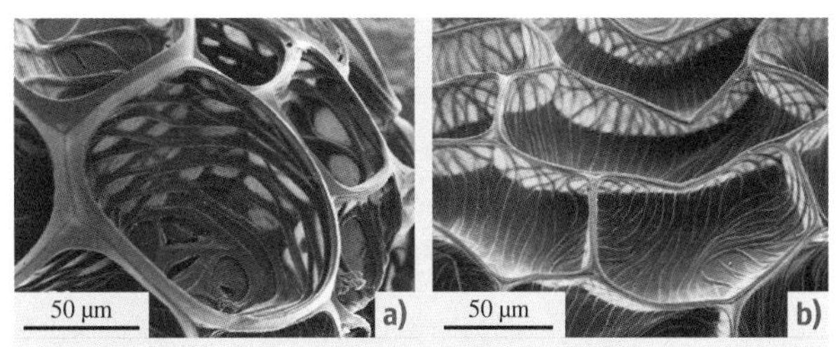

Fig. 41.12a,b SEM micrographs of two seed surfaces with concave cell sculpturing. Seeds of both species (**a**) *Aeginetia indica* and (**b**) *Triphora triantophora* are examples of lightweight constructions, optimized for seed dispersal by wind. The concave structure of the dead cells can be interpreted as due to shrinkage deformation during seed maturation and death. The bands which form an inner network in (**a**) and the surface pattern in (**b**) are build by cellulose

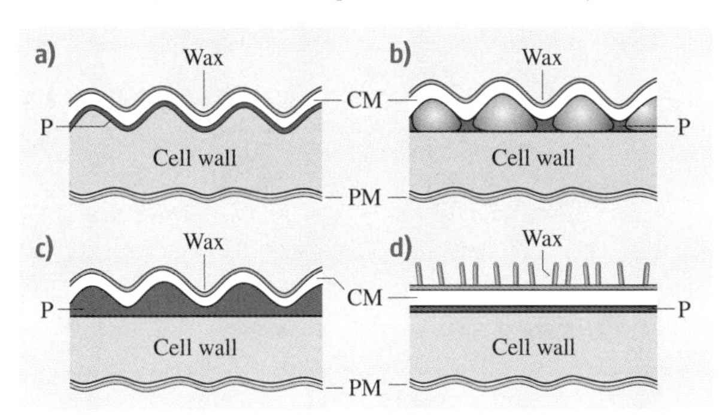

Fig. 41.13a–d Schematic cross-sections through plant epidermis cells showing different sources leading to microstructuring of cell surfaces. In (**a**) the surface profile is induced by coves of the underlying cell wall, in (**b**) by insertion of subcuticular minerals, in (**c**) by folding of the cuticle, and in (**d**) by waxes located on top of the cuticle (epicuticular wax). Wax = epicuticular waxes, CM = cuticular membrane, P = pectin, PM = plasma membrane (after [41.59])

Waxes and their structural diversity have already been introduced; thus in the following, cuticular folds and subcuticular inserts are introduced.

Cuticular patterns have been described for nearly all aboveground surfaces of plants, but are very frequently found in the leaves of flowers (petals), and on seed surfaces. They occur as folding or tubercular (verrucate) patterns, which originate due to the cuticle itself, by the expression of the bulk of the cell wall below, or by subcuticular deposits. The pattern of cuticular folds can be categorized according to the thickness (width) of the folds, distances between the folds, and by their orientation [41.59]. Additionally, the pattern of folding within a single cell can be different in the central (inner area) and anticlinal field (outer area) of a cell. Figure 41.14 shows different patterns of cuticle folding. On the leaves of *Schismatoglottis neoguinensis* (Fig. 41.14a,b) the folding is orderless and covers the central and anticlinal field of the cells. On the lower leaf side (adaxial) of *Alloca-*

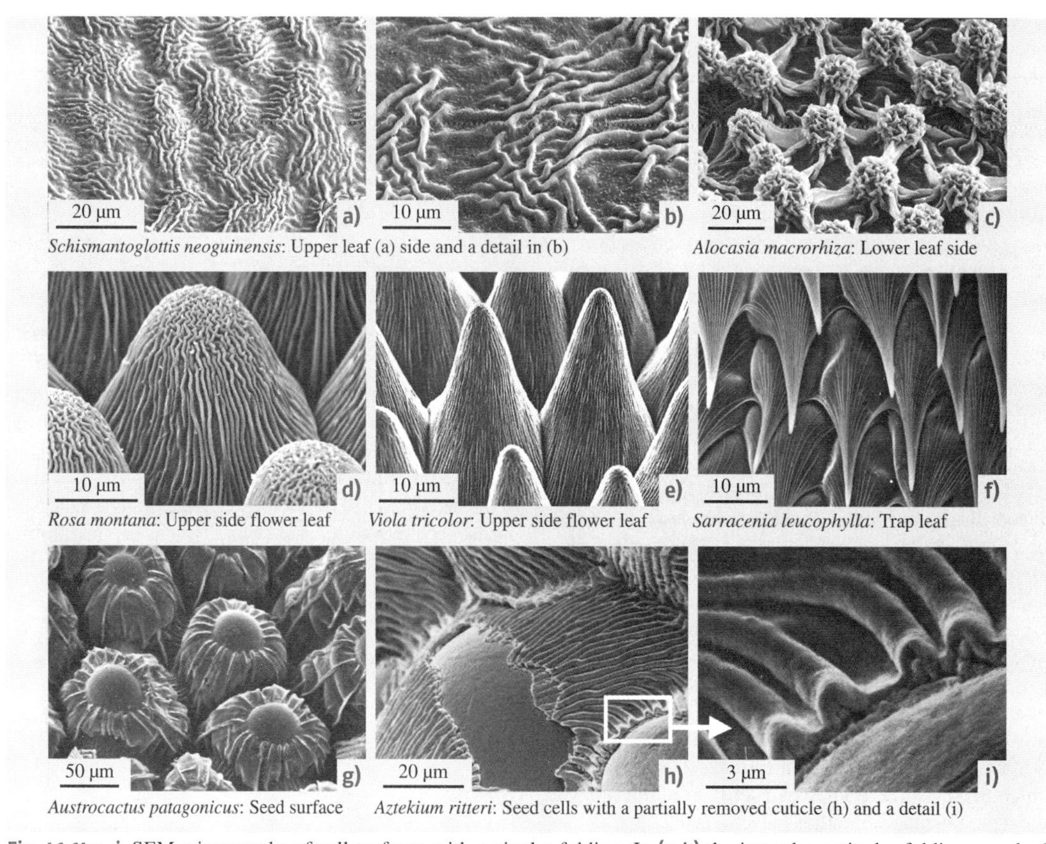

Fig. 41.14a–i SEM micrographs of cell surfaces with cuticular folding. In **(a,b)** the irregular cuticular folding on a leaf of *Schismatoglottis neoguinensis* is restricted to the central field of the cells. In **(c)** *Alocasia macrorhiza*, the cells are flat (tabular), with exposed node-like foldings in the central fields of the cells. The cells of a flower petal of *Rosa montana* with a rippled folded cuticle in the central field of the cells and parallel folding, running to the anticlinal walls of the cells, is shown in **(d)**. In **(e)** conical cells of the flower petals of *Viola tricolor* with parallel folding are shown. Cells of the inner side of a tube-like leaf of the carnivorous plant *Sarracenia leucophylla* are shown in **(f)**. These cells have a conical hair-papilla in the downward direction with parallel cuticle folding, with larger distances of the folds at the base and denser arrangement at the cell tip. In **(g–i)** seed surfaces are shown. In *Austrocactus patagonicus* **(g)** the central field of the cells is unstructured, whereas a rough folding exists in the anticlinal fields. In *Aztekium ritteri* **(h,i)** a part of the cuticle has been removed to show the cuticle folding

sia macrorhiza, shown in Fig. 41.14c, the cuticle forms node-like folding in the central part of each cell. The flower petals of *Rosa montana*, shown in Fig. 41.14d, have convex cells with a small central field with a rippled folded cuticle and parallel folds in the anticlinal field. The papilla cells of the flower petals of *Viola tricolor*, shown in Fig. 41.14e, have a parallel folding from the center to the anticlines of the cells. The cells inside the trap of the carnivorous plant *Sarracenia leucophylla*, shown in Fig. 41.14f, are hair-papillae, with a conical shape curved in a downward direction. On these, a parallel cuticle folding exists with larger distance at the base and a denser arrangement at the cell tip. The seed surface of *Austrocactus patagonicus*, shown in Fig. 41.14g, has cupolar-formed cells with unstructured central fields and broad parallel folds in the anticlinal fields. A high-magnification SEM micrograph of the seed surface of *Aztekium ritteri*, shown in Fig. 41.14h,i, shows a partially removed cuticle and demonstrates that the origin of surface folding is caused by the cuticle itself.

The functional aspects of cuticle folding are rarely investigated, but their frequent occurrence on flower petals has led to the assumption that cuticular folding forms a favorable structure for insect pollinators for walking on them. *Kevan* and *Lanet* [41.67] gave evidence that cuticular folding in flower petals is a tactile cue for bees to find the nectar source of the flowers. *Maheshwari* [41.68] reported that cuticular folding is a signal for the germination of fungi. Evidence for this hypothesis is based on in vitro investigations with structured biomimetic surfaces, microfabricated by electron-beam lithography [41.69].

Barthlott and *Ehler* [41.59] observed that cuticular folding and their characteristic pattern already exist in very young organs, and that only the amount of folding per area might increase during parthenogenesis of the organs. However, the formation and function of these complex structures are not yet completely understood.

Some microstructures on epidermis cells arise from subcuticular inserts of mineral crystals, as indicated in Fig. 41.13b. These subcuticular inserts can be solid crystals of silicon dioxide, as shown in Fig. 41.15 for the leaves of rice (*Oryza sativa*) and for tin plant or horse tail (*Equisetum*) plants. Calcium oxalate crystals are also frequently found in plants, and the presence of silicon or calcium can be verified simply by energy-dispersive x-ray (EDX) analysis, equipped in an SEM. Silicon (Si) is a bioactive element associated with beneficial effects on mechanical and physiological properties of plants. It is a common element found in plants and occurs as monosilic acid or in the polymerized form as phytoliths ($SiO_2 \cdot nH_2O$) [41.70]. In plants, Si tends to crystallize in the form of silica in cell walls, cell lumen, at intercellular spaces, and in the subcuticular layer [41.71]. Silica increases the resistance of plants to pathogenic fungi. This protective function seems to be based on a genetically controlled defense mechanism and cannot be interpreted as a function of increased mechanical stability [41.72].

Calcium oxalate crystals have been reported for more than 250 plant families of the gymnosperms and angiosperms [41.73]. Calcium oxalate crystals occur in five different morphological variations, such as raphides (needles), druses (spherical aggregates) or prisms. With some exceptions they are located within the plant body (in roots, stems, leaves, seeds). Their protective function against herbivorous animals has been discussed for a long time, but this function is still almost a question of the relation of sizes of both the herbivorous animal and the crystals [41.74].

The most common origin of cellular structuring is due to epicuticular waxes. Because of their importance as multifunctional interfaces, the morphological diversity of epicuticular waxes, their chemical diversity, and

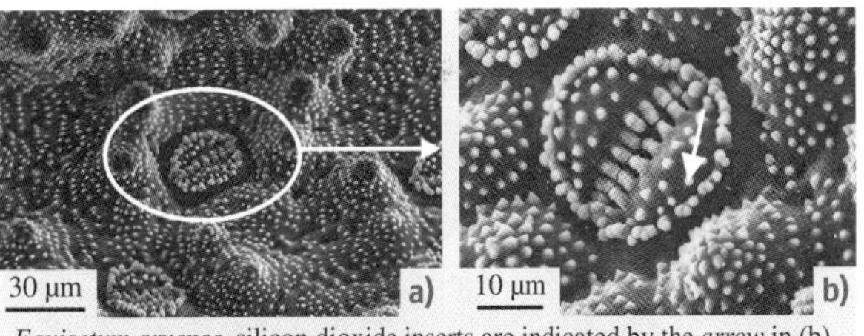

Equisetum arvense, silicon dioxide inserts are indicated by the *arrow* in (b)

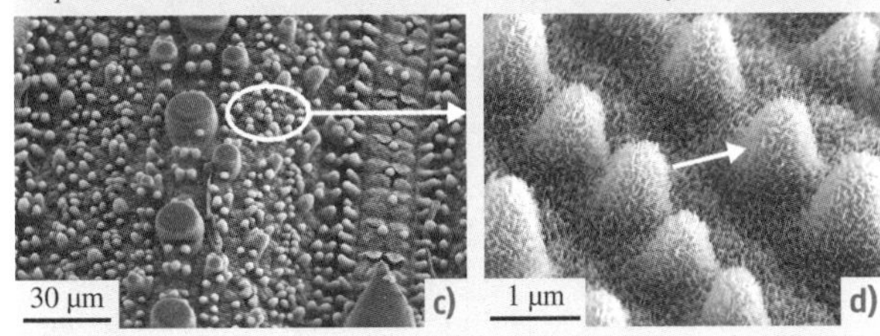

Oryza sativa, silicon dioxide inserts are indicated by the *arrow* in (d)

Fig. 41.15a–d Cell surface structuring by subcuticular silicon dioxide insertions. SEM micrographs of the common horsetail (*Equisetum arvense*) are shown in **(a,b)**. **(b)** Provides a detail of **(a)** and shows that the stomata and their surrounding cells have a micropattern of small enhanced spots formed by subcuticular inserts of silicon oxide crystals. In **(c,d)** the surface of *Oryza sativa* is shown. Here the convex sculpturing is also caused by subcuticular silicon oxide crystals

their origin by self-assembly is extensively introduced in Sect. 41.1.2.

Multicellular Surface Structures in Plants

The previous examples have shown that the large structural diversity in plants is caused by the different morphologies of the cells and their surface structures. The dimensions of these structures range from a few nanometers, e.g., wax films, to several micrometers in cells. However, on several plant surfaces, larger structures are multicellular; what that means is that more than a single cell is needed to form such structures. Several properties are correlated to multifunctional surface structures, and the most common are secreting glands and nonsecreting multicellular trichomes, which are introduced in the following.

Glands or glandular trichomes can be found on ≈ 30% of all vascular plants [41.75]. Multicellular glands include salt glands, nectaries, and the adhesive-secreting glands of some carnivorous plants [41.63]. Secretion and accumulation of toxic compounds at the plant surface allows direct contact with insects, pathogens, and herbivores, and might therefore be an effective defense strategy [41.63]. The exudates of glands are, for example, terpenoids, nicotine, alkaloids or flavonoids. The exudates of some ferns and angiosperms, in particular several members of the *Primulaceae*, are composed of flavonoids [41.76, 77]. These flavonoid exudates or *farina* are morphologically similar to waxes, but are chemically distinct from plant waxes. Other glandular trichomes, such as the glands of the carnivorous plants of the genus *Drosera* (sundew) and *Pinguicula* (butterworts), secrete adhesives and enzymes to trap and digest small insects such as mosquitoes and fruit flies.

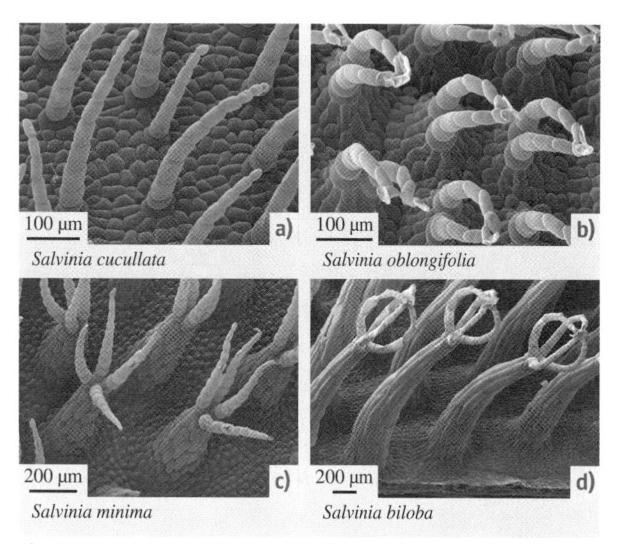

Fig. 41.16a–d SEM micrographs of four different multicellular hair types in the genus *Salvinia*. The *cucullata* type (**a**) is formed by solitary trichomes of up to 800 µm length and 50 µm in diameter. The *oblongifolia* type (**b**) is formed by groups of two trichomes, bending in the same direction and sitting on an emergence, with heights of up to 330 µm. The upper terminal (apical) cells are loosely connected. The *natans* type (**c**) is formed by four trichome branches, each elevated on a large emergence. The total height is approximately 1300 µm. The *molesta* type (**d**) is formed by four trichome branches on a large emergence, which are connected by their upper terminal cells. The height of these hairs reaches 2200 µm

Multicellular hairs are common in many flowering plant groups. Particular interesting forms occur in the floating water ferns of the genus *Salvinia*. Within this genus, morphologically different kinds of water-repellent (superhydrophobic) hairs exist. The multicellular hairs on the upper (adaxial) side of the leaves of the species of genus *Salvinia* form complex hierarchical surface structures which are able to retain an air layer at the surface, even when leaves were fixed underwater for several days [41.78, 79]. The hairs are multicellular, and their sizes are in the range of several hundred micrometers. Four different hair types have been described for the genus *Salvinia* [41.80]. The SEM micrographs in Fig. 41.16 show their morphological variability. Based on these morphological types, four groups, each with several species, have been formed. The *Cucullata* type (Fig. 41.16a) is characterized by solitary and slightly bent trichomes and occurs in *S. cucullata* and *S. hastate*. The *Oblongifolia* type (Fig. 41.16b) forms groups of two trichomes, which bend in the same direction and sit on an emergence. This type occurs on *S. oblongifolia*. The *Natans* type, shown in Fig. 41.16c, has four trichome branches, each elevated on a large multicellular base and in total has a height of up to 1300 µm. The heights of the trichome groups decrease towards the leaf margins. This type occurs in *S. natans* and *S. minima*. In the *Molesta* type (Fig. 41.16d) four trichome branches are grouped together, connected with each other by their terminal cells and sitting on a large emergence. The heights of the trichomes reach up to 2200 µm in *S. molesta*, but also decrease towards the leaf margins. This trichome type is characteristic of, e.g., *S. molesta* and *S. biloba*. In all these species, the epidermis is covered with small three-dimensional waxes in the

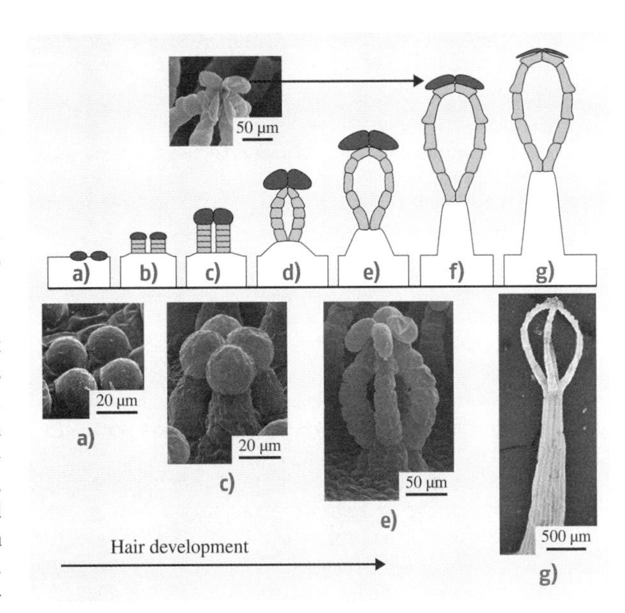

Fig. 41.17a–g Development of multicellular hairs in *S. molesta* on the upper side of the floating leaves. The schematic shows a side view (only two of four hair branches are shown) of the developmental stages during the formation of a crown-like hair. Four different developmental stages are shown in the SEM micrographs below. In this, the development starts by cell division below the four single cells shown in (**a**). Continuous growth (**b–d**) results in a crown-like structure (**e**). Further cell division leads to a multicellular hair (**g**) ▶

form of transversely ridged rodlets. The development of these complex structures, shown in Fig. 41.17, has been studied in *S. biloba* by *Barthlott* et al. [41.80]. In an early stage of leaf development, hair formation starts with a grouping of four cells, shown schematically in Fig. 41.17a. During the ontogeny of the leaf, four branches develop from these initial cells and form a crown-like structure (Fig. 41.17b–e) in which the single branches are connected with each other. Later, the base grows by cell division and cell expansion to develop a large base below the crown structure (Fig. 41.17f,g). These complex multicellular surface structures have been replicated by *Cerman* et al. [41.79], and replicas and leaves have been used to study the coherence of superhydrophobicity and air-retaining surfaces for drag reduction in underwater use (Sect. 41.2.4).

41.2 Multifunctional Plant Surfaces

Since the first plants moved from their aqueous environment to the drier atmosphere on land, plants have developed specific features for successful survival in nearly all conceivable habitats on Earth. Their adaptations have been selected by millions of years of plant evolution and include morphological, physiological, and structural specifications. Those features which were advantageous for the survival of individuals in their environment have been passed on from one generation to the next. Through natural selection, characteristics which are constitutionally present in the various groups of plants are amplified, intensified, and combined in various ways, yielding adaptation(s) to a particular habitat. However, some specific functional surface structures have evolved several times in nature; thus they exist in several groups of plants which are not closely related. Plant surfaces play an important role as interfaces to the biotic (living) and nonbiotic environment, thus it is not surprising that, in different plants and environments, a huge variety of functional surface structures have evolved.

41.2.1 Reflection and Absorption of Spectral Radiation

Temperature of plants in hot areas increases when the uptake of direct and indirect solar radiation is greater than heat transfer from the plant to the environment [41.81]. Plants usually reduce their temperature significantly by evaporation (transpiration) of water, but this mechanism of cooling requires regular uptake of water, which is limited in dry habitats. In dry areas, convective cooling is an important mechanism of heat transfer [41.82]. The latent heat flux from the vegetation to the environment depends largely on convection, which is correlated to leaf shape and size, and wind speed. Higher wind speeds cause turbulence in the plant boundary layer and force convection with mass flow [41.81, 83–85]. Most nonsucculent desert plants have small leaves which are expected to develop a small boundary-layer thickness and therefore have a high boundary-layer conductance for heat and mass transfer [41.83,86]. However, many leaves and stem surfaces are larger and therefore need a different mechanism of temperature control. In this regard, reduction of energy uptake is the most important way to reduce heat stress. The most common reflective structures on desert plants are three-dimensional waxes and a dense covering with air-filled hairs.

Spectral Properties of Plant Waxes

Waxes in many plants of dry habitats cause a bluish and white appearance of the plant surfaces. Leaves of some eucalyptus trees, as shown in Fig. 41.1g,h, show such a bluish surface structure and a dense layer of epicuticular waxes. The coloration results from the reflection of visible light, caused by surface structures in the dimensions of the wavelength of short-wave visible light, and by reflection of UV radiation [41.87–91]. Waxes on plant surfaces can reduce the uptake of radiation energy by reflection, and thereby control the temperature of the organisms [41.92]. Radiation emitted or reflected by plant surfaces occurs in the range from 280 nm to the infrared, including the wavelengths from 400 to 700 nm visible to the human eye [41.93]. Absorption of visible light is essential for photosynthesis, thus all leaf surfaces are translucent for the photosynthetically active wavelengths of solar radiation. However, intense solar radiation can depress the activity of photosynthesis, thus reflection of visible light is advantageous for desert plants to extend their photosynthesis [41.92, 94].

Three-dimensional waxes reflect a large part of the visible light, but can also play a significant role in the reflection of shortwave ultraviolet (UV) radiation [41.90]. The UV reflection of epicuticular waxes is caused by the dimensions of the waxes, which is similar to the wavelength of UV light (UV-B 280–320 nm and UV-A 320–400 nm) [41.87]. Furthermore, the epicuticular waxes of some species have been reported to contain UV-B-absorbing compounds [41.95]. Harmful ultraviolet radiation (280–400 nm) appears to be attenuated by phenols, in particular by cinnamic acids and by lipoidic flavonoids associated with the cuticular matrix or with the cuticular waxes [41.96, 97]. External flavonoid aglycones are found in various families throughout the higher plants, but occur very frequently in plants of (semi)arid habitats [41.98]. UV absorption is also important for plants growing in high altitudes in mountains, where significant UV levels can be found. UV-absorbing waxes of *Pinus mugo* subsp. *mugo* (dwarf mountain pine) contain chromophores that absorb UV radiation in such a way that it removes the most harmful UV-B and UV-A from the solar spectrum [41.99]. These fluorescent cuticular waxes remove the most damaging part of the UV radiation and convert it into harmless blue light. *Jacobs* et al. [41.99] suggest that the UV absorption and transformation found in plant waxes of *Pinus mugo* can be used for the development of new technological methods of protection against this harmful radiation.

Spectral Properties of Hairs

Besides three-dimensional waxes, the most common reflective structures on desert plants are a dense coverage with air-filled hairs. *Ehleringer* and *Björkman* [41.100] studied the spectral properties of *Encelia farinosa* leaves, which occur in different habitats. They showed that plants grown in drier areas have a denser and thicker hair layer. In *Encelia farinosa*, the hairs preferentially reflect near-infrared radiation (700–3000 nm) over photosynthetically useful solar radiation (400–700 nm). Such reflective properties have been found for several further species with a dense coverage of hairs, such as shown in Fig. 41.11a, by *Holmes* and *Keiller* [41.90]. Additionally such dense coverage of hairs may possibly increase the thickness of the leaf boundary layer, thus reducing the rate of water loss in water-limited habitats [41.85, 101]. Especially, the surfaces of plants from dry and hot environments show different microstructural adaptations to protect the plant from harmful radiation. The pubescence (presence of hairs) and glaucousness (presence of a thick epicuticular wax layer) are the two surface characteristics which have a marked effect on the total reflectance of plant surfaces in species of dry and hot habitats. However, the structural characteristics typical of plants of arid regions are frequently occurring structures, and therefore not exclusive to desert species.

41.2.2 Slippery Plant Surfaces

Many plants have evolved special structured surfaces which hinder the attachment of animals, especially insects, to protect themselves against herbivores [41.102]. Most insects possess two different types of attachment structures: claws and adhesive pads [41.103, 104]. Whereas the former are used to cling to rough surfaces, the latter enable them to stick to perfectly smooth substrates. One strategy to reduce the attachment of insects is the secretion of epicuticular waxes which assemble into three-dimensional microstructures. The other strategy is the development of a slippery surface by inducing aquaplaning.

Wax-Induced Slippery Surfaces

On a microrough wax layer, the use of both insect attachment structures mentioned above is impeded. On the one hand the wax crystals are too small and too fragile for the claw tips of insects to be inserted, and on the other hand the surface structure is too rough for the adhesive pads to develop sufficient contact area. Thus three-dimensional waxes can prevent walking of insects on the plant by the development of a slippery surface. The stems of many *Macaranga* plants (*Euphorbiaceae*) are covered by epicuticular wax crystals, rendering the surface very slippery for most insects, but specialist ants, which nest inside the hollow twigs of the plant, are able to walk on the waxy stem surfaces without any difficulty. The waxy surface is composed of long thin threads, which make the surface slippery for other ants and insects [41.105]. On these plants the wax acts as a selective barrier, protecting the symbiotic ant partners against competition from other ants.

Wax crystals on flower petals are extremely rare, because most pollinators must attach to the petal surfaces for pollination. However, in a special group of trap-flowers, insects are temporarily trapped into a flower trap, usually formed as a saccate structure (utricle) by the flower petals (perianths), for pollination. In this group, several species show waxy and slippery surface structures near the entrance of the trap-flower, and in some flowers the inner trap surface is also covered by wax, preventing the insect from climbing out. Such a waxy surface with large wax platelets is shown in Fig. 41.4f for *Aristolochia albida*. A comprehensive study of the surface structure in trap-flowers has recently been performed by *Poppinga* [41.106].

Some species of the carnivorous pitcher plants of the genus *Nepenthes* (Fig. 41.18a–e) use slippery wax layers in order to capture and retain insects. Carnivorous plants in the genus *Nepenthes* have pitcher-like leaves, formed as traps for catching and digesting insects. In these species, a layer of three-dimensional wax platelets creates a slippery zone inside the tube, above the digestive zone [41.107]. The wax plays a crucial role in animal trapping and prey retention. Insects are not able to attach to and walk on these waxy surfaces. They slide into the digestive fluid and are restrained from further escape. In *N. alata* the wax coverage consists of two overlapping layers; the upper layer is shown in the SEM micrograph in Fig. 41.18d. These layers differ in their structure, chemical composition, and mechanical properties. The upper layer is the one that comes into contact with the feet of insects. In this layer the waxes are less mechanically stable, and break easily at their smaller stalked bases into small pieces when insects slide over them. These broken wax particles adhere to the adhesive pads of insect feet (tarsi). The remaining lower wax layer provides an already rough surface structure, which reduces the adhesion area for the insect pads. Thus insect sliding is induced by a decrease in the attachment force of insects by two different mechanisms: first, by contamination of insects' attachment organs, and second, by reduction of the real contact area [41.108].

Aquaplaning on Plant Surfaces

There are several *Nepenthes* species which do not possess a waxy layer, but are nevertheless fully functional insect traps. It was found that *Nepenthes* evolved another capture mechanism which is based on special surface properties of the pitcher rim (peristome) (Fig. 41.18a,b). The peristome is characterized by a regular microstructure with radial ridges of smooth, overlapping epidermal cells, which form a series of steps towards the pitcher inside. The peristome ridges (Fig. 41.18c) mostly extend into tooth-like structures at the inner edge, in between which large glands (ex-

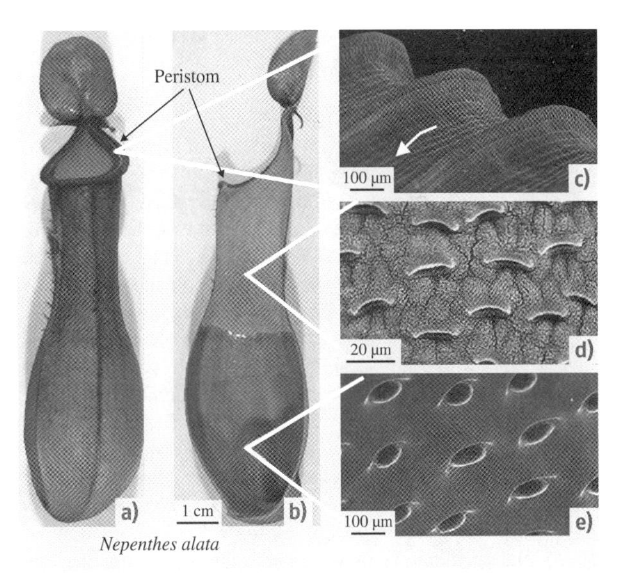

Fig. 41.18a–e Pitcher traps of the carnivorous plant *Nepenthes alata*. In (a) the complete pitchers trap of *N. alata* and in (b) a longitudinal cut through the trap are shown. (c) Shows parallel ridges of the hydrophilic peristome. The *arrow* indicates the direction toward the inside of the pitcher (courtesy Holger Bohn). In (d) the waxy and slippery surface inside the trap, with inactive stomata, is shown. In (e) glands located in the digestive zone at the lower part of the trap are shown

trafloral nectaries) are situated. Secretion of nectar by these glands attracts small insects, but leads also to a hydrophilic coverage of the surface. The plant surface microstructure, combined with hydrophilic surface chemistry, renders the pitcher rim completely wettable. Water droplets spread rapidly and form homogeneous thin films, which make the peristome extremely slippery for insects. When the peristome is wet, the fluid films prevent the insects' tarsal adhesive pads from making close contact with the surface, similar to the aquaplaning of a car tire on a wet road. In addition, the anisotropic microstructure of the peristome surface allows interlocking of claws only while the insect is running towards the pitcher inside, but not on the way out [41.109]. Under natural conditions the slippery water films are caused by rain, condensation, and nectar secretion. In contrast to this, dry peristomes are not slippery for insects. This weather-dependent variation of peristome slipperiness leads to intermittent and unpredictable activation of *Nepenthes* pitcher traps, which might make the evolution of specific avoidance behaviors more difficult [41.110].

41.2.3 Wettability and Self-Cleaning of Plant Surfaces

Basics of Surface Wetting

Wetting is the fundamental process of liquid interaction at solid–gaseous interfaces. It describes how a liquid comes into contact with a solid surface. Wetting is important in many everyday situations, for example, liquid painting on walls, printing of texts, and in the transport of fluids (water, oil, blood, and many others) through pipe systems, and it is the basis of several cleaning procedures. However, there are many situations where it is desirable to minimize wetting; water droplets adhering to window glass and car windshields reduce visibility and leave residuals after evaporation, rainwear should stay dry even during heavy showers, and movement of boats in water costs extra energy because of friction force at the interfaces. Wetting is also important for many biological processes such as the germination of seeds or microorganisms such as fungi, and reproduction of bacteria, and is essential for water uptake in soils. Under suitable conditions, microorganism proliferation results in the formation of larger populations called biofilms. Biofilms induce apparent defects in technical materials, and also their acidic excretions are damaging to buildings and technical materials [41.111]. Thus, it is not surprising that the basics of surface wetting processes have been of scientific interest for several decades and a large amount of scientific work has been carried out to understand and produce surfaces which are extremely water repellent. The basics of surface wetting are summarized here. For a deeper study, specific literature such as the books by *Israelachvili* [41.112], *Bhushan* [41.113], *De Gennes* et al. [41.114], and *Bhushan* [41.115] are recommended.

Surface Wetting and Contact Angle. A droplet on a solid surface wets the surface to a greater or lesser extent that is dependent on the contact angle (CA). A high contact angle describes surfaces on which a water droplet forms a spherical shape; thus the real contact between the adhering droplet and the surface is very small compared with wettable surfaces, on which an applied drop of water tends to spread, and the contact angle is low. The CA of a liquid on a surface depends on the surface tension (molecular forces) of the involved liquid, solid surface, and the surrounding vapor. Thus, wetting depends on the ratio between the energy necessary for the enlargement of the surface and the gain of energy due to adsorption [41.112, 116]. The basis for studying equilibrium wetting on rough surfaces was established many years ago by *Wenzel* [41.117] and *Cassie* and *Baxter* [41.118]. The Wenzel equation expresses a general amplification of the wettability induced by roughness and applies to a CA where droplets are in equilibrium, but not to advancing and receding angles of a droplet on a rough solid surface that give rise to contact-angle hysteresis (CAH) as shown in Fig. 41.19. Hysteresis is responsible for the sticking of liquids on a surface, and is defined as the difference of the advancing and receding angles of a moving or evaporating water droplet ($\mathrm{CAH} = \mathrm{CA}_{\mathrm{adv}} - \mathrm{CA}_{\mathrm{rec}}$). If additional liquid is added to a sessile drop, the contact line advances; if liquid is removed from the drop, the CA decreases to a receding value before the contact retreats. If a droplet moves over a solid surface, the CA at the front of the droplet (advancing CA) is greater than that at the back of the droplet (receding CA). However, if the droplet rolls with little resistance, the contact angle hysteresis is small.

On water-repellent surfaces, an applied droplet starts to roll off the surface when it is tilted to a specific angle. This tilt angle (TA) is simply defined as the tilting angle of a surface on which an applied drop of water starts to move. Low TA ($< 10°$) is characteristic of superhydrophobic and self-cleaning surfaces. Another important phenomenon related to wetting behavior is the bouncing of droplets. When a droplet hits a surface, it can bounce, spread or stick. In practical applications of superhydrophobic surfaces, surfaces should maintain their ability to repel penetrating droplets under dynamic conditions [41.119, 120].

Definition of Surface Wetting Contact Angle and Hysteresis. Contact angle measurement is the main method for characterization of the hydrophobicity of surfaces, and the CA is a measuring unit for the wettability of surfaces. The wetting behavior of solid surfaces can be divided into four classes, defined by their CA, and for superhydrophobic ones, also by their hysteresis. On wettable surfaces with low contact angles, a fluid will spread and cover a larger area of the surface. Such surfaces are termed superhydrophilic when the CA is $< 10°$. Surfaces with CAs of $\geq 10°$ and $< 90°$ are termed hydrophilic surfaces. Unwettable surfaces have high contact angles, meaning the liquid on the surface forms a hemispherical or spherical droplet. Surfaces on which the CA is $\geq 90°$ and $< 150°$ are termed hydrophobic. A superhydrophobic surface is defined as one that has a static CA of $\geq 150°$, and if those superhydrophobic surfaces have a low hysteresis or a low tilting angle of less than $< 10°$ they are superhydrophobic and can provide self-cleaning properties. This definition of superhydrophobic surfaces has been used in most recent reviews [41.120–124] and is the preferred one which might be used to overcome the existing variety of definitions.

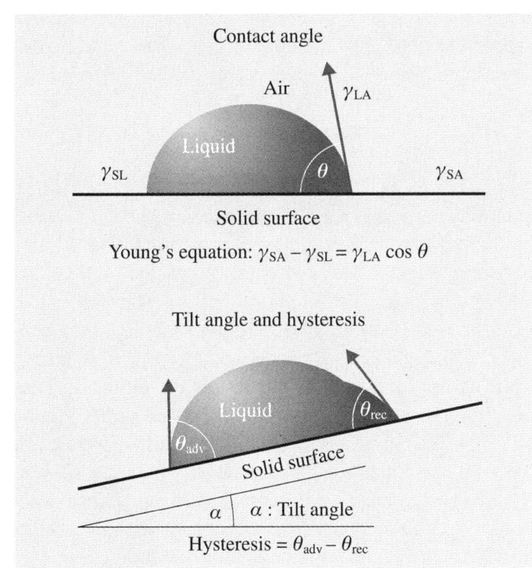

Fig. 41.19 Wetting of a solid surface with water, with air as the surrounding medium; γ_{LA}, γ_{LS}, and γ_{SA} are the interfacial tensions at the corresponding boundaries between liquid (L), solid (S), and air (A), which determine the CA of an applied water droplet, described by Young's equation. The hysteresis of a water droplet on a tilted surface represents the adhesion of the liquid on the surface and can be determined by measuring the tilting angle or the advancing and receding angle of a water droplet

Wetting of Plant Surfaces

The plant cuticle with its integrated and exposed waxes is in general a hydrophobic material, but structural and chemical modifications induce variations in surface wetting, ranging from superhydrophilicity to superhydrophobicity. Plant surface wetting is influenced by the sculptures of the cells, such as the formation of hairs, and by the fine structure of the surfaces, such as folding of the cuticle, or by epicuticular waxes. Representative plants and their structures and wettability are given for the four defined classes of surface wettability (hydrophobic, hydrophilic, superhydrophobic, and superhydrophilic). The different structural characteristics of plant surfaces and their wetting behavior are summarized in Fig. 41.20, and examples are given below.

Hydrophilic and Superhydrophilic Surfaces. On hydrophilic surfaces, a drop of water has a contact angle between $\geq 10°$ and $< 90°$. Hydrophilic surfaces are known from many leaves which have a papilla cell morphology and cuticular folding, but also from leaves with flat, tabular cells. Those leaves have only smooth wax films on their surface and no or only isolated three-dimensional wax crystals. Examples are shown in Fig. 41.14. Other hydrophilic surfaces have trichomes (hairs and glands) on their surface. On hydrophilic surfaces hairs are not covered with three-dimensional (3-D) waxes, which would make the surfaces hydrophobic. Examples are shown in Fig. 41.11. Glands influence the wettability of surfaces by their secretions, which can be very hydrophilic, as mentioned above for the nectar-secreting glands in *Nepenthes* and the water-spreading leaves of *Ruellia devosiana* [41.3]. However, why are the petals of most flowers hydrophilic? The explanation for this phenomenon is given by the functions of flower petals. In most plants, petals are developed to attract pollinators, in most cases small insects, and these pollinators must be able to walk on the petal surfaces. Thus, a coverage with three-dimensional waxes, which reduces the adhesion of most insects, would be disadvantageous. Additionally, it is important to notice that hydrophobic leaves might become hydrophilic by

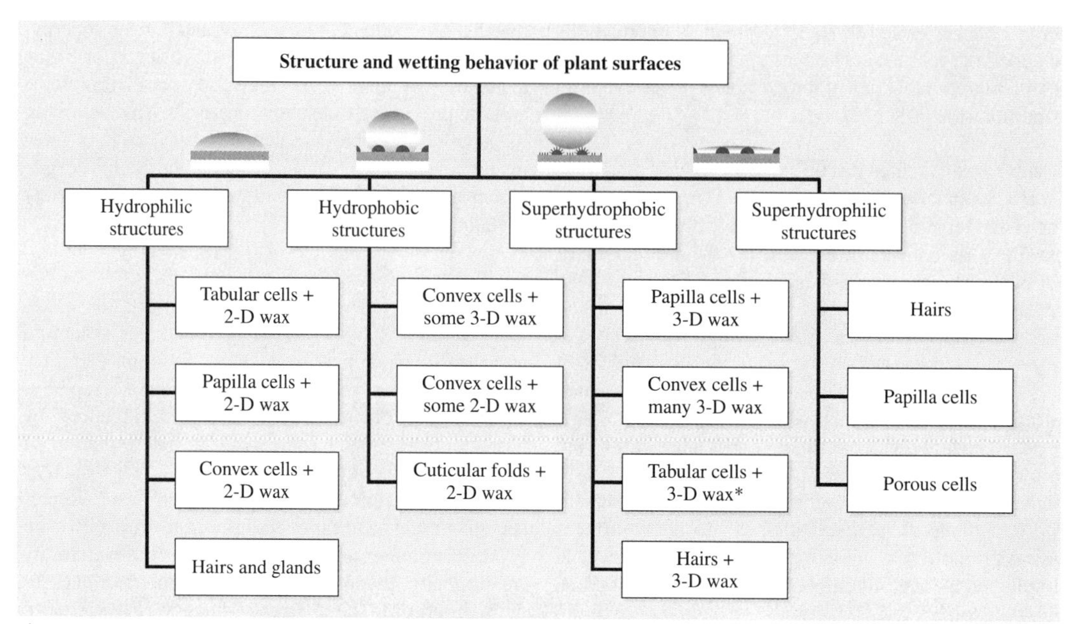

Fig. 41.20 Four groups of wettability of plant surfaces and the possible surface structures and structure combinations. The wetting drawings used for the four groups are correlated to specific contact angles. Both the hydrophobic and superhydrophobic surfaces can be built by convex cells with three-dimensional waxes on them, but only a dense layer (many wax crystals per area) leads to superhydrophobic surfaces (**Neinhuis* and *Barthlott* [41.17] defined these as temporary superhydrophobic surface structures, because erosion of 3-D waxes by environmental impacts can damage the waxes and decrease the contact angle)

the accumulation of environmental, hydrophilic contaminations such as spores, bacteria, dirt particles, and chemical aerosols on their surfaces. Atmospheric particles are mostly a conglomerate of different salts and organic materials [41.125]. These salts dissolve in rain, fog, and dew, or may become dissolved by stomatal transpiration of the leaf [41.126] and increase the hydrophilicity of plant surfaces [41.127].

Superhydrophilic surfaces are characterized by the spreading of water on the surface. Contact angles of such surfaces are, if measurable, 10° or less. Superhydrophilicity is based on different morphological structures, and by the chemistry of the surface, e.g., through secretion of hydrophilic compounds by epidermal glands. Optimized organs for the uptake of water are the roots of plants. Most roots are characterized by papillae and hairy cells, but porous surface structures have also been described for air roots of epiphytic plants. Superhydrophilicity is a great advantage for lower plants (a collective term for three groups of plants: the mosses, liverworts, and lichens), which have no roots for water uptake and no vascular system for water transport. Some species have not even developed a waxy cuticle as a transpiration and stabilization element. Water uptake and the uptake of nutrients in many lower plants occurs over the complete surface. Peat moss (*Sphagnum*) belongs to the lower plants and is common in wet swamps and mires, where it might represent the dominant biomass. The complete plant surface is free of stomata, which function in higher plants for water transpiration control. Water uptake and gas exchange occurs via pores, shown in Fig. 41.21a, which are spread over the plant body. Pores are relatively large openings of 10–20 μm in diameter, whereas the openings of a porous surface texture with openings of 0.2 μm and less are much smaller. An example of a water-absorbing porous cell surface structure is given by the leaf surface of the moss *Rhacocarpus purpurescens*. The superhydrophilic surface of *R. purpurescens* is shown in the SEM micrographs presented in Fig. 41.21b. The functional characteristic of the surface structure of the leaves of this moss is rapid absorption of fog, dew or rain [41.128, 129]. Water absorption via the plant surface is not limited to lower plants. In higher plants, all specimens of the *Bromelia* family, e.g., pineapple (*Ananas comosus*) or Spanish moss (*Tillandsia usneoides*) (Fig. 41.21c,d), have water-absorbing, multicellular absorptive trichomes on the epidermis cells. Some genera in this group form funnels by arranging their leaves in the form of a rosette. In these funnels, water can be stored after a rain shower and organic litter can accumulate and decay, so hairs at the funnel surface also absorb nutrients dissolved in the water. Further water-absorbing structures are porous thorns in some cacti, which absorb water condensed at the surface [41.1]. Some leaves of tropical plants, such as as those of *Ruellia devosiana* [41.3], *Marantha leuconeura*, and *Calathea zebrina*, are also superhydrophilic. The latter ones show strong convex sculpturing of their epidermis cells, but little information about the physicochemical basis of superhydrophilicity and the biological advantage of these properties is available.

It can be summarized that, in some lower plants and the species of the *Bromeliaceae*, superhydrophilic surfaces have evolved for water and nutrient uptake. However, a water film on the surface can reduce uptake of CO_2, which is required for photosynthesis [41.62]. On wet surfaces the growth of most pathogen microorganisms, including bacteria and fungi, is provided by permanent or temporary water availability. Additionally, lensing effects of water droplets on leaf surfaces can increase incident sunlight by over 20-fold directly beneath individual droplets [41.62], which may have important implications for processes such as photosynthesis and transpiration for a large variety of plant species. Thus, there may be strong selective pressure for increased water repellency in terrestrial plant leaves.

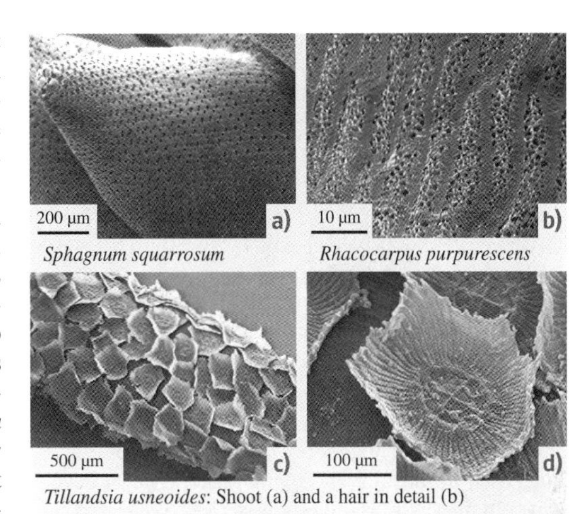

Fig. 41.21a–d SEM micrographs of superhydrophilic plant surfaces. In (**a**) the water uptake pores of a *Sphagnum squarrosum* moss and in (**b**) the water-adsorbing porous surface of the moss *Rhacocarpus purpurescens* are shown. In (**c**) the epiphytic-growing Spanish moss (*Tillandsia usneoides*) is shown. Figure (**d**) shows a higher magnification of the water-absorbing hairs of *T. usneoides*

Hydrophobic and Superhydrophobic Surfaces. The chemical constituents of most plant surfaces are hydrophobic, meaning that a water droplet applied to the surface has a static contact angle between $\geq 90°$ and $< 150°$. This phenomenon is based on the hydrophobic waxy nature of the plant cuticle, the primary function of which is as a barrier against water loss. The micromorphological characteristics of hydrophobic plant surfaces are tabular cells or only slightly convex epidermal cells, covered with a thin wax film (2-D waxes) or by a relatively low amount of 3-D waxes. Additionally, tabular cells might be structured by cuticular folding, and then a 2-D wax layer is present on the surface. However, for most plants, these wax layers have not been observed, because they are not or are only rarely visible in SEM. It might also be that a hairy leaf surface forms hydrophobic structures, but evidence with given contact angles has not been found; thus, we forego adding these structures to the more frequent surface morphologies.

Many leaf surfaces are superhydrophobic, as demonstrated by *Neinhuis* and *Barthlott* [41.17], who investigated over 200 water-repellent plant species. As defined above, a surface is termed superhydrophobic when the static CA is equal to or above 150° and low hysteresis lets a water droplet roll off at surface inclinations below 10°. The morphological characteristics of superhydrophobic leaves in most cases are a hierarchical surface structure, formed by convex to papillose epidermal cells, and a very dense arrangement of three-dimensional epicuticular waxes of different shapes. In Fig. 41.22 three species with such hierarchical surface structures are shown. All known crystal forms were found on superhydrophobic leaves, and their size ranged from 0.5 to about 20 µm in length [41.6]. Based on these investigations, it has been concluded that the superhydrophobicity of most plant surfaces is achieved by hierarchical structures of convex or papilla epidermal cells with three-dimensional wax structures on top.

It has also been found that tabular cells with a dense arrangement of wax crystals are superhydrophobic (e.g., *Brassica oleracea* and *Crambe maritima*), but those

Fig. 41.22a–c SEM micrographs of the hierarchical structures of superhydrophobic leaves of (**a**) *Colocasia esculenta*, *Euphorbia myrsinites*, and *Nelumbo nucifera* (lotus). Leaves of all three species are characterized by convex (**a**),(**b**) to papillose (**c**) cells, with coverage of three-dimensional wax crystals

leaves are water repellent for only a limited period of time. This, in most cases, is when leaves are in development and wax biosynthesis is active, but on mature leaves damage or erosion of the waxes can influence the wax structure and result in a less hydrophobic surface.

Besides the superhydrophobic leaf structures described before, a second method of water repellency has been developed in plants. Hairy leaf surfaces, such as those on the leaves of the lady's mantle (*Alchemilla vulgaris* L.) can very efficiently repel water. On such surfaces, a deposited drop bends the fibers (hairs), but the stiffness of the hairs prevents contact with the substrate, and promotes a fakir state of the water droplet [41.130]. The hydrophobic hairs of the water fern *Salvinia*, which have been introduced before and are shown in Figs. 41.16 and 41.17, are also superhydrophobic [41.79]. The use of these biological structures for the development of biomimetic air-retaining surfaces will be discussed in detail later. Superhydrophobic hairy surface structures are also known from animals, for example, water beetles and the water spider. These hairy systems may also be extremely useful for underwater systems because they minimize the wetted area of immersed surfaces and therefore may greatly reduce drag, as well as the rate of biofouling. Therefore, underwater superhydrophobicity is of great interest in biomimetics and is discussed in more detail later.

Hierarchical Structures for Superhydrophobicity and Self-Cleaning

The self-cleaning ability of microstructured, superhydrophobic plant surfaces was first described by *Barthlott* and *Ehler* [41.59], and the large shield-shaped leaves of the sacred lotus plant (*Nelumbo nucifera*) show this phenomenon to perfection [41.6]. Without knowledge of the physicochemical basics of self-cleaning, lotus has been a symbol of purity in Asian religions for over 2000 years [41.131]. Even emerging from muddy waters it unfolds its leaves unblemished and untouched by pollution. The hierarchical (double-structured) surface is characteristic of the lotus leaf and several other superhydrophobic leaves [41.17]. On lotus leaves the composite or hierarchical surface structure (Fig. 41.23) is built by convex papilla cells and a much smaller superimposed layer of hydrophobic three-dimensional wax tubules. The arrangement of the papilla cells is irregular, and orientation of the wax tubules is random. Wetting of such surfaces is minimized, because air is trapped in the cavities of the convex cell sculptures, and the hierarchical roughness enlarges the water–air interface while the solid–water interface is reduced. Water on such a surface gains very little energy through adsorption and forms a spherical droplet (Fig. 41.23), and both the contact area and the adhesion to the surface are dramatically reduced [41.6]. For lotus leaves the static contact angle is about 160° and the tilting angle $< 4°$.

The leaves of lotus (*Nelumbo nucifera*) are superhydrophobic and anti-adhesive with respect to particulate contaminations; thus contaminant particles are carried away by water droplets, resulting in a cleaned surface, as shown in Fig. 41.23a–c. A particle on a structured surface is like a fakir on his bed of nails, and the contact area and physical adhesion forces between a particle and the underlying leaf surface are considerably reduced. If water rolls over such a structured hydrophobic surface, contaminating particles are picked up by the water droplets, or they adhere to the surface of the droplets and are then removed with the droplets as they roll off. The reason for this is that only weak van der Waals forces bind the particle to the surface [41.132], whereas much stronger capillary forces between the particle and an adhering water droplet occur [41.119, 133, 134]. Thus, self-cleaning occurs on superhydrophobic leaves on which water moves over the surface to remove particles. Self-cleaning was found to be a result of an intrinsic hierarchical surface structure built by randomly oriented small hydrophobic wax tubules on the top of convex cell papillae (Fig. 41.23d–f). This self-cleaning process is independent of the chemistry of the adhering particles, i. e., whether they are hydrophilic or hydrophobic, and results in a smart protection against particle accumulation and is also a protection against plant pathogens such as fungi and bacteria [41.6, 17, 135]. In 2000, the trademark Lotus-effect was registered to label self-cleaning products based on the model of the lotus.

The lotus plant (*Nelumbo nucifera*) (a): Removal of dirt particles by water (b,c)

SEM micrographs of the lotus leaf (d) show the papillose cells (e) and wax tabules (f) on it

Fig. 41.23a–f Superhydrophobic and self-cleaning surface of lotus (*Nelumbo nucifera*). A flowering plant of lotus (**a**), a lotus leaf contaminated with clay (**b**) and removal of the adhering particles by water (**c**). The SEM micrographs (**d–f**) show the lotus leaf surface at different magnifications: (**d**) shows randomly distributed cell papilla, (**e**) shows a detail of the cell papilla, and in (**f**) the epicuticular wax tubules on the cells are shown

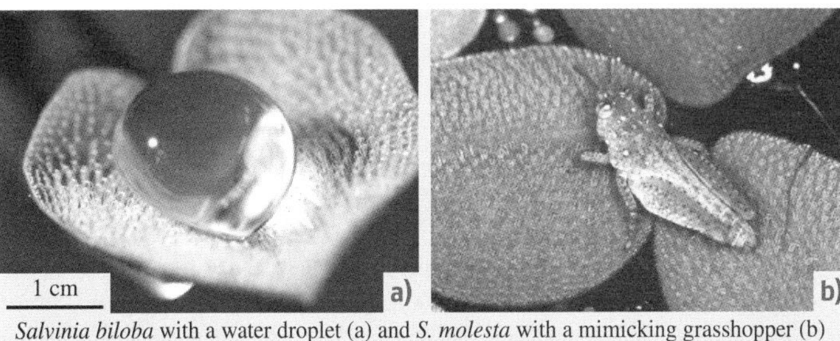

Salvinia biloba with a water droplet (a) and *S. molesta* with a mimicking grasshopper (b)

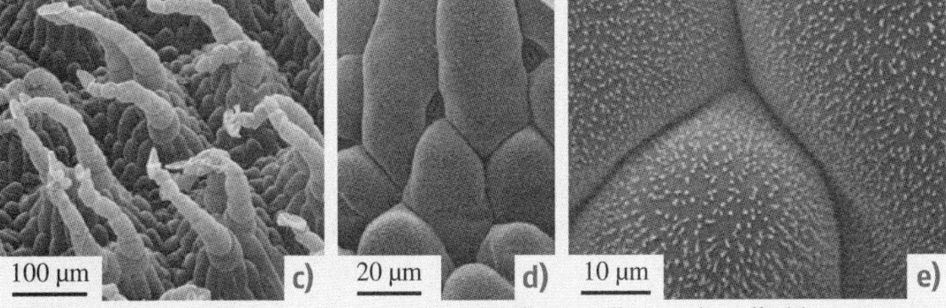

Salvina minima: Hairs (c), epidermis cells (d), and wax rodlets on the cells (e)

Fig. 41.24a–e Air-retaining surfaces of the water fern *Salvinia*. In (**a**) a water droplet on the upper leaf side of *Salvinia biloba* and in (**b**) a mimicking grasshopper on the leaf surface are shown. The SEM micrographs (**c–e**) of the leaf surface of *Salvinia minima* show the multicellular hairs (**c**), and at higher magnification the epidermis cells (**d**) and the wax rodlets (**e**) on the epidermis cells

41.2.4 Superhydrophobic Air-Retaining Surfaces

Trichomes in general have a strong influence on leaf wettability. *Neinhuis* and *Barthlott* [41.17] found that the wettability of hairy leaves strongly depends on the presence or absence of wax crystals on the trichome surface. Leaves with nonwaxy trichomes were only water repellent for a short time after a water droplet had been applied. In contrast, leaves with waxy trichomes were extremely water repellent, for example, the leaves of *Salvinia auriculata* and *Pistia stratiotes*. The different types of hairs of *Salvinia* leaves have been introduced before. These surfaces are superhydrophobic (Fig. 41.24a), but the water droplets do not penetrate between the hairs; thus small particles from the leaf surface will not be removed by rinsing with water. However, a study by *Cerman* [41.79] showed that *Salvinia* surfaces are able to retain an air film for up to 17 days when positioned underwater. The crucial factor for superhydrophobicity in *Salvinia* leaves is the hairs, which are several hundred micrometers high and covered by hydrophobic wax crystals (Fig. 41.24c–d). These leaves, as well as their technical replicas, are air retaining and stay dry for several days when placed underwater.

Water plants developed such air-retaining surfaces to float on the water. As a result of coevolution, also insects such as younger nymphs of the semiaquatic grasshopper *Paulinia acuminata* (Fig. 41.24b) show similar structures by mimicking the effective surface texture of the leaves [41.136]. The surface adaptation of the grasshopper is an optical and functional mimicry, which means that the coloration and pattern of colors of the grasshopper mimic the fern's surface texture, and its waxy surface provides water repellency [41.136].

41.3 Technical Uses of Superhydrophobicity

Biological surfaces are evolutionary optimized interfaces and provide a large diversity of structures and functions. Wetting phenomena, as described for lotus leaves, and several other biological surface phenomena are based on physicochemical factors and therefore are transferable to technical surfaces.

41.3.1 Biomimetic Surfaces

Bionics or *biomimicry describes a process in which the ideas and concepts developed by nature are taken and implemented into technology* [41.137]. Biomimetic research deals with the analysis, extraction, and transformation of biological structures, materials, processes, and principles into technical use, and is of great interest for the design of modern functional innovative materials. Bionics contains a wide spectra of research fields, for example, lightweight constructions, fluid dynamics, robotics, micro- and nanoelectromechanical systems (MEMS, NEMS), and sensors. The dimensions of interest encompass the range from the molecular level up to the function of complex organisms. However, even if nature has successful solutions, these are not necessarily optimal for technical performance. It is very important to get a profound understanding of the principles of nature's solutions, by analyzing the functions and boundary conditions, in order to transfer them into artificial systems.

The first prominent example of successful transfer of biological surface structures is the drag-reducing surface structure of shark skin and the artificial surfaces (rippled foils) developed after this model [41.138]. Recently, the structure of shark skin has been used as a model for the development of swimming suits with reduced surface drag when diving in water [41.139]. The description of superhydrophobic and self-cleaning plant surfaces by *Barthlott* and *Neinhuis* in 1997 [41.6] can be interpreted as a stimulating moment for many scientists to focus research on functional biological surfaces [41.140, 141]. Prominent examples are the development of superhydrophobic surfaces after the model plant lotus [41.54, 55, 122, 123]. Other biological models are the feet of several arthropods and some vertebrates with their remarkable ability of reversible attachment to varying surfaces [41.142–144]. Remarkable adhesion systems can also be found in plants. The climbing shoots of ivy (*Hedera helix*) adhere to smooth and rough surfaces by the secretion of organic glue, which forms nanoparticles that allow aerial roots to affix to a surface [41.145]. Another bioinspired attachment system is the hook-and-loop fastener, which plants use for the dispersal of their seeds by attaching fruits to animals. Self-repairing processes in plants that seal fissures served as concept generators for the development of biomimetic coatings for membranes of pneumatic structures [41.146]. These are only a few biomimetic examples; comprehensive overviews are given by *Benyus* [41.48], *Forbes* [41.141], and *Bar-Cohen* [41.147].

41.3.2 Structural Basics of Biomimetic Superhydrophobic Surfaces

Much theoretical and experimental work has been done to understand the physicochemical basics of superhydrophobicity. In this, several plants and their surface structures have been investigated. The structural parameters of the hierarchical surfaces and their influence on water and methanol repellency for 33 water-repellent plants has been studied by *Wagner* et al. [41.148]. This study showed that wax crystals of sizes from 0.5 and 5 μm had no significant influence on the hydrophobicity of the surfaces, but repellency increased when the aspect ratio of the cells increased. *Wagner* et al. [41.148] also characterized the number, height, and average lateral distance of papillae per unit area of six species with superhydrophobic surfaces. Even when papillae density varied between 737 and 3431 mm^{-2} and cell aspect ratios ranged from 0.8 to 0.5, no statistically significant variations in superhydrophobicity were found. However, a higher amount of smaller papillae seems to be more effective in terms of water repellency compared with surfaces with larger but fewer papillae. In this study also AFM and confocal light microscopy were used to calculate the solid–liquid contact area during wetting, and it was found that roughening on two hierarchical levels results in a reduction of the contact area between a water droplet and the plant surface of more than 95%. The influence of the wax nanostructures on the wettability of lotus leaves has been investigated by *Cheng* et al. [41.149]. In this, removal of the wax structure reduced the static contact angle from 146° to 126° (dry lotus leaf with molten wax), and hysteresis allowed water drops to adhere, even at tilting angles of 90°. This promising experimental approach did not change the chemistry of the surface, but led to a dramatic reduction of the cell aspect ratio by shrinkage of the cells.

Superhydrophobicity of hierarchical plant surface structures has been theoretically postulated earlier by *Otten* and *Herminghaus* [41.130] and was also confirmed by *Nosonovsky* and *Bhushan* [41.6, 150]. *Extrand* [41.151] investigated the interaction of capillary forces and gravity for a single asperity to suspend a liquid drop. In this, a microstructured surface was modified with a secondary structure by adding notches with sharp edges. It was stated that, if an asperity has a hierarchical structure, the secondary features can greatly enhance the surface repellency. *Marmur* [41.152] carried out theoretical analysis of the underlying mechanisms of superhydrophobicity and indicated that nature uses metastable wetting states as the key to superhydrophobicity. He concluded that the specific shape of the lotus leaf protrusions (cell papilla) lower the sensitivity of the superhydrophobic state to the protrusion distances. In these calculations simple parabolic shapes were used to simulate the morphology of the lotus cell papilla, but the influence of the secondary-level structure on the wetting behavior of the surfaces was not addressed.

For self-cleaning of superhydrophobic surfaces, a low hysteresis of less than 10° tilting angle is required. The degree of hysteresis is correlated with the wetting state of a liquid on a surface. On rough surfaces, an applied water droplet in the Wenzel state creates a wet contact mode with high hysteresis. In the Cassie state, the droplet sits on top of the structure asperities, hysteresis is low, and liquids roll off at low inclination angles. The intermediate wetting state occurs when the droplet sinks partially into the surface structures, but cavities of the surfaces are still air filled. Further wetting states with a high static contact angle, but also a high adhesive force (high hysteresis), have been described and called the Gecko wetting state [41.153] and petal effect [41.154]. However, these phenomena still seem to need more explanation before they can be accepted as new classifications of wetting states of superhydrophobic surfaces. *Li* and *Amirfazli* [41.155] combined theoretical calculations with experimental data by analyzing the thermodynamics of surface wetting of different surface textures (pillar height, width, and spacing). These calculations showed that small pillar spacing is needed for composite wetting states (Cassie state), whereas large pillar spacing is needed for large stable CA. Thus for achievement of large CA and small CAH simultaneously, a compromise between pillar spacing and other geometrical parameters such as pillar height, and spacing is necessary. *Nosonovsky* and *Bhushan* [41.150, 156–158] demonstrated that the effect of roughness on wetting and also the mechanisms that lead to destabilization of a composite interface are scale dependent. They concluded that multiscale roughness based on tightly packed convex papillae (asperities) with *nanobumps* as well as a hydrophobic nature of the material is optimum for superhydrophobicity. Optimal parameters for the development of stable, superhydrophobic rough surfaces have been calculated, and their requirements for optimal superhydrophobic surfaces fit well with the already realized biological water-repellent lotus leaf surfaces with hierarchical structures.

41.3.3 Generation of Biomimetic Superhydrophobic Surfaces

Plant surfaces and their properties stimulated the development of new functional biomimetic surfaces and materials, and many investigations have been made into creating superhydrophobic surfaces. In the past, several new methods for surface functionalization have been developed, and special interest has been given to techniques for the development of superhydrophobic surfaces [41.123]. Hierarchical surfaces are the preferred structures for highly water-repellent surfaces [41.151, 155, 159]. The generation of superhydrophobic surfaces includes bottom-up techniques, in which the desired structures are built up from small units such as single molecules, and top-down techniques, in which a structure is applied to a material by, e.g., lithography or molding. Templating, a top-down technique, is a simple and fast technique for replication of surface structures. In this, a master surface is used as the template and is pressed or printed onto a material or a liquid is filled in. Lithography is a technique to produce precise geometric structures with different aspect ratios (height to width) on scales from a few nanometers to several micrometers.

Bottom-up approaches to fabrication use chemical or physical forces operating at the nanoscale to assemble basic units into larger structures. Thus, the synthesis of a micro- or nanostructured material starts at the atomic or molecular level. In this procedure the precursor particles grow in size by chemical reactions or self-assembly [41.160]. Chemical self-assembly is a technique to exploit selective attachment of molecules to specific surfaces. Nanostructures of a wide range of materials, including organic and biological compounds, inorganic oxides, and metals, can be processed using chemical self-assembly techniques. In chemical self-assembly the binding between the units is covalent, whereas the binding in self-assembly is noncovalent. Materials which have been used to create superhydrophobic surfaces by self-assembly of the material are metals and metal oxides such as silver aggregates, cobalt hydroxide, CuS-coated copper oxide, and ZnO nanorods. For references and further examples, see *Roach* et al. [41.123]. For the development of nano- and microstructures it is important to control structure sizes, shapes, and the distribution of structures, e.g., single particles or the formation of particle agglomerations.

In plant surfaces, the secondary level of superhydrophobic hierarchical structures is built up by three-dimensional waxes, whose different morphologies originate by self-assembly. *Koch* et al. [41.161] used self-assembled wax platelets on artificial substrates and a lost wax casting process to transfer the platelet structures into artificial resin (Spurr's resin) surfaces. Another fast and low-cost technique to develop superhydrophobic surfaces is molding of hierarchical leaf surfaces, based on which a fast nanostructure replication method for water-containing biomaterial has been developed. The technique is based on three production steps:

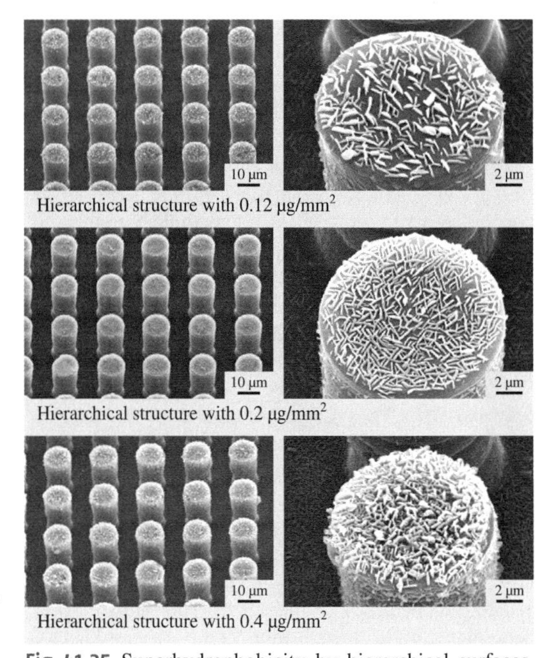

Fig. 41.25 Superhydrophobicity by hierarchical surfaces. SEM micrographs show replicated microstructured silicon surfaces with pillars of 14 μm diameter, 30 μm height, and 23 μm pitch, modified with self-assembled alkanes (hexatriacontane). From top to bottom an increase in crystal density is shown. The highest water repellency and lowest hysteresis has been found for the structures in the center, where 20 $\mu g/cm^2$ hexatriacontane was applied to the surfaces. These surfaces have been used for detailed study of wetting and adhesion (after [41.55])

1. Molding of a master surface
2. Removal of residues, such as wax crystals, from the mold
3. Filling of the mold with epoxy resin [41.162]

With this technique hierarchical structures of lotus leaves have been replicated and used for microscopy analysis and characterization of different properties, such as wettability and adhesion [41.4].

Superhydrophobic hierarchical surface structures can also be made by a combination of top-down and bottom-up techniques. One example is molding and multiple replication of artificial microstructured master surfaces and subsequent modification of the surfaces with self-assembling wax compounds or complete mixtures of plant waxes. For surface modification and homogenous covering of substrate surfaces with small three-dimensional crystalline nanostructures, thermo-evaporation of long-chain hydrocarbons has been used [41.162]. With this technique different molecules can be used to develop different nanostructure morphologies. Crystal sizes can be varied by changing the time of crystal growth, and the density of crystals on the surface can be controlled by the amount of evaporated material. We used this method to modify the nanostructures of replicated microstructured silicon surfaces for detailed wetting and adhesion analysis. Optimized hierarchical surface structures for self-cleaning approaches have been developed by replication of microstructured silicon surfaces and alkane and wax tubule self-assembly (Fig. 41.25). The optimized surfaces show contact angles above 170° and very low hysteresis of $\approx 2°$ tilting angle and therefore comply with the requirements for self-cleaning surfaces. The results of these studies are shown in [41.54, 55, 163].

41.3.4 Existing and Potential Use of Superhydrophobic Surfaces

Self-Cleaning Superhydrophobic Surfaces

Inspired by the self-cleaning behavior of lotus leaves, various artificial superhydrophobic self-cleaning surfaces have been prepared by creating appropriate surface morphology and roughness. A patent on technical micro- and nanostructured self-cleaning surfaces was assigned to *Barthlott* [41.164]. Based on this the trademark lotus-effect has been introduced, and several industrial manufactures have developed first products, labeled with the lotus-effect trademark. The first product available was a façade paint named lotusan, which has been successfully on the market since 1999. Sprays for temporary covering of surfaces create a superhydrophobic film on artificial surfaces and can, if no longer required, simply be removed by wiping. Self-cleaning glasses have been installed in sensors of traffic control units on German autobahns, and the introduction of building textiles, including awnings, tents, and flags, is to be expected. The list of applications for external surfaces includes lacquers for vehicles [41.165, 166], waterproofing of clothes [41.167] and other textiles [41.168, 169], plastics [41.170], roof tiles [41.171], temporary coatings [41.172], and plastics for microfluidics. However, some existing products are only easy to clean. These surfaces have very low surface energies and are smooth on the microscale. They are superhydrophobic, but the movement of water cannot remove adhering particles from them.

There are only two mechanisms which lead to the self-cleaning property of surfaces. First is the described superhydrophobic double-structured surface, as described for the lotus leaf. The second is a photocatalytic hydrophilic surface. In this, titanium dioxide is present in the outermost surface layer, which disintegrates any adhering organic materials. An application of these functional surfaces is, for example, roof tiles or glasses.

The use of self-cleaning surfaces is limited by the micro- and nanoscale of the surface structures, which requires low wear for the maintenance of the material functionality; thus materials must either be very wear resistant or uses must involve low friction [41.173]. For self-cleaning, movement is required and exterior applications where surfaces are exposed to rain or where surfaces can be artificially sprayed with water are preferred. Self-cleaning surfaces have limited success in glass materials, because surface microstructures affect the diffraction of light and thus change the optical properties of, e.g., lenses. For integration of superhydrophobicity and transparency within the same surface, the dimensions of the roughness should be lower than the wavelength of visible light (ca. 380–760 nm) [41.124].

Superhydrophobicity for Underwater Use

The surfaces of a number of floating plants and semi-aquatic animals provide technical solutions for the design of underwater air-trapping surfaces, and some have recently been successfully transferred to technical prototypes [41.78, 79].

Solga et al. [41.78] and *Cerman* et al. [41.79] studied the morphology and capacity for air retention of the surfaces of several species of the floating water fern

Salvinia. These leaves are characterized by multicellular hairs with coverage of small three-dimensional waxes (Figs. 41.16 and 41.17) and are able to retain air films underwater for up to 17 days. For technical surfaces, the advantage of staying dry underwater is the reduction of drag during movement. A small layer of air on a superhydrophobic surface reduces friction drag by 80% at a speed of 4 m/s and by 55% at 8 m/s [41.174, 175]. *Solga* et al. [41.78] list five surface characteristics as crucial for a stable, long-lasting underwater air film:

1. Hydrophobicity
2. Hairs with lengths of a few micrometers to several millimeters
3. Additional fine structures such as ridges, hairs or waxes
4. Micro- and nanocavities
5. Elasticity of the structures

Based on these, a textile prototype which stays dry for about 4 days when submerged in water has been developed. *Cerman* et al. [41.79] performed experiments with biological templates of different *Salvinia* species and showed that these biomimetic surfaces are able to retain an air film underwater for 4 days. A patent for air-retaining surfaces, outlining different fields of application, such as textiles, varnishes, and coatings, has been submitted by *Cerman* et al. [41.176]. However, *Cerman* et al. [41.79] discuss that it is obvious that even hairy superhydrophobic surfaces alone cannot stay permanently dry underwater and suggest that the solution will lie in a combination of microbubble technology (a technique in which the air layer is permanently refilled) and optimization of air-retaining superhydrophobic surfaces.

Underwater superhydrophobicity might also be a solution for the reduction of accumulation of algae, bacteria, and other sessile marine organisms on underwater surfaces. Whether underwater superhydrophobic surfaces are a solution against marine fouling has recently been discussed in a review by *Genzer* and *Efimenko* [41.177]. They present the implications of superhydrophobicity for marine fouling and potential designs of coatings and roughness on multiple length scales, which are assumed to represent a new and promising platform for fabrication-efficient foul-release marine coatings. However, it seems that some work still needs to be done to fully understand the phenomenon of the different mechanisms of adhesion by different marine organisms.

41.4 Conclusions

The diversity of plant surface structures is a result of several million years of evolutionary processes, through which nature has developed and proved surface structures and their functions. As a result of this, different highly functional surfaces have been developed. Plants developed surfaces with dense layers of hairs and surface waxes for light reflection and the absorption of harmful UV radiation. Anti-adhesive surfaces have been developed by some carnivorous plants to catch small insects. Such anti-adhesive surfaces are built either by epicuticular waxes or by a hydrophilic surface, covered with a water film. In plants, superhydrophobic surfaces are formed by microstructured cells with three-dimensional waxes on them, or by multicellular hairs, which are also covered by three-dimensional waxes. Hierarchical roughness in the first type of plant surface leads to self-cleaning, however droplets sit on top of hairs in the second type of plants and do not provide self-cleaning of particles trapped between the hair, although air can be retained by this structures effectively. Some plants developed superhydrophilic, hairy or porous surfaces, for the uptake of water and nutrients. The given examples show that the cuticle and its hydrophobic waxes represent a multifunctional boundary layer. In this, the surface functionality is predominantly influenced by the occurrence of epicuticular waxes. These waxes represent a hydrophobic covering for the primary surfaces of nearly all land-living plants. Waxes mainly function as a transport barrier, but on many plants waxes form small three-dimensional crystals, which provide the structural basis for superhydrophobicity, light refection, adsorption of UV radiation, and the creation of anti-adhesive surfaces. The three-dimensional wax crystals originate by self-assembly of mainly alkane molecules and their derivates (alcohols, esters, and others). These molecules can be used to develop biomimetic surfaces with superhydrophobic, self-cleaning, and anti-adhesive properties. However, impressive new techniques and materials have been developed in recent years to develop superhydrophobic and self-cleaning surfaces after the model plant lotus (*Nelumbo nucifera*). The superhydrophobic surfaces of the water fern *Salvinia* retain an air layer under water over several days. This surface property has been used to develop air-retaining biomimetic surfaces with potential use for underwater materials, e.g., for drag reduction on surfaces moving in water. It is to be expected that more biomimetic materials will be developed in the future, and the functional plant surface structures presented here might further stimulate research in biomimicry.

References

41.1 K. Koch, B. Bhushan, W. Barthlott: Diversity of structure, morphology and wetting of plant surfaces, Soft Matter **4**, 1799–1804 (2008)

41.2 K. Koch, B. Bhushan, W. Barthlott: Multifunctional surface structures of plants: An inspiration for biomimetics, Prog. Mater. Sci. **54**, 137–178 (2009)

41.3 K. Koch, I.C. Blecher, G. König, S. Kehraus, W. Barthlott: The superhydrophilic and superoleophilic leaf surface of *Ruellia devosiana* (*Acanthaceae*): A biological model for spreading of water and oil on surfaces, Funct. Plant Biol. **36**, 339–350 (2009)

41.4 K. Koch, B. Bhushan, Y.C. Jung, W. Barthlott: Fabrication of artificial Lotus leaves and significance of hierarchical structure for superhydrophobicity and low adhesion, Soft Matter **5**, 1386–1393 (2009)

41.5 M. Riederer, L. Schreiber: Protecting against water loss: analysis of the barrier properties of plant cuticles, J. Exp. Bot. **52**, 2023–2032 (2001)

41.6 W. Barthlott, C. Neinhuis: The purity of sacred lotus or escape from contamination in biological surfaces, Planta **202**, 1–8 (1997)

41.7 H. Bargel, K. Koch, Z. Cerman, C. Neinhuis: Structure-function relationships of the plant cuticle and cuticular waxes – A smart material?, Funct. Plant. Biol. Evans Rev. Ser. **3**, 893–910 (2006)

41.8 P. Kenrick, P.R. Crane: The origin and early evolution of plants on land, Nature **389**, 33–39 (1997)

41.9 P.J. Holloway: Section I – Reviews. Plant cuticles: Physiochemical characteristics and biosynthesis. In: *Air Pollutants and the Leaf Cuticle*, ed. by K.E. Percy, J.N. Cape, R. Jagels, C.J. Simpson (Springer, Berlin Heidelberg 1994)

41.10 P.E. Kolattukudy: Polyesters in higher plants, Adv. Biochem. Eng. Biotechnol. **71**, 4–49 (2001)

41.11 C.E. Jeffree: The fine structure of the plant cuticle. In: *Biology of the Plant Cuticle*, ed. by M. Riederer, C. Müller (Blackwell, Oxford 2006) pp. 11–125

41.12 J.T. Martin, B.E. Juniper: *The Cuticles of Plants* (Edward Arnold, London 1970)

41.13 D.F. Cutler, K.L. Alvin, C.E. Price (Eds.): *The Plant Cuticle* (Academic Press, London 1982)

41.14 G. Kerstiens: *Plant cuticles: an integrated functional approach* (BIOS Scientific Publishers, Oxford 1996)

41.15 M. Riederer, C. Müller: *Biology of the Plant Cuticle* (Blackwell, Oxford 2006)

41.16 M. Riederer, L. Schreiber: Waxes – The transport barriers of plant cuticles. In: *Waxes: Chemistry, Molecular Biology and Functions*, ed. by R.J. Hamilton (The Oily Press, Dundee 1995) pp. 131–156

41.17 C. Neinhuis, W. Barthlott: Characterization and distribution of water-repellent, self-cleaning plant surfaces, Ann. Bot. **79**, 667–677 (1997)

41.18 R. Fürstner, W. Barthlott, C. Neinhuis, P. Walzel: Wetting and self-cleaning properties of artificial superhydrophobic surfaces, Langmuir **21**, 956–961 (2005)

41.19 L.Q. Ren, S.J. Wang, X.M. Tian, Z.W. Han, L.N. YanQiu, Z.M. Qiu: Non-smooth morphologies of typical plant leaf surfaces and their anti-adhesion effects, J. Bionic Eng. **4**, 33–40 (2007)

41.20 H. Bargel, C. Neinhuis: Tomato (*Lycopersicon esculentum* Mill.) fruit growth and ripening as related to the biomechanical properties of fruit skin and isolated cuticle, J. Exp. Bot. **56**, 1049–1060 (2005)

41.21 H.G. Edelmann, C. Neinhuis, H. Bargel: Influence of hydration and temperature on the rheological properties of plant cuticles and their impact on plant organ integrity, J. Plant Growth Regul. **24**, 116–126 (2005)

41.22 M. Riederer, K. Markstädter: Cuticular waxes: A critical assessment of current knowledge. In: *Plant Cuticles – An Integrated Functional Approach* (BIOS Scientific, Oxford 1996) pp. 189–198

41.23 L. Kunst, A.L. Samuels: Biosynthesis and secretion of plant cuticular wax, Prog. Lipid Res. **42**, 51–80 (2003)

41.24 R. Jetter, S. Schäffer: Chemical composition of the Prunus laurocerasus leaf surface. Dynamic changes of the epicuticular wax film during leaf development, Plant Phys. **126**, 1725–1737 (2001)

41.25 R. Jetter, L. Kunst, A.L. Samuels: Composition of plant cuticular waxes, Annu. Plant Rev. **23**, 145–175 (2006)

41.26 E.A. Baker: Chemistry and morphology of plant epicuticular waxes. In: *The Plant Cuticle*, ed. by D.F. Cutler, K.L. Alvin, C.E. Price (Academic, London 1982) pp. 139–165

41.27 T. Shepherd, D.W. Griffiths: The effects of stress on plant cuticular waxes, New Phytol. **171**, 469–499 (2006)

41.28 K. Koch, K.D. Hartmann, L. Schreiber, W. Barthlott, C. Neinhuis: Influence of air humidity on epicuticular wax chemical composition, morphology and wettability of leaf surfaces. Env. Exp. Bot. **56**, 1–9 (2006)

41.29 C. Markstädter, W. Federle, R. Jetter, M. Riederer, B. Hölldobler: Chemical composition of the slippery epicuticular wax blooms on *Macaranga* (*Euphorbiaceae*) ant-plants, Chemoecology **10**, 33–40 (2000)

41.30 M. Riedel, A. Eichner, R. Jetter: Slippery surfaces of carnivorous plants: Composition of epicuticular wax crystals in *Nepenthes alata* Blanco pitchers, Planta **218**, 87–97 (2003)

41.31 M. Wen, C. Buschhaus, R. Jetter: Nanotubules on plant surfaces: Chemical composition of epicuticular wax crystals on needles of *Taxus baccata* L, Phytochemistry **67**, 1808–1817 (2007)

41.32 H. Ensikat, C. Neinhuis, W. Barthlott: Direct access to plant epicuticular wax crystals by a new mechanical isolation method, Int. J. Plant Sci. **161**, 143–148 (2000)

41.33 H.J. Ensikat, M. Boese, W. Mader, W. Barthlott, K. Koch: Crystallinity of plant epicuticular waxes: Electron and X-ray diffraction studies, Chem. Phys. Lipids **144**, 45–59 (2006)

41.34 W. Barthlott, C. Neinhuis, D. Cutler, F. Ditsch, I. Meusel, I. Theisen, H. Wilhelmi: Classification and terminology of plant epicuticular waxes, Bot. J. Linn. Soc. **126**, 237–260 (1998)

41.35 N.D. Hallam, B.E. Juniper: The anatomy of the leaf surface. In: *The Ecology of Leaf Surface Microorganisms*, ed. by T.F. Preece, C.H. Dickinson (Academic, London 1971) pp. 3–37

41.36 C.E. Jeffree, E.A. Baker, P.J. Holloway: Ultrastructure and recrystallization of plant epicuticular waxes, New Physiol. **75**, 539–549 (1975)

41.37 K. Koch, C. Neinhuis, H.J. Ensikat, W. Barthlott: Self assembly of epicuticular waxes on plant surfaces investigated by atomic force microscopy (AFM), J. Exp. Bot. **55**, 711–718 (2004)

41.38 K. Koch, H.J. Ensikat: The hydrophobic coatings of plant surfaces: epicuticular wax crystals and their morphologies, crystallinity and molecular self-assembly, Micron **39**, 759–772 (2008)

41.39 P.J. Holloway, C.E. Jeffree, E.A. Baker: Structural determination of secondary alcohols from plant epicuticular waxes, Phytochemistry **15**, 1768–1770 (1976)

41.40 R. Jetter, M. Riederer: In vitro reconstitution of epicuticular wax crystals: Formation of tubular aggregates by long chain secondary alkendiols, Bot. Acta **108**, 111–120 (1995)

41.41 W. Barthlott, I. Theisen, T. Borsch, C. Neinhuis: Epicuticular waxes and vascular plant systematics: Integrating micromorphological and chemical data. In: *Deep Morphology: Toward a Renaissance of Morphology in Plant Systematics*, Regnum Vegetabile, Vol. 141, ed. by T.F. Stuessy, V. Mayer, E. Hörandl (Gantner Verlag Ruggell/Liechtenstein 2003) pp. 189–206

41.42 I. Meusel, C. Neinhuis, C. Markstädter, W. Barthlott: Chemical composition and recrystallization of epicuticular waxes: coiled rodlets and tubules, Plant Biol. **2**, 1–9 (2000)

41.43 I. Meusel, C. Neinhuis, C. Markstädter, W. Barthlott: Ultrastructure, chemical composition and recrystallisation of epicuticular waxes: transversely ridged rodlets, Can. J. Bot. **77**, 706–720 (1999)

41.44 International Union of Crystallography: Report of the executive committee for 1991, Acta Crystalogr. A **48**, 922–946 (1992)

41.45 R. Jetter, M. Riederer: Epicuticular crystals of nonacosan-10-ol: In-vitro reconstitution and factors influencing crystal habits, Planta **195**, 257–270 (1994)

41.46 K. Koch, W. Barthlott, S. Koch, A. Hommes, K. Wandelt, W. Mamdouh, S. De-Feyter, P. Broekmann: Structural analysis of wheat wax (*Triticum aestivum*): From the molecular level to three dimensional crystals, Planta **223**, 258–270 (2006)

41.47 K. Koch, A. Dommisse, W. Barthlott: Chemistry and crystal growth of plant wax tubules of Lotus (*Nelumbo nucifera*) and Nasturtium (*Tropaeolum majus*) leaves on technical substrates, Cryst. Growth Des. **6**, 2571–2578 (2006)

41.48 J.M. Benyus: *Biomimicry: Innovation Inspired by Nature*, 2nd edn. (H. Collins Pub., New York 2002)

41.49 G.M. Whitesides, M. Boncheva: Beyond molecules: Self-assembly of mesoscopic and macroscopic components, Proc. Natl. Acad. Sci. USA **99**(8), 4769–4774 (2002)

41.50 J. Zhang, W. Zhong-Lin, J. Liu, C. Shaowei, G. Liu: *Self Assembled Nanostructures* (Kluwer Academic, New York 2003)

41.51 N. Boden, P.J.B. Edwards, K.W. Jolley: Self-assembly and self-organization in micellar liquid crystals. In: *Structure and Dynamics of Strongly Interacting Colloids and Supermolecular Aggregates in Solutions*, ed. by S.H. Chen, J.S. Huang, P. Tartaglia (Kluwer, Dordrecht 1992)

41.52 C. Neinhuis, K. Koch, W. Barthlott: Movement and regeneration of epicuticular waxes through plant cuticles, Planta **213**, 427–434 (2001)

41.53 D. Dorset: Development of lamellar structures in natural waxes – An electron diffraction investigation, J. Phys. D **32**, 1276–1280 (1999)

41.54 B. Bhushan, K. Koch, Y.C. Jung: Biomimetic hierarchical structure for self-cleaning, Appl. Phys. Lett. **93**, 093101 (2008)

41.55 B. Bhushan, K. Koch, Y.C. Jung: Nanostructures for superhydrophobicity and low adhesion, Soft Matter **4**, 1799–1804 (2008)

41.56 S. De Feyter, F.C. De Schryver: Self-assembly at the liquid/solid interface: STM reveals, J. Phys. Chem. B **109**, 4290–4302 (2005)

41.57 F.C. Meldrum, S. Ludwigs: Template-directed control of crystal morphologies, Macromol. Biosci. **7**, 152–162 (2007)

41.58 A. Dommisse: Self-assembly and pattern formation of epicuticular waxes on plant surfaces. Ph.D. Thesis (Rheinische Friedrich-Wilhelms Universität, Bonn 2007)

41.59 W. Barthlott, N. Ehler: *Rasterelektronenmikroskopie der Epidermis-Oberflächen von Spermatophyten, Tropische und subtropische Pflanzenwelt* (Akademie der Wissenschaften und Literatur/Franz Steiner, Mainz/Wiesbaden 1977), in German

41.60 W. Barthlott: Scanning electron microscopy of the epidermal surface in plants. In: *Scanning Electron Microscopy in Taxonomy and Functional Morphology*, ed. by D. Claugher (Clarendon Press, Oxford 1990) pp. 69–94

41.61 C. Martin, B.J. Glover: Functional aspects of cell patterning in aerial epidermis, Curr. Opin. Plant Biol. **10**, 70–82 (2007)

41.62 C.A. Brewer, W.K. Smith, T.C. Vogelmann: Functional interaction between leaf trichomes, leaf wettability and the optical properties of water droplets, Plant Cell Environ. **14**, 955–962 (1991)

41.63 G.J. Wagner, E. Wang, R.W. Shephers: New approaches for studying and exploiting an old protuberance, the plant trichome, Ann. Bot. **93**, 3–11 (2004)

41.64 E. Rodriguez, P.L. Healey, I. Mehta: *Biology and Chemistry of Plant Trichomes* (Plenum, New York 1984)

41.65 H.D. Behnke: Plant trichomes – structure and ultrastructure: General terminology, taxonomic applications, and aspects of trichome bacterial interaction in leaf tips of *Dioscorea*. In: *Biology and chemistry of plant trichomes*, ed. by E. Rodriguez, P.L. Healey, I. Mehta (Plenum Press, New York 1984) pp. 1–21

41.66 W. Barthlott, D. Hunt: Seed-diversity in *Cactaceae* subfam. *Cactoideae*. In: *Succulent Plant Research*, Vol. 5, ed. by D. Hundt (Milborne Port, England 2000)

41.67 P.G. Kevan, M.A. Lanet: Flower petal microtexture is a tactile cue for bees, Proc. Natl. Acad. Sci. USA **82**, 4750–4752 (1985)

41.68 R. Maheshwari: A scourge of mankind: From ancient times into the genomics era, Curr. Sci. **93**, 1249–1256 (2007)

41.69 H.C. Hoch, R.C. Staples, B. Whitehead, J. Comeau, E.D. Wolf: Signaling for growth orientation and cell differentiation by surface topography, Uromyces Sci. **239**, 1659–1663 (1987)

41.70 L.H.P. Jones, K.A. Handreck: Silica in soils, plants, and animals, Adv. Agron. **19**, 107–149 (1967)

41.71 A.G. Sangster, M.J. Hodson, H.J. Tubb: Silicon deposition in higher plants. In: *Silicon in Agriculture*, ed. by L.E. Datnoff, G.H. Snyder, G.H. Korndörfer (Elsevier, Amsterdam 2001) pp. 85–114

41.72 F. Fauteux, W. Remus-Borel, J.G. Menziesb, R.R. Belanger: Silicon and plant disease resistance against pathogenic fungi, FEMS Microbiol. Lett. **249**, 1–6 (2005)

41.73 R.K. Saeedur: *Calcium Oxalate in Biological Systems* (CRC, Boca Raton 1995) p. 375

41.74 V.R. Franceschi, P.A. Nakata: Calcium oxalate in plants: Formation and function, Annu. Rev. Plant Biol. **56**, 41–71 (2005)

41.75 A. Fahn: Structure and function of secretory cells, Adv. Bot. Res. **31**, 37–75 (2000)

41.76 E. Wollenweber: The distribution and chemical constituents of the farinose exudates in gymnogrammoid ferns, Am. Fern. J. **68**, 13–28 (1978)

41.77 W. Barthlott, E. Wollenweber: Zur Feinstruktur, Chemie und Taxonomischen Signifikanz Epicuticularer Wachse und ähnlicher Sekrete, Trop. Subtrop. Pflanzenwelt **32**, 7–67 (1981), in German

41.78 A. Solga, Z. Cerman, B.F. Striffler, M. Spaeth, W. Barthlott: The dream of staying clean: Lotus and biomimetic surfaces, Bioinspir. Biomimetics **2**, 1–9 (2007)

41.79 Z. Cerman, B.F. Striffler, W. Barthlott: Dry in the water: The superhydrophobic water fern *Salvinia* – A model for biomimetic surfaces. In: *Functional Surfaces in Biology: Little Structures with Big Effects*, Vol. I, ed. by S.N. Gorb (Springer, Berlin Heidelberg 2009)

41.80 W. Barthlott, S. Wiersch, Z. Colic, K. Koch: Classification of trichome types within the water ferns *Salvinia* and ontogeny of the eggbeater trichomes, Botany **87**, 830–836 (2009)

41.81 H.G. Jones, E. Rotenberg: Energy, radiation and temperature regulation in plants. In: *Encyclopedia of Life Sciences* (Wiley, New York 2001) pp. 1–8 (2001)

41.82 D.M. Gates: Transpiration and leaf temperature, Ann. Rev. Plant Phys. **19**, 211–238 (1968)

41.83 D.M. Gates: Energy exchange and transpiration. In: *Water and Plant Life*, Ecological Studies, Vol. 19, ed. by O.L. Lange, L. Kappen, E.D. Schulze (Springer, New York 1976) pp. 137–147

41.84 P.H. Schuepp: Model experiments on free convection heat and mass transfer of leaves and plant elements, Bound.-Layer Meteorol. **3**, 454–457 (1973)

41.85 P.H. Schuepp: Leaf boundary layers, New Phytol. **125**, 477–507 (1993)

41.86 P.S. Nobel: *Physicochemical and Environmental Plant Physiology* (Academic, San Diego 1991)

41.87 M. Riederer: The cuticles of conifers: Structure, composition and transport properties. In: *Forest Decline and Air Pollution: A study of spruce (Picea abies) on acid soil*, Ecological Studies, Vol. 77, ed. by E.D. Schulze, O.L. Lange, R. Oren (Springer, Berlin Heidelberg 1989), pp. 157–192

41.88 R.H. Grant, G.M. Heisler, W. Gao, M. Jenks: Ultraviolet leaf reflectance of common urban trees and the prediction of reflectance from leaf surface characteristics, Agric. For. Meteorol. **120**, 127–139 (2003)

41.89 J.D. Barnes, J. Cardoso-Vilhena: Interactions between electromagnetic radiation and the plant cuticle. In: *Plant Cuticles, an integrated approach*, ed. by G. Kerstiens (BIOS Scientific, Oxford 1996) pp. 157–170

41.90 M.G. Holmes, D.R. Keiller: Effects of pubescence and waxes on the reflectance of leaves in the ultraviolet and photosynthetic wavebands: A comparison of a range of species, Plant Cell Env. **25**, 85–93 (2002)

41.91 E.E. Pfündel, G. Agati, Z.G. Cerovic: Optical properties of plant surfaces, Annu. Plant Rev. **3**, 216–239 (2006)

41.92 S. Robinson, C.E. Lovelock, C.B. Osmond: Wax as a mechanism for protection against photoinhibition: A study of *Cotyledon orbiculata*, Bot. Acta **106**, 307–312 (1993)

41.93 C. Müller, M. Riederer: Plant surface properties in chemical ecology, Chem. Ecol. **3**, 2621–2651 (2005)

41.94 R. Sinclair, D.A. Thomas: Optical properties of leaves of some species in arid South Australia, Austral. Bot. **18**, 261–273 (1970)

41.95 P. Krauss, C. Markstädter, M. Riederer: Attenuation of UV radiation by plant cuticles from woody species, Plant Cell Env. **20**, 1079–1085 (1997)

41.96 A.E. Stapleton, V. Walbot: Flavonoids can protect maize DNA from the induction of ultraviolet radiation damage, Plant Physiol. **105**, 881–889 (1994)

41.97 L.C. Olsson, M. Veit, J.F. Bornman: Epidermal transmittance and phenolics composition of atrazine-tolerant and atrazine-sensitive cultivars of *Brassica napus* grown under enhanced UV-B radiation, Phys. Plant **107**, 259–266 (1999)

41.98 E. Wollenweber, U.H. Dietz: Occurrence and distribution of free flavonoid aglycones in plants, Phytochemistry **20**, 869–932 (1981)

41.99 J.F. Jacobs, G.J.M. Koper, W.N.J. Ursem: UV protective coatings: A botanical approach, Prog. Org. Coat. **58**, 166–171 (2007)

41.100 J.R. Ehleringer, O. Björkman: Pubescence and leaf spectral characteristics in a desert shrub *Encelia farinosa*, Oecologia **36**, 151–162 (1978)

41.101 P.S. Nobel: *Biophysical Plant Physiology and Ecology* (Freeman, San Francisco 1983)

41.102 S.D. Eigenbrode: Plant surface waxes and insect behaviour. In: *Plant Cuticles: An Integrated Functional Approach*, ed. by G. Kerstiens (BIOS Scientific Publishers, Oxford 1996) pp. 201–222

41.103 R.G. Beutel, S.N. Gorb: Ultrastructure of attachment specializations of hexapods (*Arthropoda*): Evolutionary patterns inferred from a revised ordinal phylogeny, J. Zool. Syst. Evol. Res. **39**, 177–207 (2001)

41.104 S. Gorb: *Attachment Devises of Insect Cuticles* (Kluwer, Dortrecht 2001)

41.105 W. Federle, U. Maschwitz, B. Fiala, M. Riederer, B. Hölldobler: Slippery ant-plants and skilful climbers: Selection and protection of specific ant partners by epicuticular wax blooms in *Macaranga* (*Euphorbiaceae*), Oeco **112**, 217–224 (1997)

41.106 S. Poppinga: Pflanzen fangen Tiere: Mikroskopische Charakteristika von Gleitfallen. Diploma Thesis (Nees Institut for Biodiversity of Plants, Universtität Bonn 2007), in German

41.107 B.E. Juniper, R.J. Robins, D.M. Joel: *The Carnivorous Plants* (Academic, London 1989)

41.108 E. Gorb, K. Haas, A. Henrich, S. Enders, N. Barbakadze, S. Gorb: Composite structure of the crystalline epicuticular wax layer of the slippery zone in the pitchers of the carnivorous plant *Nepenthes alata* and its effect on the insect attachment, J. Exp. Biol. **208**, 4651–4662 (2005)

41.109 H. Bohn, W. Federle: Insect aquaplaning: *Nepenthes* pitcher plants capture prey with the peristome, a fully wettable water-lubricated anisotropic surface, Proc. Natl. Acad. Sci. USA **39**, 14138–14143 (2004)

41.110 U. Bauer, H.F. Bohn, W. Federle: Harmless nectar source or deadly trap: *Nepenthes* pitchers are activated by rain, condensation and nectar, Proc. R. Soc. B **275**, 259–265 (2008)

41.111 H.C. Flemming: Auswirkungen mikrobieller Materialzerstörung. In: *Mikrobielle Materialzerstörung*, ed. by H. Brill (Georg Fischer, Stuttgart 1995) pp. 15–23, in German

41.112 J.N. Israelachvili: *Intermolecular and Surface Forces*, 2nd edn. (Academic Press, London 1992)

41.113 B. Bhushan: *Introduction to Tribology* (Wiley, New York 2002)

41.114 P.G. De Gennes, F. Brochard-Wyart, D. Quere: *Capillarity and Wetting Phenomena: Drops, Bubbles, Pearls, Waves* (Springer, New York 2004)

41.115 B. Bhushan (Ed.): *Nanotribology and Nanomechanics – An Introduction*, 2nd edn. (Springer, Heidelberg 2008)

41.116 A.V. Adamson: *Physical Chemistry of Surfaces* (Wiley, New York 1990)

41.117 R.N. Wenzel: Resistance of solid surfaces to wetting by water, Ind. Eng. Chem. **28**, 988 (1936)

41.118 A.B.D. Cassie, S. Baxter: Wettability of porous surfaces, Trans Faraday Soc. **40**, 546 (1944)

41.119 M. Reyssat, J.M. Yeomans, D. Quere: Impalement of fakir drops, Europhys Lett. **81**, 26006 (2008)

41.120 Y.C. Jung, B. Bhushan: Wetting behavior during evaporation and condensation of water microdroplets on superhydrophobic patterned surfaces, J. Microsc. **229**, 127–140 (2008)

41.121 C.W. Extrand: Model for contact angle and hysteresis on rough and ultraphobic surfaces, Langmuir **18**, 7991–7999 (2002)

41.122 B. Bhushan, Y.C. Jung: Wetting study of patterned surfaces for superhydrophobicity, Ultramicroscopy **107**, 1033–1041 (2007)

41.123 P. Roach, N.J. Shirtcliffe, M.I. Newton: Progress in superhydrophobic surface development, Soft Matter **4**, 224–240 (2008)

41.124 X. Zhang, F. Shi, J. Niu, Y. Jiang, Z. Wang: Superhydrophobic surfaces: from structural control to functional application, J. Mater. Chem. **18**, 621–633 (2008)

41.125 J. Heintzenberg: Fine particles in the global troposphere – A review, Tellus B **41**, 149–160 (1989)

41.126 J. Burkhardt, H. Kaiser, H. Goldbach, L. Kappen: Measurements of electrical leaf surface conductance reveal re-condensation of transpired water vapour on leaf surfaces, Plant Cell Environ. **22**, 189–196 (1999)

41.127 J. Burkhardt, K. Koch, H. Kaiser: Deliquescence of deposited atmospheric particles on leaf surfaces, Water Air Soil Pollut. **1**, 313–321 (2001)

41.128 W. Barthlott, W. Schultze-Motel: Zur Feinstruktur der Blattoberflächen und Systematischen Stellung der Laubmoosgattung *Rhacocarpus* und anderer *Hedwigiaceae*, Willdenowia **11**, 3–11 (1981), in German

41.129 H.G. Edelmann, C. Neinhuis, M. Jarvis, B. Evans, E. Fischer, W. Barthlott: Ultrastructure and chemistry of the cell wall of the moss *Rhacocarpus purpurascens* (*Rhacocarpaceae*): A puzzling architecture among plants, Planta **206**, 315–321 (1998)

41.130 A. Otten, S. Herminghaus: How plants keep dry: A physicist's point of view, Langmuir **20**, 2405–2408 (2004)

41.131 W.E. Ward: The lotus symbol: Its meaning in Buddhist art and philosophy, J. Aesthet. Art Crit. **11**, 135–146 (1952)

41.132 T.S. Chow: Nanoscale surface roughness and particle adhesion on structures substrates, Nanotechnology **18**, 1–4 (2007)

41.133 O. Pitois, X. Chateau: Small particle at a fluid interface: effect of contact angle hysteresis on force and work of detachment, Langmuir **18**, 9751–9756 (2002)

41.134 B. Bhushan, Y.C. Jung: Wetting, adhesion and friction of superhydrophobic and hydrophilic leaves and fabricated micro/nanopatterned surfaces, J. Phys.: Condens. Matter **20**, 225010 (2008)

41.135 A.K. Stosch, A. Solga, U. Steiner, C. Oerke, W. Barthlott, Z. Cerman: Efficiency of self-cleaning properties in wheat (*Triticum aestivum* L.), Appl. Bot. Food Qual. **81**, 49–55 (2007)

41.136 W. Barthlott, K. Riede, M. Wolter: Mimicry and ultrastructural analogy between the semi-aquatic grasshopper *Paulinia acuminata* (*Orthoptera: Pauliniidae*) and its foodplant, the water-fern *Salvinia auriculata* (*Filicatae: Salviniaceae*), Amazoniana **13**, 47–58 (1994)

41.137 M. Ayre: *Biomimicry – A review. Work package report, European Space Research & Technology Centre (ESTEC)* (European Space Agency (ESA), Noordwijk 2003)

41.138 D.W. Bechert, M. Bruse, W. Hage, R. Meyer: Fluid mechanics of biological surfaces and their technological application, Naturwissenschaften **87**, 157–171 (2000)

41.139 S.J. Abbott, P.H. Gaskell: Mass production of bio-inspired structured surfaces, Proc. Inst. Mech. Eng, Part C: J. Mech. Eng. Sci. **221**, 1181–1191 (2007)

41.140 H.C. Von Baeyer: The Lotus effect, The Sciences **40**, 12–15 (2000)

41.141 P. Forbes: *The Gecko's Foot* (Fourth Estate, London 2005)

41.142 A.K. Geim, S.V. Dubonos, I.V. Grigorieva, K.S. Novoselov, A.A. Zhukov, S.Y. Shapoval: Microfabricated adhesive mimicking gecko foot-hair, Nat. Mater. **2**, 461–463 (2003)

41.143 S. Gorb, M. Varenberg, A. Peressadko, J. Tuma: Biomimetic mushroom-shaped fibrillar adhesive microstructures, J. R. Soc. Interface **4**, 271–275 (2007)

41.144 B. Bhushan, R.A. Sayer: Surface characterization and friction of a bio-inspired reversible adhesive tape, Microsyst. Technol. **13**, 71–78 (2007)

41.145 M. Zhang, M. Liu, H. Prest, S. Fischer: Nanoparticles secreted from ivy rootlets for surface climbing, Nano Lett. **8**, 1277–1280 (2008)

41.146 T. Speck, T. Masselter, B. Prüm, O. Speck, R. Luchsinger, S. Fink: Plants as concept generators for biomimetic light-weight structures with variable stiffness and self-repair mechanisms, J. Bionics Eng. **1**, 199–205 (2004)

41.147 Y. Bar-Cohen: *Biomimetics: Biologically Inspired Technologies* (CRC Press Book, New York 2006)

41.148 P. Wagner, R. Fürstner, W. Barthlott, C. Neinhuis: Quantitative assessment to the structural basis of water repellency in natural and technical surfaces, J. Exp. Bot. **54**, 1295–1303 (2003)

41.149 Y.T. Cheng, D.E. Rodak, C.A. Wong, C.A. Hayden: Effects of micro- and nano-structures on the self-cleaning behaviour of lotus leaves, Nanotechnology **17**, 1359–1362 (2006)

41.150 M. Nosonovsky, B. Bhushan: Hierarchical roughness optimization for biomimetic superhydrophobic surfaces, Ultramicroscopy **107**, 969–979 (2007)

41.151 C.W. Extrand: Modeling of ultralyophobicity: Suspension of liquid drops by a single asperity, Langmuir **21**, 10370–10374 (2005)

41.152 A. Marmur: Wetting on hydrophobic rough surfaces: To be heterogeneous or not to be?, Langmuir **19**, 8343–8348 (2003)

41.153 S. Wang, L. Jiang: Definition of superhydrophobic states, Adv. Mater. **19**, 3423–3424 (2007)

41.154 L. Feng, Y. Zhang, J. Xi, Y. Zhu, N. Wang, F. Xia, L. Jiang: Petal effect: A superhydrophobic state with high adhesive force, Langmuir **24**, 4114–4119 (2008)

41.155 W. Li, A.A. Amirfazli: Hierarchical structures for natural superhydrophobic surfaces, Soft Matter **4**, 462–466 (2008)

41.156 M. Nosonovsky, B. Bhushan: Hierarchical roughness makes superhydrophobic states stable, Microelectron. Eng. **84**, 382–386 (2007)
41.157 M. Nosonovsky, B. Bhushan: Multiscale friction mechanisms and hierarchical surfaces in nano and bio-tribology, Mater. Sci. Eng. R **58**, 162–193 (2007)
41.158 M. Nosonovsky, B. Bhushan: Roughness induced superhydrophobicity: A way to design non adhesive surfaces, J. Phys. D **20**, 225009 (2008)
41.159 M. Nosonovsky, B. Bhushan: Biologically inspired surfaces: Broadening the scope of roughness, Adv. Func. Mater. **18**, 843–855 (2008)
41.160 C.J. Brinker: Evaporation-induced self-assembly: Functional nanostructures made easy, Mater. Res. Bull. **29**, 631–640 (2004)
41.161 K. Koch, A. Dommisse, W. Barthlott, S. Gorb: The use of plant waxes as templates for micro- and nanopatterning of surfaces, Acta Biomater. **3**, 905–909 (2007)
41.162 K. Koch, A.J. Schulte, A. Fischer, S. Gorb, W. Barthlott: A fast, precise and low cost replication technique for nano- and high aspect ratio structures of biological and artifical surfaces, Bioinspir. Biomimetics **3**, 046002 (2008)
41.163 B. Bhushan, Y.C. Jung, A. Niemietz, K. Koch: Lotus-like biomimetic hierarchical structures developed by self-assembly of tubular plant waxes, Langmuir **25**, 1659–1666 (2009)
41.164 W. Barthlott: Self-cleaning surfaces of objects and process for producing same. Patent, EP 0772514 B1, 8, Germany (1998)
41.165 A. Born, J. Ermuth, C. Neinhuis: Fassadenfarbe mit Lotus-Effekt: Erfolgreiche übertragung bestätigt, Phänomen Farbe **2**, 34–36 (2000)
41.166 W. Ming, D. Wu, R. van Benthem, G. de With: Superhydrophobic films from raspberrylike particles, Nano Lett. **5**, 2298–2301 (2005)
41.167 H. Höcker: Plasma treatment of textile fibres, Pure Appl. Chem. **74**, 423–427 (2002)
41.168 L. Gao, T.J. McCarthy: "Artificial Lotus Leaf" prepared using a 1945 patent and a commercial textile, Langmuir **22**, 5998–6000 (2006)
41.169 M. Pociute, B. Lehmann, A. Vitkauskas: Wetting behaviour of surgerical polyester woven fabrics, Mater. Sci. **9**, 410–413 (2003)
41.170 E. Nun, M. Oles, B. Schleich: Lotus-Effect®-surfaces, Macromol. Symp. **187**, 677–682 (2002)
41.171 P. Dendl, J. Interwies: Method for imparting a self-cleaning feature to a surface, and an object provided with a surface of this type. Patent WO 2001/079141, Germany (2001)
41.172 F. Müller, P. Winter: Clean surfaces with the lotus-effect, J. Com. Esp. Deterg. **34**, 103–111 (2004)
41.173 M. Nosonovsky, B. Bhushan: *Multiscale Dissipative Mechanisms and Hierarchical Surfaces* (Springer, Berlin Heidelberg 2008)
41.174 J. Tokunaga, M. Kumada, Y. Sugiyama, N. Watanabe, Y.B. Chong, N. Matsubara: Method of forming air film on submerged surface of submerged part-carrying structure, and film structure on submerged surface, European Patent EP 0616940 (1993) pp. 1–14
41.175 K. Fukuda, J. Tokunaga, T. Nobunaga, T. Nakatani, T. Iwasaki, Y. Kunitake: Frictional drag reduction with air lubricant over a super-water-repellent surface, J. Marine Sci. Tech. **5**, 123–130 (2000)
41.176 Z. Cerman, B.F. Striffler, W. Barthlott, T. Stegmeier, A. Scherrieble, V. von Arnim: Superhydrophobe Oberflächen für Unterwasseranwendungen. Patent, DE 10 2006 009, (2006), in German
41.177 J. Genzer, K. Efimenko: Recent developments in superhydrophobic surfaces and their relevance to marine fouling: A review, Biofouling **22**, 339–360 (2006)

42. Lotus Effect: Surfaces with Roughness-Induced Superhydrophobicity, Self-Cleaning, and Low Adhesion

Bharat Bhushan, Yong Chae Jung, Michael Nosonovsky

Superhydrophobic surfaces exhibit extreme water-repellent properties. These surfaces with high contact angle and low contact angle hysteresis also exhibit a self-cleaning effect and low drag for fluid flow. These surfaces are of interest in various applications, including self-cleaning windows, exterior paints for buildings, navigation ships, textiles, solar panels, and applications requiring antifouling and a reduction in fluid flow, e.g., in micro/nanochannels. Superhydrophobic surfaces can also be used for energy conservation and energy conversion, such as in the development of a microscale capillary engine. Superhydrophobic surfaces prevent the formation of menisci at a contacting interface and can be used to minimize adhesion and stiction. Certain plant leaves, notably lotus leaves, are known to be superhydrophobic and self-cleaning due to hierarchical roughness and the presence of wax tubules on the leaf surface. This phenomenon is known as the *lotus effect*. Superhydrophobic and self-cleaning surfaces can be produced by using roughness combined with hydrophobic coatings. In this chapter, the theory of roughness-induced superhydrophobicity and self-cleaning is presented, followed by the characterization data of natural leaf surfaces. Micro-, nano-, and hierarchical patterned structures have been fabricated, and the wetting properties and adhesion have been characterized to validate models and provide design guidelines for superhydrophobic and self-cleaning surfaces. In addition, a model of contact angle for oleophilic/phobic surfaces is presented. The wetting behavior of fabricated surfaces is investigated. Fundamental physical mechanisms of wetting responsible for the transition between various wetting regimes, contact angle, and contact angle hysteresis are also discussed.

42.1 Background

The primary parameter that characterizes wetting is the static contact angle, which is defined as the angle that a liquid makes with a solid. The contact angle depends on several factors, such as surface energy, surface roughness, and its cleanliness [42.1–5]. If the liquid wets the surface (referred to as wetting liquid or hydrophilic surface), the value of the static contact angle is $0° \le \theta \le 90°$, whereas if the liquid does not wet the surface (referred to as a nonwetting liquid or hydrophobic surface), the value of the contact angle is $90° < \theta \le 180°$. The term hydrophobic/philic, which was originally applied only to water (*hydro* means water in Greek), is often used to describe the contact of a solid surface with any liquid. The term *oleophobic/philic* is sometimes used with regard to wetting by oil and organic liquids. The term *amphiphobic/philic* is used for surfaces that are both hydrophobic/philic and oleophobic/philic. Surfaces with high energy, formed by polar molecules, tend to be hydrophilic, whereas those with low energy and built of nonpolar molecules tend to be hydrophobic.

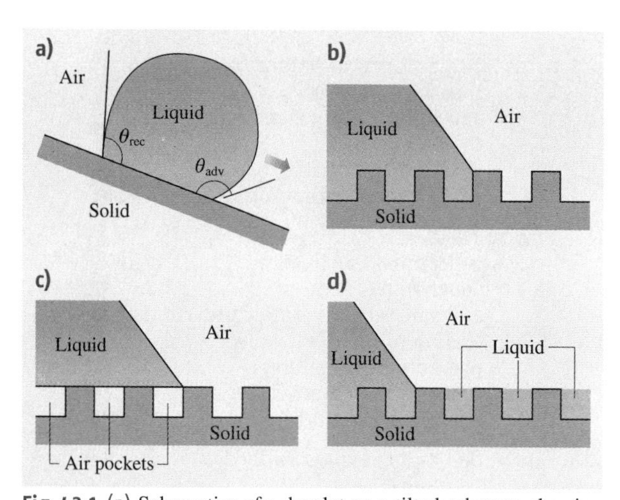

Fig. 42.1 **(a)** Schematics of a droplet on a tilted substrate showing advancing (θ_{adv}) and receding (θ_{rec}) contact angles. The difference between these angles constitutes the contact angle hysteresis. Configurations described by **(b)** the Wenzel equation for the homogeneous interface (42.6), **(c)** the Cassie–Baxter equation for the composite interface with air pockets (42.9), and **(d)** the Cassie equation for the homogeneous interface (42.10)

Surfaces with a contact angle between 150° and 180° are called superhydrophobic. In fluid flow, in order to have low drag and for applications requiring the self-cleaning feature, in addition to the high contact angle, superhydrophobic surfaces should also have very low water contact angle hysteresis. Water droplets roll off (with some slip) on these surfaces and take contaminants with them, providing the self-cleaning ability known as the *lotus effect*. The contact angle at the front of the droplet (advancing contact angle) is greater than that at the back of the droplet (receding contact angle), resulting in contact angle hysteresis (CAH) (Fig. 42.1a), which is the difference between the advancing and receding contact angles, representing two stable values. It occurs due to surface roughness and surface heterogeneity. Contact angle hysteresis reflects the irreversibility of the wetting/dewetting cycle. It is a measure of energy dissipation during the flow of a droplet along a solid surface. At a low value of CAH, the droplets may roll in addition to sliding, which facilitates removal of contaminant particles. Surfaces with low contact angle hysteresis have a low water roll-off (tilt) angle, which denotes the angle to which a surface must be tilted for roll off of water drops [42.5–12]. Surfaces with CAH or a low tilting angle of $< 10°$ are generally referred to as self-cleaning surfaces. Self-cleaning surfaces are of interest in various applications, including self-cleaning windows, windshields, exterior paints for buildings and navigation ships, utensils, roof tiles, textiles, solar panels, and applications requiring antifouling and a reduction of drag in fluid flow, e.g., in micro-/nanochannels.

Superhydrophobic surfaces can also be used for energy conservation and conversion [42.13–15]. Recent advances in superhydrophobic surfaces make such applications possible. Several concepts can be used. First, the hydrophobic/philic properties of a surface significantly affect the capillary adhesion force that, in turn, affects friction and energy dissipation during the sliding contact of solid surfaces. Selection of a proper superhydrophobic surface allows the reduction of energy dissipation. Second, superhydrophobic and superoleophobic surfaces can be used for fuel economy. Third, the recently discovered effect of reversible superhydrophobicity provides potential for new ways of energy conversion such as the microscale capillary engine.

Wetting may lead to the formation of menisci at the interface between solid bodies during sliding contact, which increases adhesion and friction. In some cases, the wet friction force can be greater than the dry friction force, which is usually undesirable [42.3, 4, 16, 17]. On the other hand, high adhesion is desirable in some applications, such as adhesive tapes and adhesion of cells to biomaterial surfaces; therefore, enhanced wetting would be desirable in these applications. Numerous applications, such as magnetic storage devices and micro-/nanoelectromechanical systems (MEMS/NEMS), require surfaces with low adhesion and stiction [42.16–21]. As the size of these devices decreases, surface forces tend to dominate over the volume forces, and adhesion and stiction constitute a challenging problem for proper operation of these devices. This makes the development of nonadhesive surfaces crucial for many of these emerging applications.

42.1.1 Natural Superhydrophobic and Self-Cleaning Surfaces

Starting in the 1990s, materials scientists paid attention to natural superhydrophobic surfaces. Among them are the leaves of water-repellent plants such as *Nelumbo nucifera* (lotus) and *Colocasia esculenta*, which have high contact angles with water [42.22–29]. First, the surface of the leaves is usually covered with a range of different waxes made from a mixture of hydrocarbon compounds that have a strong phobia of being wet. Second, the surface is very rough due to so-called papillose epidermal cells, which form asperities or papillae. In addition to the microscale roughness due to the papillae, the surface of the papillae is also rough with submicron-sized asperities composed of three-dimensional (3-D) epicuticular waxes, which occur in different morphologies such as tubules or platelets [42.28, 29]. Thus, they have hierarchical micro- and nanostructures, which have been studied by *Burton* and *Bhushan* [42.25] and *Bhushan* and *Jung* [42.26]. The water droplets on these surfaces readily sit on the apex of the nanostructures because air bubbles fill in the valleys of the structure under the droplet. Therefore, these leaves exhibit considerable superhydrophobicity. The water droplets on the leaves remove any contaminant particles from their surfaces when they roll off, leading to self-cleaning [42.22]. It has been reported that all superhydrophobic and self-cleaning leaves consist of an intrinsic hierarchical structure [42.28, 29]. The static contact angle and contact angle hysteresis of a lotus leaf are $\approx 164°$ and $3°$, respectively [42.29, 30]. Other examples of biological objects include water striders (*Gerris remigis*) [42.31] and mosquito (*Culex pipiens*) eyes [42.32]. Their hierarchical structures are responsible for the superhydrophobicity. Duck feathers and butterfly wings also provide superhydrophobicity. Their corrugated surfaces provide air pockets that prevent water from completely touching the surface.

A model surface for superoleophobicity and self-cleaning is provided by fish, which are known to be well protected from contamination by oil pollution although they are wetted by water [42.14]. Fish scales have a hierarchical structure consisting of sector-like scales with diameters of 4–5 mm covered by papillae 100–300 μm in length and 30–40 μm in width [42.33]. Shark skin, which is a model from nature for a low-drag surface, is covered by very small individual tooth-like scales called dermal denticles (little skin teeth), ribbed with longitudinal grooves (aligned parallel to the local flow direction of the water). These grooved scales reduce the formation of vortices present on a smooth surface, resulting in water moving efficiently over their surface [42.27, 34]. The water surrounding these complex structures can lead to protection from marine fouling and play a role in the defense against adhesion and growth of marine organisms, e.g., bacteria and algae [42.29, 35]. If oil is present on the surfaces in air or water, surfaces are known to be oleophobic and may provide self-cleaning and antifouling. Many sea animals including fishes and sharks are known to be oleophobic under water. Superoleophobic surfaces can also reduce significant losses of residual fuel in fuel tanks and pipes [42.13].

Study and simulation of biological objects with desired properties is referred to as *biomimetics*. Biomimetics involves taking engineering solutions from nature, mimicking them, and implementing them in an application. The term "biomimetics" is derived from the Greek word *biomimesis*, meaning to mimic life. Several other names are used as well, such as bionics and biognosis. Mimicking the lotus effect falls into the field of biomimetics.

42.1.2 Roughness-Induced Superhydrophobicity and Self-Cleaning

One of the ways to increase the hydrophobic or hydrophilic properties of a surface is to increase its surface roughness, so roughness-induced hydrophobicity or hydrophilicity has become the subject of extensive investigations. *Wenzel* [42.36] suggested a simple model predicting that the contact angle of a liquid with a rough surface is different from that with a smooth surface. *Cassie* and *Baxter* [42.37] showed that a gaseous phase including water vapor, commonly referred to as *air* in the literature, may be trapped in the cavities of a rough surface, resulting in a composite solid–liquid–air interface, as opposed to the homogeneous solid–liquid interface. These two models describe two possible wetting regimes or states: the homogeneous (Wenzel) and the composite (Cassie–Baxter) regimes. *Johnson* and *Dettre* [42.38] showed that the homogeneous and composite interfaces correspond to the two equilibrium states of a droplet. Many authors have investigated the stability of artificial superhydrophobic surfaces and showed that whether the interface is homogeneous or composite may depend on the history of the system, in particular whether the liquid was applied from the top or condensed at the bottom [42.39–44]. *Extrand* [42.6] pointed out that whether the interface is homogeneous or composite depends on droplet size. It has also been suggested that so-called two-tiered (or double) roughness, composed of a superposition of two roughness patterns at different length scales [42.45–47], and fractal roughness [42.48] may enhance superhydrophobicity.

Herminghaus [42.45] showed that certain self-affine profiles may result in superhydrophobic surfaces even for wetting liquids, if the local equilibrium condition for the triple line (the line of contact between solid, liquid, and air) is satisfied. *Nosonovsky* and *Bhushan* [42.49, 50] pointed out that such configurations, although formally possible, are likely to be unstable. *Nosonovsky* and *Bhushan* [42.50,51] proposed a probabilistic model for wetting of rough surfaces with a certain probability associated with every equilibrium state. According to their model, the overall contact angle with a two-dimensional roughness profile is calculated by assuming that the overall configuration of a droplet occurs as a result of superposition of numerous metastable states. The probability-based concept is consistent with the experimental data [42.8, 30, 52–62], which suggests that the transition between the composite and homogeneous interfaces is gradual, rather than instant. *Nosonovsky* and *Bhushan* [42.5, 9–11, 63–68] have identified mechanisms which led to the destabilization of the composite interface, namely the capillary waves, condensation and accumulation of nanodroplets, and surface inhomogeneity. These mechanisms are scale dependent, with different characteristic length scales. To effectively resist these scale-dependent mechanisms, a multiscale (hierarchical) roughness is required. High asperities resist the capillary waves, while nanobumps prevent nanodroplets from filling the valleys between asperities and pin the tripe line in case of a hydrophilic spot.

Various criteria have been formulated to predict the transitions from a metastable composite state to a wetted state [42.5, 11, 43, 53, 59–70]. *Extrand* [42.69] formulated the transition criterion referred to as the contact line density criterion, which was obtained by balancing the droplet weight and the surface forces along the contact line. *Patankar* [42.43] proposed a transition criterion based on energy balance. There is an energy barrier in going from a higher-energy Cassie–Baxter droplet to a lower-energy Wenzel droplet. The most probable mechanism is that the decrease in the gravitational potential energy during the transition helps to overcome this energy barrier. This energy barrier was estimated by considering an intermediate state in which the water fills the grooves below the contact area of a Cassie–Baxter droplet but the liquid–solid contact is yet to be formed at the bottom of the valleys. These criteria were tested on selected experiments from the literature [42.44, 71–73]. *Bhushan* et al. [42.53] and *Nosonovsky* and *Bhushan* [42.64, 70] found that the transition occurs at a critical value of the spacing factor, a nondimensional parameter which is defined as the diameter of the pillars divided by the pitch distance between them for patterned surfaces and its ratio to the droplet size. *Bhushan* and *Jung* [42.8, 52] and *Jung* and *Bhushan* [42.59–61] proposed the transition criterion based on the pitch distance between the pillars and the curvature of the droplet governed by the Laplace equation, which relates pressure inside the droplet to its curvature. In addition, the transition can occur by applying external pressure to the droplet, or by the impact of a droplet on the patterned surfaces [42.42, 61, 68, 74–76].

It has been demonstrated experimentally that roughness changes contact angle in accordance with the Wenzel model or Cassie–Baxter model, dependent upon whether the surface is hydrophilic or hydrophobic. *Yost* et al. [42.77] found that roughness enhances wetting of a copper surface with Sn–Pb eutectic solder, which has a contact angle of 15–20° for a smooth surface. *Shibuichi* et al. [42.48] measured the contact angle of various liquids (mixtures of water and 1,4-dioxane) on alkylketene dimer (AKD) substrate (contact angle not > 109° for a smooth surface). They found that, for wetting liquids, the contact angle decreases with increasing roughness, whereas for nonwetting liquids it increases. *Semal* et al. [42.78] investigated the effect of surface roughness on contact angle hysteresis by studying a sessile droplet of squalane spreading dynamically on multilayer substrates (behenic acid on glass) and found that an increase in microroughness slows the rate of droplet spread. *Erbil* et al. [42.79] measured the contact angle of polypropylene (contact angle of 104° for smooth surface) and found that the contact angle increases with increasing roughness. *Burton* and *Bhushan* [42.80] measured contact angle with roughness of patterned surfaces and found that, in the case of hydrophilic surfaces, it decreases with increasing roughness, whereas for hydrophobic surfaces, it increases with increasing roughness. *Bhushan* and *Jung* [42.8, 52], *Jung* and *Bhushan* [42.59–61, 81], *Bhushan* et al. [42.53–58], and *Koch* et al. [42.30] studied the influence of micro-, nano-, and hierarchical structures, created by replication of micropatterns and by self-assembly of hydrophobic alkanes and plant wax, on static contact angle, contact angle hysteresis, tilt angle and air pocket formation, adhesive force as well as efficiency of self-cleaning. They showed that, for micro-, nano-, and hierarchical structures, introduction of roughness increased the hydrophobicity of the surfaces. Hierarchical structure composed of a microstructure with a superimposed nanostructure of hydrophobic waxes led to superhydrophobicity with static contact angles of 173° and low contact angle hysteresis of 2°.

Contact angle on selected patterned surfaces has been measured to understand how the transition between the Cassie–Baxter regime and Wenzel regime occurs. Evaporation studies are useful in characterizing wetting behavior because droplets with various sizes can be created to evaluate the transition criterion on a given patterned surface [42.9, 10, 54, 56–60, 65, 66, 70, 82–85]. It has been reported that the wetting state changes from the Cassie–Baxter to Wenzel state as the droplet becomes smaller than a critical value on patterned surfaces during evaporation. Another important phenomenon related to wetting behavior is the bouncing of droplets. When a droplet hits a surface, it can bounce, spread or stick. In practical applications of superhydrophobic surfaces, surfaces should maintain their ability to repel penetrating droplets under dynamic conditions. *Jung* and *Bhushan* [42.61] and *Nosonovsky* and *Bhushan* [42.68] showed that the transition can occur by the impact of a droplet on a given patterned surface at a critical velocity with a critical geometric parameter.

An environmental scanning electron microscope (ESEM) can be used to condense or evaporate water droplets on surfaces by adjusting the pressure of the water vapor in the specimen chamber and the temperature of the cooling stage. Transfer of the water droplet has been achieved by a specially designed microinjector device on wool fibers and then imaged at room temperature in ESEM [42.86]. Images of water droplets show strong topographic contrast in ESEM such that reliable contact angle measurements can be made on the surfaces [42.87]. Water condensation and evaporation studies on patterned surfaces were carried out by *Jung* and *Bhushan* [42.60] and *Nosonovsky* and *Bhushan* [42.9, 10, 65, 66, 70], where the change of static contact angle and contact angle hysteresis was related with the surface roughness.

42.1.3 Scope of the Chapter

In this chapter, numerical models which provide relationships between roughness and contact angle, and contact angle hysteresis as well as the Cassie–Baxter and Wenzel regime transition are discussed. The role of microbumps and nanobumps is examined by analyzing surface characterization of hydrophobic and hydrophilic leaves on the micro- and nanoscale. Along with measuring and characterizing surface roughness, the contact angle and adhesion and friction properties of these leaves are also considered. The knowledge gained by examining these properties of the leaves and by quantitatively analyzing the surface structure has helped in the design of superhydrophobic and self-cleaning surfaces. Next, the techniques of producing superhydrophobic surfaces are described. Micro-, nano-, and hierarchical patterned structures have been fabricated using soft lithography, photolithography, and a technique which involves the replication of micropatterns and self-assembly of hydrophobic alkanes and plant waxes. They have been characterized to validate the models and to provide design guidelines for superhydrophobic and self-cleaning surfaces. To further examine the effect of meniscus force and real area of contact, scale dependence is considered with the use of atomic force microscope (AFM) tips of various radii. To investigate how the effects of droplet size and impact velocity influence the transition, evaporation and bouncing studies are conducted on silicon surfaces patterned with pillars of two different diameters and heights and with varying pitch values and deposited with a hydrophobic coating. In order to generate submicron droplets, an AFM-based technique using a modified nanoscale dispensing probe is presented. An ESEM study on the wetting behavior for a microdroplet with $\approx 20\,\mu$m radius on the patterned Si surfaces is presented. Furthermore, a model for predicting the oleophobic/philic nature of the surfaces is discussed. To validate the model, it is investigated how the water and oil droplets in three-phase interfaces influence the wetting behavior on micropatterned surfaces with varying pitch values and the nano- and hierarchical structures as well as the shark-skin replica as an example of aquatic animal.

42.2 Modeling of Contact Angle for a Liquid in Contact with a Rough Surface

42.2.1 Contact Angle Definition

The surface atoms or molecules of liquids or solids have fewer bonds with neighboring atoms, and therefore they have higher energy than similar atoms and molecules in the interior. This additional energy is characterized quantitatively by the surface tension or free surface energy γ, which is equal to the work required to create unit area of the surface at constant pressure and temperature. The unit of γ is J/m^2 or N/m, and it can be interpreted either as energy per unit surface area or as tension force per unit length of a line at the surface. When a solid is in contact with liquid, the molecular attraction will reduce the energy of the system below that for the two separated surfaces. This is expressed by the Dupré equation

$$W_{SL} = \gamma_{SA} + \gamma_{LA} - \gamma_{SL} , \tag{42.1}$$

where W_{SL} is the work of adhesion per unit area, γ_{SA} and γ_{SL} are the surface energies of the solid against air and liquid, and γ_{LA} is the surface energy of liquid against air [42.1–3].

If a liquid droplet is placed on a solid surface, the liquid and solid surfaces come together under equilibrium at a characteristic angle called the static contact angle θ_0 (Fig. 42.1a). This contact angle can be determined by minimizing the net surface free energy of the system [42.1–3]. The total energy E_{tot} is given by

$$E_{tot} = \gamma_{LA}(A_{LA} + A_{SL}) - W_{SL} A_{SL} , \tag{42.2}$$

where A_{SL} and A_{LA} are the contact areas of the liquid with the solid and air, respectively. It is assumed that the droplet of density ρ is smaller than the capillary length $(\gamma_{LA}/\rho g)^{1/2}$, so that the gravitational potential energy can be neglected. It is also assumed that the volume and pressure are constant, so that the volumetric energy does not change. At the equilibrium $dE_{tot} = 0$, which yields

$$\gamma_{LA}(dA_{LA} + dA_{SL}) - W_{SL}\, dA_{SL} = 0 . \tag{42.3}$$

For a droplet of constant volume, it is easy to show using geometrical considerations, that

$$\frac{dA_{LA}}{dA_{SL}} = \cos\theta_0 . \tag{42.4}$$

Combining (42.1), (42.3), and (42.4), the well-known Young equation for the contact angle is obtained

$$\cos\theta_0 = \frac{\gamma_{SA} - \gamma_{SL}}{\gamma_{LA}} . \tag{42.5}$$

Equation (42.5) provides an expression for the static contact angle for given surface energies. Note that although we use the term *air*, the analysis does not change in the case of another gas, such as water vapor.

42.2.2 Heterogeneous Interfaces and the Wenzel and Cassie–Baxter Equations

In this section, we will introduce and discuss the equations that govern the contact angle of liquid with a rough surface and heterogeneous interface.

The Wenzel equation, which can be derived using the surface force balance and empirical considerations, relates the contact angle of a water droplet upon a rough solid surface θ with that upon a smooth surface θ_0,

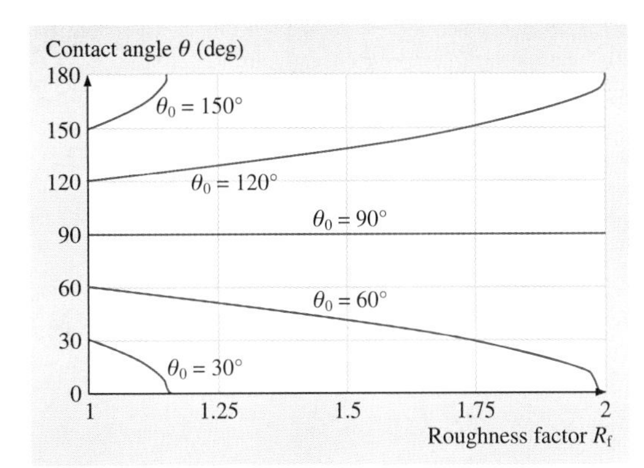

Fig. 42.2 Contact angle for a rough surface (θ) as a function of the roughness factor (R_f) for various contact angles for smooth surface (θ_0)

for a homogeneous interface (Fig. 42.1b), through the nondimensional surface roughness factor $R_f > 1$ equal to the ratio of the surface area A_{SL} to its flat projected area A_F

$$\cos\theta = \frac{dA_{LA}}{dA_F} = \frac{A_{SL}}{A_F}\frac{dA_{LA}}{dA_{SL}} = R_f \cos\theta_0 \,, \quad (42.6)$$

where

$$R_f = \frac{A_{SL}}{A_F} \,. \quad (42.7)$$

The dependence of the contact angle on the roughness factor is presented in Fig. 42.2 for different values of θ_0. The Wenzel model predicts that a hydrophobic surface ($\theta_0 > 90°$) becomes more hydrophobic with an increase in R_f, while a hydrophilic surface ($\theta_0 < 90°$) becomes more hydrophilic with an increase in R_f [42.49, 81].

In a similar manner, for a surface composed of two fractions, one with fractional area f_1 and contact angle θ_1 and the other with f_2 and θ_2, respectively (so that $f_1 + f_2 = 1$), the contact angle for the heterogeneous interface is given by the Cassie equation

$$\cos\theta = f_1 \cos\theta_1 + f_2 \cos\theta_2 \,. \quad (42.8)$$

For the case of a composite interface (Fig. 42.1c), consisting of the solid–liquid fraction ($f_1 = f_{SL}$, $\theta_1 = \theta_0$) and liquid–air fraction ($f_2 = f_{LA} = 1 - f_{SL}$, $\cos\theta_2 = -1$), combining (42.7) and (42.8) yields the Cassie–Baxter equation

$$\cos\theta = R_f f_{SL} \cos\theta_0 - 1 + f_{SL} \quad \text{or}$$
$$\cos\theta = R_f \cos\theta_0 - f_{LA}(R_f \cos\theta_0 + 1) \,. \quad (42.9)$$

The opposite limiting case of $\cos\theta_2 = 1$ ($\theta_2 = 0°$ corresponds to the water-on-water contact) yields the Cassie equation

$$\cos\theta = 1 + f_{SL}(\cos\theta_0 - 1) \,. \quad (42.10)$$

Equation (42.10) is sometimes used for a homogeneous interface instead of (42.6), if the rough surface is covered by holes filled with water [42.88] (Fig. 42.1d).

Two situations in wetting of a rough surface should be distinguished: the homogeneous interface without any air pockets shown in Fig. 42.1b (called the Wenzel interface, since the contact angle is given by the Wenzel equation or (42.6)) and the composite interface with air pockets trapped between the rough details as shown in Fig. 42.1c or Fig. 42.3a (called the Cassie or Cassie–Baxter interface, since the contact angle is given by (42.9)).

While (42.9) for the composite interface was derived using (42.6) and (42.8), it could also be obtained independently. For this purpose, two sets of interfaces are considered: a liquid–air interface with the ambient and a flat composite interface under the droplet, which itself involves solid–liquid, liquid–air, and solid–air interfaces. For fractional flat geometrical areas of the solid–liquid and liquid–air interfaces under the droplet, f_{SL} and f_{LA} ($f_{SL} = 1 - f_{LA}$), the flat area of the composite interface is

$$A_F = f_{SL} A_F + f_{LA} A_F = R_f A_{SL} + f_{LA} A_F \,. \quad (42.11)$$

In order to calculate the contact angle in a manner similar to the derivation of (42.6), the differential area of the liquid–air interface under the droplet $f_{LA}\,dA_F$ should be subtracted from the differential of the total liquid–air area dA_{LA}, which yields the Cassie–Baxter equation (42.9),

$$\begin{aligned}\cos\theta &= \frac{dA_{LA} - f_{LA}\,dA_F}{dA_F} = \frac{A_{SL}}{A_F}\frac{dA_{LA}}{dA_{SL}} - f_{LA}\\ &= R_f f_{SL}\cos\theta_0 - f_{LA}\\ &= R_f \cos\theta_0 - f_{LA}(R_f \cos\theta_0 + 1) \,.\end{aligned}$$

The dependence of the contact angle on the roughness factor and fractional liquid–air area for hydrophilic and hydrophobic surfaces is presented in Fig. 42.3b.

According to (42.9), even for a hydrophilic surface, the contact angle increases with increasing f_{LA}. At a high value of f_{LA}, a surface can become hydrophobic; however, the value required may be unachievable, or the formation of air pockets may become unstable. Using the Cassie–Baxter equation, the value of f_{LA} at which

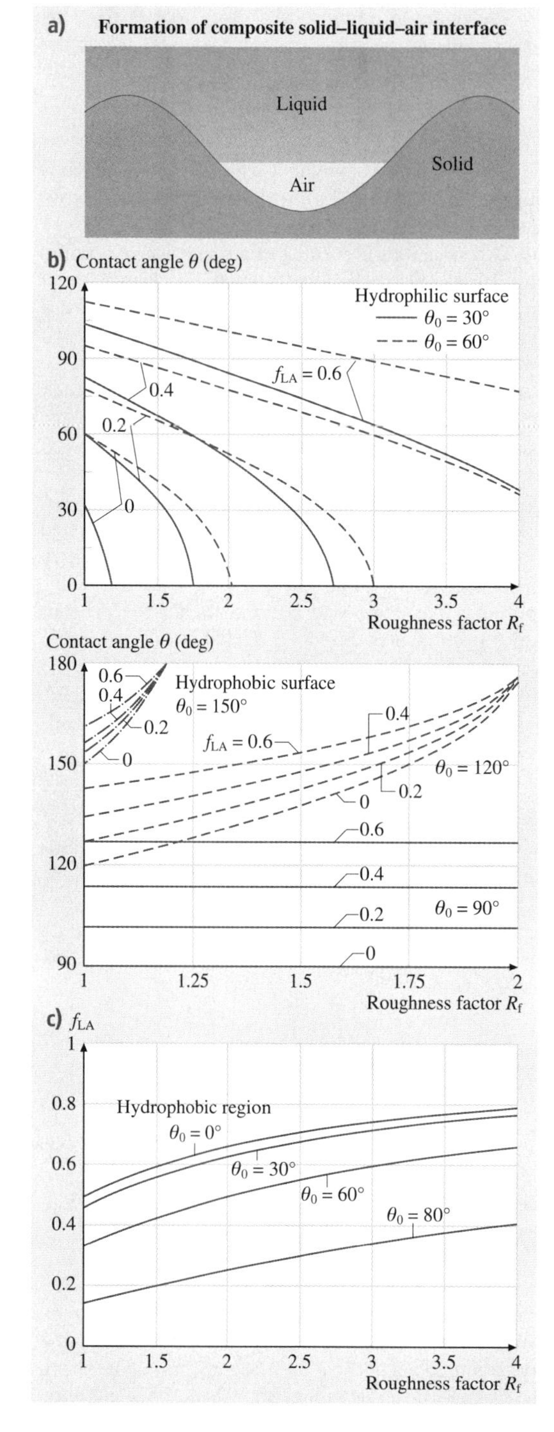

Fig. 42.3 **(a)** Schematic of the formation of a composite solid–liquid–air interface for rough surface, **(b)** the contact angle for rough surface (θ) as a function of the roughness factor (R_f) for various f_{LA} values on the hydrophilic surface and the hydrophobic surface, and **(c)** f_{LA} requirement for a hydrophilic surface to be hydrophobic as a function of the roughness factor (R_f) and θ_0 (after [42.81]) ◄

a hydrophilic surface could turn into a hydrophobic one is given as [42.81]

$$f_{LA} \geq \frac{R_f \cos\theta_0}{R_f \cos\theta_0 + 1} \quad \text{for} \quad \theta_0 < 90^\circ . \tag{42.12}$$

Figure 42.3c shows the value of f_{LA} requirement as a function of R_f for four surfaces with different contact angles θ_0. Hydrophobic surfaces can be achieved above a certain f_{LA} value, as predicted by (42.12). The upper part of each contact angle line is the hydrophobic region. For the hydrophobic surface, contact angle increases with an increase in f_{LA} for both smooth and rough surfaces.

Spreading of a liquid over a rough solid surface continues until simultaneously (42.5) (the Young equation) is satisfied locally at the triple line and the surface area is minimum over the entire liquid–air interface. The minimal surface area condition states that the sum of the inverse of principal radii of curvature, R_1 and R_2 of the liquid surface, along the two mutually orthogonal planes (mean curvature), is constant at any point, which governs the shape of the liquid–air interface. The same condition is also a consequence of the Laplace equation, which relates pressure change through an interface ΔP with its mean curvature

$$\frac{1}{R_1} + \frac{1}{R_2} = \frac{\Delta P}{\gamma_{LA}} . \tag{42.13}$$

Note that, at the thermodynamic equilibrium (when condensation and evaporation occur at the same speed), ΔP is dependent on the partial vapor pressure. For contact with saturated vapor the mean curvature of the liquid–air interface is zero at equilibrium. A convex interface ($1/R_1 + 1/R_2 > 0$) results in evaporation prevailing over condensation; this is why small droplets tend to evaporate. However, a concave interface ($1/R_1 + 1/R_2 < 0$) results in condensation of saturated vapor prevailing over evaporation. Since condensation prevails, a concave interface may be in thermodynamic equilibrium with undersaturated vapor. This is why concave menisci tend to condense even when the relative humidity is $< 100\%$.

Limitations of the Wenzel and Cassie Equations

The Cassie equation (42.8) is based on the assumption that the heterogeneous surface is composed of well-separated distinct patches of different material, so that the free surface energy can be averaged. It has been argued also ,that when the size of the chemical heterogeneities is very small (of atomic or molecular dimensions), the quantity that should be averaged is not the energy but the dipole moment of a macromolecule [42.89], and (42.8) should be replaced by

$$(1+\cos\theta)^2 = f_1(1+\cos\theta_1)^2 + f_2(1+\cos\theta_2)^2 . \tag{42.14}$$

Experimental studies of polymers with different functional groups showed good agreement with (42.14) [42.90].

Later investigations put the Wenzel and Cassie equations into a thermodynamic framework; however, they also showed that there is no one single value of the contact angle for a rough or heterogeneous surface [42.38, 40, 91]. The contact angle can be in a range of values between the receding contact angle θ_{rec} and the advancing contact angle θ_{adv}. The system tends to achieve the receding contact angle when liquid is removed (for example, at the rear end of a moving droplet), whereas the advancing contact angle is achieved when liquid is added (for example, at the front end of a moving droplet) (Fig. 42.4a). When the liquid is neither added nor removed, the system tends to have a static or *most stable* contact angle, which is given approximately by (42.5–42.10).

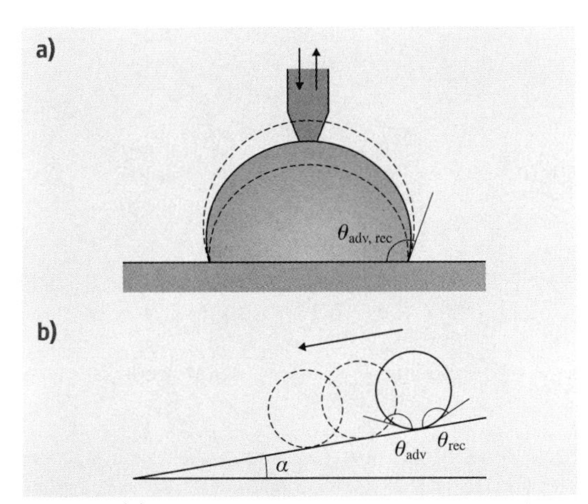

Fig. 42.4 (**a**) A liquid droplet in contact with a rough surface (advancing and receding contact angles are θ_{adv} and θ_{rec}, respectively) and (**b**) tilted surface profile (tilt angle α with a liquid droplet)

It is emphasized that the contact angle provided by (42.5–42.10) is a macroscale parameter, so it is sometimes called *the apparent contact angle*. The actual angle under which the liquid–air interface comes into contact with the solid surface at the micro- and nanoscale can be different. There are several reasons for this. First, water molecules tend to form a thin layer upon the surfaces of many materials. This is because of a long-distance van der Waals adhesion force that creates the so-called disjoining pressure [42.92]. This pressure is dependent upon the liquid layer thickness and may lead to the formation of stable thin films. In this case, the shape of the droplet near the triple line (the line of contact of the solid, liquid, and air, to be shown later in Fig. 42.6) transforms gradually from the spherical surface into a precursor layer, and thus the nanoscale contact angle is much smaller than the apparent contact angle. In addition, adsorbed water monolayers and multilayers are common for many materials. Second, even carefully prepared atomically smooth surfaces exhibit certain roughness and chemical heterogeneity. Water tends first to cover the hydrophilic spots with high surface energy and low contact angle [42.93]. The tilt angle due to roughness can also contribute to the apparent contact angle. Third, the very concept of the static contact angle is not well defined. For practical purposes, the contact angle, which is formed after a droplet is gently placed upon a surface and stops propagating, is considered the static contact angle. However, depositing the droplet involves adding liquid while leaving it involves evaporation, so it is difficult to avoid dynamic effects. Fourth, for small droplets and curved triple lines, the effect of the contact line tension may be significant. Molecules at the surface of a liquid or solid phase have higher energy because they are bonded to fewer molecules than those in the bulk. This leads to surface tension and surface energy. In a similar manner, molecules at the concave surface and, especially, at the edge have fewer bonds than those at the surface, which leads to line tension and curvature dependence of the surface energy. This effect becomes important when the radius of curvature is comparable with the so-called Tolman's length, normally of the molecular size [42.94]. However, the triple line at the nanoscale can be curved so that line tension effects become important [42.95]. The contact angle, taking into account the contact line effect, for a droplet with radius R is given by $\cos\theta = \cos\theta_0 + 2\tau/(R\gamma_{LA})$, where τ is the contact line tension, and θ_0 is the value given by the

Table 42.1 Wetting of a superhydrophobic surface as a multiscale process (after [42.11, 70])

Scale level	Characteristic length	Parameters	Phenomena	Interface
Macroscale	Droplet radius (mm)	Contact angle, droplet radius	Contact angle hysteresis	2-D
Microscale	Roughness detail (μm)	Shape of the droplet, position of the liquid–air interface (h)	Kinetic effects	3-D solid surface, 2-D liquid surface
Nanoscale	Molecular heterogeneity (nm)	Molecular description	Thermodynamic and dynamic effects	3-D

Young equation [42.96]. Thus, while the contact angle is a convenient macroscale parameter, wetting is governed by interactions at the micro- and nanoscale, which determine the contact angle hysteresis and other wetting properties. Table 42.1 shows various scale levels which affect wetting of a superhydrophobic surface.

Range of Applicability of the Wenzel and Cassie Equations

Gao and *McCarthy* [42.98] showed experimentally that the contact angle of a droplet is defined by the triple line and does not depend upon the roughness under the bulk of the droplet. A similar result for chemically heterogeneous surfaces was obtained by *Extrand* [42.99]. *Gao* and *McCarthy* [42.98] concluded that the Wenzel and Cassie–Baxter equations *should be used with the knowledge of their fault.* The question remained, however, under what circumstances the Wenzel and Cassie–Baxter equations can be safely used and under what circumstances they become irrelevant.

For a liquid front propagating along a rough two-dimensional profile (Fig. 42.5a,b), the derivative of the free surface energy (per liquid front length) W with respect to the profile length t yields the surface tension force $\sigma = \mathrm{d}W/\mathrm{d}t = \gamma_{SL} - \gamma_{SA}$. The quantity of practical interest is the component of the tension force that corresponds to the advancing of the liquid front in the horizontal direction for $\mathrm{d}x$. This component is given by $\mathrm{d}W/\mathrm{d}x = (\mathrm{d}W/\mathrm{d}t)(\mathrm{d}t/\mathrm{d}x) = (\gamma_{SL} - \gamma_{SA})\mathrm{d}t/\mathrm{d}x$. It is noted that the derivative $\mathrm{d}t/\mathrm{d}x$ is equal to Wenzel's roughness factor (R_f) in the case when the roughness factor is constant throughout the surface. Therefore, the Young equation (42.5), which relates the contact angle to the solid, liquid, and air interface tensions, is modified to [42.97]

$$\gamma_{LA}\cos\theta = R_f(\gamma_{SA} - \gamma_{SL}) . \qquad (42.15)$$

The empirical Wenzel equation (42.6) is a consequence of (42.15) combined with the Young equation.

Nosonovsky [42.97] showed that, for the more complicated case of nonuniform roughness given by the profile $z(x)$, the local value of the derivative, $r(x) = \mathrm{d}t/\mathrm{d}x = [1 + (\mathrm{d}z/\mathrm{d}x)^2]^{1/2}$, matters. In the cases that were studied experimentally by *Gao* and *McCarthy* [42.98] and *Extrand* [42.99], the roughness was present ($r > 1$) under the bulk of the droplet, but there was no roughness ($r = 0$) at the triple line, and the contact angle was given by (42.6) (Fig. 42.5c). In the general case of a 3-D rough surface $z(x, y)$, the roughness factor can be defined as a function of the coordinates $r(x, y) = [1 + (\mathrm{d}z/\mathrm{d}x)^2 + (\mathrm{d}z/\mathrm{d}y)^2]^{1/2}$.

Whereas (42.6) is valid for uniformly rough surfaces, that is, surfaces with $r = \text{const}$, for nonuniformly

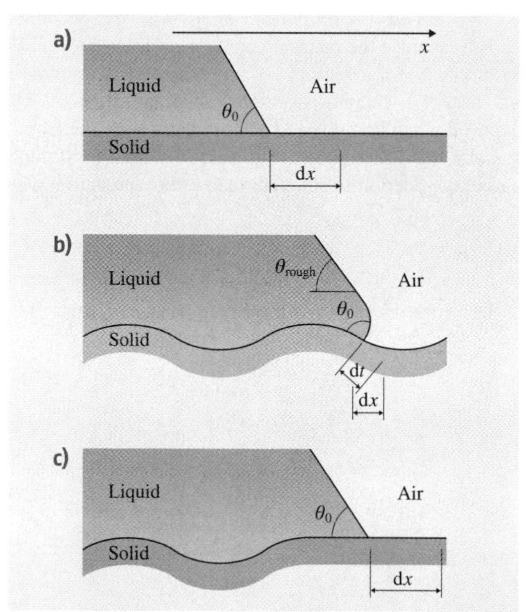

Fig. 42.5a–c Liquid front in contact with **(a)** a smooth solid surface and **(b)** a rough solid surface; propagation for a distance $\mathrm{d}t$ along the curved surface corresponds to the distance $\mathrm{d}x$ along the horizontal surface. **(c)** Surface roughness under the bulk of the droplet does not affect the contact angle (after [42.97])

rough surfaces the generalized Wenzel equation is formulated to determine the local contact angle (a function of x and y) with rough surfaces at the triple line [42.5, 10, 97]

$$\cos\theta = r(x, y)\cos\theta_0 . \quad (42.16)$$

The difference between the Wenzel equation (42.6) and the Nosonovsky–Bhushan equation (42.16) is that the latter is valid for nonuniform roughness with the roughness factor a function of the coordinates. Equation (42.16) is consistent with the experimental results of the scholars who showed that roughness beneath the droplet does not affect the contact angle, since it predicts that only roughness at the triple line matters. It is also consistent with the results of the researchers who confirmed the Wenzel equation (for the case of the uniform roughness) and of those who reported that only the triple line matters (for nonuniform roughness). A summary of experimental results for uniform and nonuniform rough and chemically heterogeneous surfaces is presented in Table 42.2.

The Cassie equation for the composite surface can be generalized in a similar manner by introducing the spatial dependence of the local densities, f_1 and f_2, of the solid–liquid interface with the contact angle as a function of x and y, given by

$$\cos\theta_{\text{composite}} = f_1(x, y)\cos\theta_1 + f_2(x, y)\cos\theta_2 , \quad (42.17)$$

where $f_1 + f_2 = 1$, and θ_1 and θ_2 are the contact angles of the two components [42.97].

The important question remains: what should be the typical size of roughness/heterogeneity details in order for the generalized Wenzel and Cassie equations (42.16) and (42.17) to be valid? Some scholars have suggested that roughness/heterogeneity details should be comparable with the thickness of the liquid–air interface and thus *the roughness would have to be of molecular dimensions to alter the equilibrium conditions* [42.101], whereas others have claimed that roughness/heterogeneity details should be small compared with the linear size of the droplet [42.8, 38, 52, 53, 59–61, 81, 91, 100]. The interface in our analysis is an idealized two-dimensional (2-D) object, which has no thickness. In reality, the triple line zone has two characteristic dimensions: the thickness (of the order of molecular dimensions) and the length (of the order of the droplet size).

The apparent contact angle, given by (42.16) and (42.17), may be viewed as the result of averaging of the local contact angle at the triple line over its length, and thus the size of the roughness/heterogeneity details should be small compared with the length (and not the thickness) of the triple line. A rigorous definition of the generalized equation requires the consideration of several length scales. The length dx needed for averaging of the energy gives the length over which the averaging is performed to obtain $r(x, y)$. This length should be larger than roughness details. However, it is still smaller than the droplet size and the length scale at which the apparent contact angle is observed (at which local variations of the contact angle level out). Since, of these lengths (roughness size, dx, droplet size), the first and the last are of practical importance, we conclude that the roughness details should be smaller than the droplet size. When the liquid–air interface is studied at the length scale of roughness/heterogeneity details, the local contact angle θ_0 is given by (42.6–42.10).

Table 42.2 Summary of experimental results for uniform and nonuniform rough and chemically heterogeneous surfaces. For nonuniform surfaces, the results shown are for droplets larger than the islands of nonuniformity. Detailed quantitative values of the contact angle in various sets of experiments can be found in the cited sources (after [42.97])

Experiment	Roughness/ hydrophobicity at the triple line and at the rest of the surface	Roughness at the bulk (under the droplet)	Experimental contact angle (compared with that at the rest of the surface)	Theoretical contact angle, Wenzel–Cassie equations	Theoretical contact angle, generalized Wenzel–Cassie (42.16,42.17)
Gao and *McCarthy* [42.98]	Hydrophobic	Hydrophilic	Not changed	Decreased	Not changed
	Rough	Smooth	Not changed	Decreased	Not changed
	Smooth	Rough	Not changed	Increased	Not changed
Extrand [42.99]	Hydrophilic	Hydrophobic	Not changed	Increased	Not changed
	Hydrophobic	Hydrophilic	Not changed	Decreased	Not changed
Bhushan et al. [42.53]	Rough	Rough	Increased	Increased	Increased
Barbieri et al. [42.100]	Rough	Rough	Increased	Increased	Increased

The liquid–air interface at that scale has perturbations caused by the roughness/heterogeneity, and the scale of the perturbations is the same as the scale of the roughness/heterogeneity details. However, when the same interface is studied at a larger scale, the effect of the perturbation vanishes, and apparent contact angle is given by (42.16) and (42.17) (Fig. 42.5c). This apparent contact angle is defined at the length scale for which the small perturbations of the liquid–air interface vanish, and the interface can be treated as a smooth surface. The values of $r(x, y)$, $f_1(x, y)$, and $f_2(x, y)$ in (42.16) and (42.17) are values averaged over an area (x, y) with a size larger than a typical roughness/heterogeneity detail size. Therefore, the generalized Wenzel and Cassie equations can be used at the scale at which the effect of the interface perturbations vanish, or in other words, when the size of the solid surface roughness/heterogeneity details is small compared with the size of the liquid–air interface, which is of the same order as the size of the droplet.

Nosonovsky and *Bhushan* [42.65] used the surface energy approach to find the domain of validity of the Wenzel and Cassie equations (uniformly rough surfaces) and generalized it to a more complicated case of nonuniform surfaces. The generalized equations explain a wide range of existing experimental data, which could not be explained by the original Wenzel and Cassie equations.

42.2.3 Contact Angle Hysteresis

Contact angle hysteresis is another important characteristic of a solid–liquid interface. Contact angle hysteresis occurs due to surface roughness and heterogeneity. Although for surfaces with roughness carefully controlled on the molecular scale it is possible to achieve contact angle hysteresis as low as $< 1°$ [42.102], hysteresis cannot be eliminated completely, since even atomically smooth surfaces have a certain roughness and heterogeneity. Contact angle hysteresis is a measure of energy dissipation during the flow of a droplet along a solid surface. Low contact angle hysteresis results in a very low water roll-off angle, which denotes the angle to which a surface must be tilted for roll-off of water drops (i. e., very low water contact angle hysteresis) [42.6–8,52,59, 60] (Fig. 42.4b). Low water roll-off angle is important in liquid flow applications such as in micro-/nanochannels and surfaces with self-cleaning ability.

There is no simple expression for contact angle hysteresis as a function of roughness; however, certain conclusions about the relationship of contact angle hysteresis with roughness can be made. It is known that the energy gained for surfaces during contact is greater than the work of adhesion for separating the surfaces, due to so-called adhesion hysteresis. Factors that affect contact angle hysteresis include adhesion hysteresis, surface roughness, and inhomogeneity. *Bhushan* et al. [42.53] and *Nosonovsky* and *Bhushan* [42.63] assumed that contact angle hysteresis is equal to the adhesion hysteresis term and a term corresponding to the effect of roughness H_r. They further noted that adhesion hysteresis can be assumed to be proportional to the fractional solid–liquid area $(1 - f_{LA})$. Using (42.9), the difference of cosines of the advancing and receding angles is related to the difference of those for a nominally smooth surface θ_{adv0} and θ_{rec0} as

$$\cos\theta_{adv} - \cos\theta_{rec} = R_f(1 - f_{LA})(\cos\theta_{adv0} - \cos\theta_{rec0}) + H_r\ . \quad (42.18)$$

The first term in the right-hand side of the equation, which corresponds to the inherent contact angle hysteresis of a smooth surface, is proportional to the fraction of the solid–liquid contact area $1 - f_{LA}$. The second term, H_r, is the effect of surface roughness, which is equal to the total perimeter of the asperity per unit area, or in other words, to the length density of the triple line [42.53]. Thus (42.18) involves terms proportional to both the solid–liquid interface area and the triple line length. It is observed from (42.9) and (42.18) that increasing $f_{LA} \to 1$ results in increasing contact angle ($\cos\theta \to -1, \theta \to \pi$) and decreasing contact angle hysteresis ($\cos\theta_{adv} - \cos\theta_{rec} \to 0$). In the limiting case of very small solid–liquid fractional contact area under the droplet, when the contact angle is large ($\cos\theta \approx -1 + (\pi - \theta)^2/2$, $\sin\theta \approx \theta - \pi$) and where the contact angle hysteresis is small ($\theta_{adv} \approx \theta \approx \theta_{rec}$), based on (42.9) and (42.18) [42.63]

$$\pi - \theta = \sqrt{2(1 - f_{LA})(R_f\cos\theta_0 + 1)}\ , \quad (42.19)$$

$$\theta_{adv} - \theta_{rec} = (1 - f_{LA})R_f\frac{\cos\theta_{adv0} - \cos\theta_{rec0}}{-\sin\theta} = \left(\sqrt{1 - f_{LA}}\right)R_f\frac{\cos\theta_{rec0} - \cos\theta_{adv0}}{\sqrt{2(R_f\cos\theta_0 + 1)}}\ . \quad (42.20)$$

For a homogeneous interface $f_{LA} = 0$, whereas for composite interface f_{LA} is nonzero. It is observed from (42.19) and (42.20) that, for a homogeneous interface, increasing roughness (high R_f) leads to increasing contact angle hysteresis (high values of $\theta_{adv} - \theta_{rec}$), while for a composite interface, an approach to unity of f_{LA} provides both high contact angle and small contact angle hysteresis [42.53, 63, 64, 81]. Therefore, the composite interface is desirable for self-cleaning.

A sharp edge can pin the line of contact of the solid, liquid, and air (also known as the *triple line*) at a position far from stable equilibrium, i. e., at contact angles different from θ_0 [42.103]. This effect is illustrated in the bottom sketch of Fig. 42.6, which shows a droplet propagating along a solid surface with grooves. At the edge point, the contact angle is not defined and can have any value between the values corresponding to contact with the horizontal and inclined surfaces. For a droplet moving from left to right, the triple line will be pinned at the edge point until it will be able to proceed to the inclined plane. As observed from Fig. 42.6, the change of the surface slope (α) at the edge is the cause of the pinning. Because of the pinning, the value of the contact angle at the front of the droplet (the dynamic maximum advancing contact angle or $\theta_{adv} = \theta_0 + \alpha$) is greater than θ_0, whereas the value of the contact angle at the back of the droplet (the dynamic minimum receding contact angle or $\theta_{rec} = \theta_0 - \alpha$) is smaller than

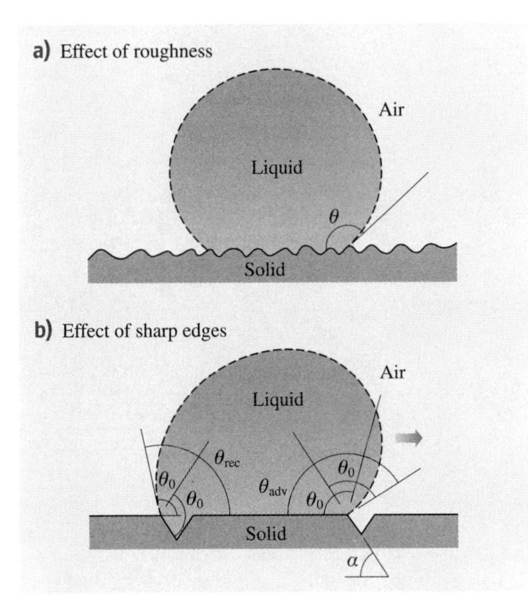

Fig. 42.6a,b A droplet of liquid in contact with a solid surface: rough surface, contact angle θ; and a surface with sharp edges. For a droplet moving from left to right at the sharp edge (*arrow*), the contact angle at the edge may take any value between the contact angle with the horizontal and the inclined planes. This effect results in the difference between the advancing ($\theta_{adv} = \theta_0 + \alpha$) and receding ($\theta_{rec} = \theta_0 - \alpha$) contact angles (after [42.49])

θ_0. A hysteresis domain of the dynamic contact angle is thus defined by the difference $\theta_{adv} - \theta_{rec}$. The liquid can travel easily along the surface if the contact angle hysteresis is small. It is noted that the static contact angle lies within the hysteresis domain; therefore, increasing the static contact angle up to the values of a superhydrophobic surface (approaching 180°) will also result in a reduction of the contact angle hysteresis. In a similar manner, contact angle hysteresis can also exist even if the surface slope changes smoothly, without sharp edges. There is an analogy between the two mechanisms leading to contact angle hysteresis (energy dissipation at the solid–liquid interface and pinning of the triple line) and dissipation mechanisms of dry friction (adhesion and deformation) [42.104].

42.2.4 Stability of a Composite Interface and the Role of Hierarchical Structure

Stability of a composite interface is an important issue. Even though it may be geometrically possible for the system to become composite, it may be energetically profitable for the liquid to penetrate into the valleys between asperities and form a homogeneous interface. *Marmur* [42.40] formulated geometrical conditions for a surface under which the energy of the system has a local minimum and the composite interface may exist. *Patankar* [42.43] pointed out that whether the homogeneous or composite interface exists depends on the system's history, i. e., on whether the droplet was formed at the surface or deposited. However, the above-mentioned analyses do not provide an answer to which of the two possible configurations, homogeneous or composite, will actually form.

Formation of a composite interface is also a multiscale phenomenon which depends upon the relative sizes of the liquid droplet and roughness details. A composite interface is fragile and can be irreversibly transformed into a homogeneous interface, thus damaging superhydrophobicity. In order to form a stable composite interface with air pockets between solid and liquid, the destabilizing factors such as capillary waves, nanodroplet condensation, surface inhomogeneity, and liquid pressure should be avoided.

1. The capillary waves at the liquid–air interface may destabilize the composite interface. Due to an external perturbation, a standing capillary wave can form at the liquid–air interface. If the amplitude of the

capillary wave is greater than the height of the asperity, the liquid can touch the valley between the asperities, and if the angle under which the liquid comes into contact with the solid is greater than θ_0, it is energetically profitable for the liquid to fill the valley [42.49, 50]. When the composite interface is destroyed and the space between the asperities is filled with water, it is highly unlikely that the composite interface will be formed again because the transition from the noncomposite solid–liquid interface to a composite interface would require a large activation energy. Such a transition has never been observed. The effect of capillary waves is more pronounced for small asperities with height comparable to the wave amplitude.

2. Nanodroplets may condense and accumulate in the valleys between asperities and eventually destroy the composite interface. *Cheng* et al. [42.105] observed condensation of submicron-size droplets on a lotus leaf surface and found that droplets tend to condense at areas adjacent to bumps (i. e., in the valleys) and have a contact angle of $< 90°$, whereas larger droplets have higher contact angle, thus demonstrating that the contact angle is scale dependent. The scale effect is observed for small droplets or at small distances near the triple line. Scale dependence of the contact angle has been reported by *Nosonovsky* and *Bhushan* [42.63]. At nanoscale distances from the triple line the liquid touches the solid under a much lower contact angle.
3. Even hydrophobic surfaces are usually not chemically homogeneous and can have hydrophilic spots. It is known from experiments that, for droplets of submicron size, the value of the contact angle is usually smaller than for droplets at the macroscale [42.42]. *Checco* et al. [42.93] suggested that surface inhomogeneity is responsible for this scale effect, since nanodroplets tend to sit at the highest free surface energy (most hydrophilic) spots and thus have lower contact angles. Their phenomenological numerical simulations showed good agreement with experimental data.

For high f_{LA}, a nanopattern is desirable because whether a liquid–air interface is generated depends upon the ratio of the distance between two adjacent asperities and the droplet radius. Furthermore, asperities can pin liquid droplets and thus prevent liquid from filling the valleys between asperities. High R_f can be achieved by both micropatterns and nanopatterns. *Nosonovsky* and *Bhushan* [42.5, 9–11, 63, 64, 67, 70] have demonstrated that a combination of microroughness and nanoroughness (multiscale roughness) with convex surfaces can help resist the destabilization by pinning the interface. It also helps in preventing the gaps between the asperities from filling with liquid, even in the case of a hydrophilic material. The effect of roughness on wetting is scale dependent, and mechanisms that lead to destabilization of a composite interface are also scale dependent. To resist these scale-dependent mechanisms effectively, it is expected that multiscale roughness is optimum for superhydrophobicity.

To guarantee the equilibrium and a local minimum of the total energy ($\mathrm{d}^2W > 0$) the following should be satisfied

$$\mathrm{d}^2W = \mathrm{d}^2A_{SL}\left(\cos\theta_0 - \frac{\gamma_{SA} - \gamma_{SL}}{\gamma_{LA}}\right)\gamma_{SL} + \mathrm{d}A_{LA}\,\mathrm{d}(\cos\theta) > 0\,. \qquad (42.21)$$

Using (42.5), which is satisfied at the equilibrium, and the fact that $\cos\theta$ decreases monotonically with θ in the domain of interest, $0 < \theta < 180°$, yields

$$\mathrm{d}A_{SL}\,\mathrm{d}\theta_0 < 0\,. \qquad (42.22)$$

Based on (42.22), for the interface to be stable, the value of the contact angle should decrease when the liquid–air interface advances, whereas for receding liquid the contact angle should increase. For a two-dimensional surface, the change of angle is equal to the change of the slope of the surface, and whether the configuration is stable or not depends on the sign of curvature of the surface. A convex surface (with bumps) leads to a stable interface, whereas a concave surface (with grooves) leads to an unstable interface. This approach was suggested for creating superoleophobic surfaces, since the surface tension of oil and organic liquids is much lower than that of water, and it is difficult to create a surface not wetted by oil [42.106]. Since oleophilic surfaces in air (solid–oil–air system) can become superoleophobic when immersed in water (solid–oil–water system), underwater superoleophobicity has potential for self-cleaning antifouling surfaces for ships.

An experiment suggesting that the sign of the curvature is indeed important for hydrophobicity was conducted by *Sun* et al. [42.47]. They produced both a positive and a negative replica of a lotus leaf surface by nanocasting using poly(dimethylsiloxane), which has a contact angle with water of $\approx 105°$. This value is close to the contact angle of the wax which covers lotus leaves ($\approx 103°$, as reported by *Kamusewitz* et al. [42.107]). The positive and negative replicas have the same roughness factor and thus should produce the same contact angle in the case of a homogeneous interface, according to (42.6); however, the values of surface curvature are opposite. The value of contact angle for the positive replica was found to be 160° (the same as for lotus leaf), while for the negative replica it was only 110°. This result suggests that the high contact angle for lotus leaf is due to the composite, rather than homogeneous interface, and that the sign of the surface curvature indeed plays a critical role for formation of the composite interface.

42.2.5 The Cassie–Baxter and Wenzel Wetting Regime Transition

Since superhydrophobicity requires a stable composite interface, it is important to understand the destabilization mechanisms for the Cassie–Baxter and Wenzel wetting transition. It is known from experimental observations that the transition from the Cassie–Baxter to Wenzel regime can be an irreversible event. Whereas such a transition can be induced, for example, by applying pressure or force to the droplet [42.61, 68], electric voltage [42.108, 109], light for a photocatalytic texture [42.110], and vibration [42.111], the opposite transition is never observed. Several approaches have been proposed for investigation of the transition between the Cassie–Baxter and Wenzel regimes, referred to as *the Cassie–Wenzel transition*. It has been suggested that the transition takes place when the net surface energy of the Wenzel regime becomes equal to that of the Cassie–Baxter regime, or in other words, when the contact angle predicted by the Cassie–Baxter equation is equal to that predicted by the Wenzel equation. *Lafuma* and *Quéré* [42.42] noticed that in certain cases the transition does not occur even when it is energetically profitable, and considered such a Cassie–Baxter state metastable. *Extrand* [42.99] suggested that the weight of the droplet is responsible for the transition and proposed the contact line density model, according to which the transition takes place when the weight exceeds the surface tension force at the triple line. *Patankar* [42.43] suggested that which of the two states is realized may depend upon how the droplet was formed, that is, upon the history of the system. *Quéré* [42.112] also suggested that the droplet curvature (which depends upon the pressure difference between inside and outside of the droplet) governs the transition. *Nosonovsky* and *Bhushan* [42.50] suggested that the transition is a dynamic process of destabilization and identified possible destabilizing factors. It has also been suggested that the curvature of multiscale roughness defines the stability of the Cassie–Baxter wetting regime [42.9, 10, 63–66, 104, 113] and that the transition is a stochastic, gradual process [42.49, 111, 114, 115]. Numerous experimental results support many of these approaches; however, it is not clear which particular mechanism prevails.

There is an asymmetry between the wetting and dewetting processes, since less energy is released during wetting than the amount required for dewetting, due to adhesion hysteresis. Adhesion hysteresis is one of the reasons that leads to contact angle hysteresis, and it also results in the hysteresis of the Wenzel and Cassie–Baxter state transition. Figure 42.7 shows the contact angle of a rough surface as a function of surface roughness parameter, given by (42.9). Here it is assumed that $R_f \approx 1$ for a Cassie–Baxter regime with a stable composite interface, and the liquid droplet sits flat over the surface. It is noted that, at a certain point, the contact angles given by the Wenzel and Cassie–Baxter equations are the same, and $R_f = (1 - f_{LA}) -$

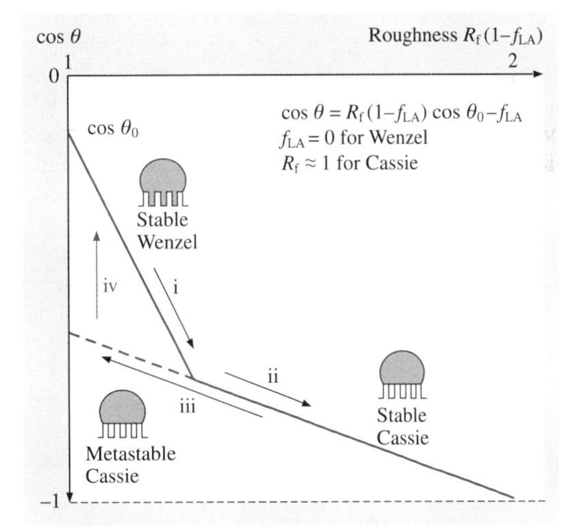

Fig. 42.7 Wetting hysteresis for a superhydrophobic surface: contact angle as a function of roughness. The stable Wenzel state (i) can transform into the stable Cassie state with increasing roughness (ii). The metastable Cassie state (iii) can abruptly transform (iv) into the stable Wenzel state. The transition (i/ii) corresponds to equal Wenzel and Cassie states free energies, whereas the transition (iv) corresponds to significant energy dissipation and thus is irreversible (after [42.65])

$f_{LA}/\cos\theta_0$. At this point, the lines corresponding to the Wenzel and Cassie–Baxter regimes intersect. This point corresponds to equal net energy of the Cassie–Baxter and Wenzel regimes. For lower roughness (e.g., larger pitch between the asperities) the Wenzel state is more energetically profitable, whereas for higher roughness the Cassie–Baxter regime is more energetically profitable.

It is observed from Fig. 42.7 that an increase of roughness may lead to the transition between the Wenzel and Cassie–Baxter regimes at the intersection point. With decreasing roughness, the system is expected to transit to the Wenzel state. However, experiments [42.8, 52, 53, 59–61, 100] show that, despite the energy in the Wenzel regime being lower than that in the Cassie–Baxter regime, the transition does not necessarily occur, and the droplet may remain in the metastable Cassie–Baxter regime. This is because there are energy barriers associated with the transition, which occurs due to destabilization by dynamic effects (such as waves and vibration).

In order to understand contact angle hysteresis and the transition between the Cassie–Baxter and Wenzel regimes, the shape of the free surface energy profile can be analyzed. The free surface energy of a droplet upon a smooth surface as a function of the contact angle has a distinct minimum which corresponds to the most stable contact angle. As shown in Fig. 42.8a, the macroscale profile of the net surface energy allows us to find the contact angle (corresponding to energy minimums); however, it fails to predict the contact angle hysteresis and Cassie–Baxter and Wenzel transition, which are governed by micro- and nanoscale effects. As soon as microscale substrate roughness is introduced, the droplet shape cannot be considered as

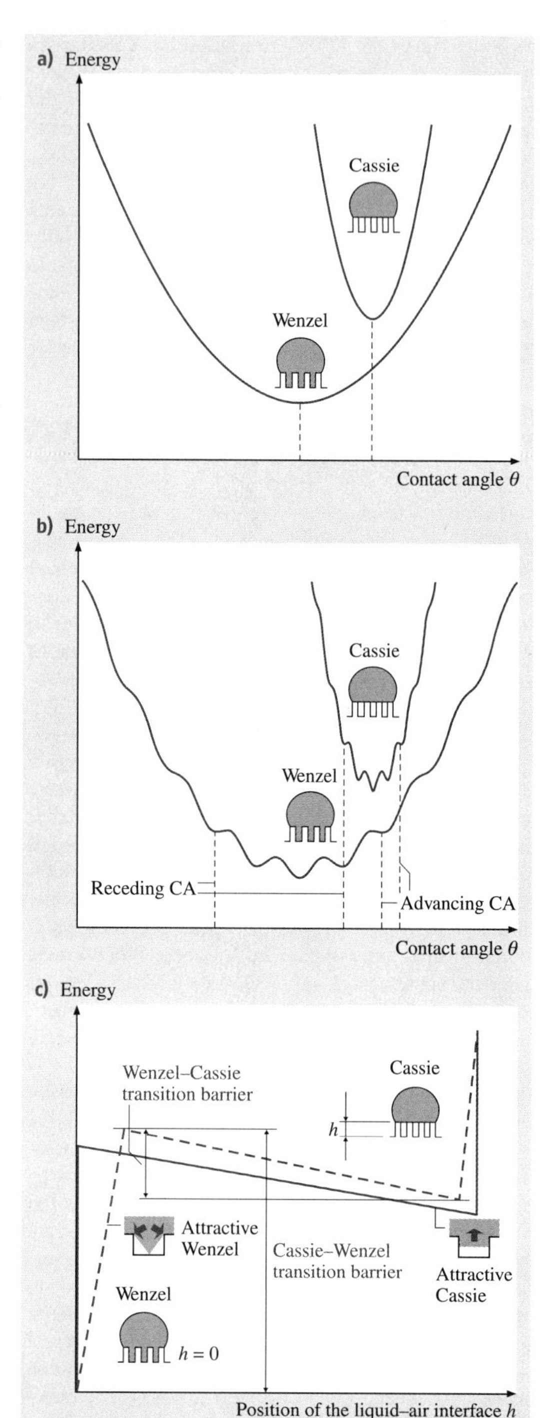

Fig. 42.8a–c Schematic net free-energy profiles. **(a)** Macroscale description; energy minimums correspond to the Wenzel and Cassie states. **(b)** Microscale description with multiple energy minimums due to surface texture. The largest and smallest values of the energy minimum correspond to the advancing and receding contact angles. **(c)** The origin of the two branches (Wenzel and Cassie) is found when the dependence of energy upon h (air-layer thickness or the vertical position of the liquid–air interface) is considered for the microscale description (solid line) or nanoscale imperfection (*dashed line*) (after [42.65]). When nanoscale imperfection is introduced, it is observed that the Wenzel state corresponds to an energy minimum and the energy barrier for the Wenzel–Cassie transition is much smaller than that for the opposite transition ▶

an ideal truncated sphere anymore, and energy profiles have multiple energy minima, corresponding to the location of the asperities (Fig. 42.8b). The microscale energy profile (solid line) has numerous energy maxima and minima due to surface asperities. While exact calculation of the energy profile for a 3-D droplet is complicated, a qualitative shape may be obtained by assuming a periodic sinusoidal dependence [42.38] superimposed upon the macroscale profile, as shown in Fig. 42.8b. Thus the advancing and receding contact angles can be identified as the maximum and minimum possible contact angles corresponding to energy minimum points. However, the transition between the Wenzel and Cassie–Baxter branches still cannot be explained. Note also that Fig. 42.8b explains qualitatively the hysteresis due to the kinetic effect of the asperities, but not the inherited adhesion hysteresis, which is characterized by the molecular length scale and cannot be captured by the microscale model.

The energy profile as a function of the contact angle does not provide information on how the transition between the Cassie–Baxter and Wenzel regimes occurs, because their two states correspond to completely isolated branches of the energy profile in Fig. 42.8a,b. However, the energy may depend not only upon the contact angle, but also upon micro-/nanoscale parameters, such as for example the vertical position of the liquid–air interface under the droplet, h (assuming that the interface is a horizontal plane) or similar geometrical parameters (assuming a more complicated shape of the interface). In order to investigate the Wenzel and Cassie–Baxter transition, the dependence of the energy upon these parameters should be studied. We assume that the liquid–air interface under the droplet is a flat horizontal plane. When such air layer thickness or the vertical position of the liquid–air interface h is introduced, the energy can be studied as a function of the droplet's shape, the contact angle, and h (Fig. 42.8c). For an ideal situation, the energy profile has an abrupt minimum at the point corresponding to the Wenzel state, which corresponds to the sudden net energy change due to the destruction of the solid–air and liquid–air interfaces [$\gamma_{SL} - \gamma_{SA} - \gamma_{LA} = -\gamma_{LA}(\cos\theta + 1)$ times the interface area] (Fig. 42.8c). In a more realistic case, the liquid–air interface cannot be considered horizontal due to nanoscale imperfectness or dynamic effects such as the capillary waves [42.50]. A typical size of the imperfectness is much smaller than the size of details of the surface texture and thus belongs to the molecular scale level. The height of the interface h can now be treated as an average height. The dependence of energy on h is now not as abrupt as in the idealized case. For example, for the *triangular* shape shown in Fig. 42.8c, the Wenzel state may become the second attractor for the system. It is seen that there are two equilibriums which correspond to the Wenzel and Cassie–Baxter regimes, with the Wenzel state corresponding to a much lower energy level. The dependence of energy on h governs the transition between the two states, and it is observed that a much larger energy barrier exists for the transition from Wenzel to Cassie–Baxter regime than for the opposite transition. This is why the former transition has never been observed experimentally [42.70].

To summarize, the contact angle hysteresis and Cassie–Baxter and Wenzel transition cannot be determined from the macroscale equations and are governed by micro- and nanoscale phenomena.

42.3 Lotus Effect Surfaces in Nature

Many biological surfaces are known to be superhydrophobic and self-cleaning. In this section, we will discuss various plant leaves and their roughness and wax coatings in relation to their hydrophobic/hydrophilic and self-cleaning properties.

42.3.1 Superhydrophobic and Self-Cleaning Plant Leaves

Hydrophobic and water-repellent abilities of many plant leaves have been known for a long time. Scanning electron microscope (SEM) studies since the 1970s have revealed that the hydrophobicity of the leaf surface is related to its microstructure. The outer cells covering a plant, especially the leaf, are called epidermis cells. The epidermis in all plant surfaces is covered by a cuticle composed of soluble lipids embedded in a polyester matrix, which makes the plant surface hydrophobic in most cases [42.23, 28, 29]. The chemical structure of the epicuticular waxes has been studied extensively by plant scientists and lipid chemists in recent decades [42.116, 117]. The epicuticular waxes can be either thin with a 2-D structure, thick with a 3-D structure or a combination thereof. The hydrophobicity of the leaves is related to another important effect: the ability to remain clean after being immersed in dirty water, known as self-cleaning. This ability is best known for the lotus (*Nelumbo nucifera*) leaf that is considered by some Asian cultures as sacred due to its purity. Not surprisingly, the ability of lotus-like surfaces for self-cleaning and water repellency was dubbed the *lotus effect*. As far as the biological implications of the lotus effect, self-cleaning plays an important role in defense against pathogens binding to the leaf surface. Many spores and conidia of pathogenic organisms – most fungi – require water for germination and can infect leaves only in the presence of water.

Neinhuis and *Barthlott* [42.22] systematically studied surfaces and wetting properties of about 200 water-repellent plants (for a comprehensive review, see *Koch* et al. [42.28,29]). Among the epidermal relief features are the papillose epidermal cells, either with every epidermal cell forming a single papilla or cells being divided into papillae. The scale of the epidermal relief ranged from 5 μm in multipapillate cells to 100 μm in large epidermal cells. Some cells are also convex (rather than having real papillae) and/or have hairs (trichomes). *Neinhuis* and *Barthlott* [42.22] also found various types and shapes of wax crystals at the surface; also see *Koch* et al. [42.28, 29]. Interestingly, the hairy surfaces with a thin film of wax exhibited water repellency for short periods (minutes), after which water penetrated between the hairs, whereas hairs with a thick film led to strong water repellency. The wax crystal creates nanoroughness, in addition to the microroughness created by the papillae. Apparently, roughness plays the dominant role in the lotus effect.

The SEM study reveals that the lotus leaf surface is covered by *bumps*, more exactly called papillae (papillose epidermal cells), which in turn are covered by an additional layer of epicuticular waxes [42.23]. The wax is present in crystalline tubules, composed of a mixture of aliphatic compounds, principally nonacosanol and nonacosanediols [42.28, 29, 118]. The wax is hydrophobic with a water contact angle of $\approx 95-110°$, whereas the papillae provide the tool to magnify the contact angle based on the Wenzel model, discussed in the preceding section. The experimental value of the static water contact angle with the lotus leaf was reported to be $\approx 160°$ [42.23, 25, 26]. Indeed, taking the papillae density of $3400\,/\mathrm{mm}^2$, the average radius of the hemispherical asperities $r = 10\,\mu\mathrm{m}$, and the aspect ratio $h/r = 1$, provides, based on (42.6), a value of the roughness factor $R_\mathrm{f} \approx 4$ [42.49]. Taking the value of the contact angle for wax, $\theta_0 = 104°$ [42.107], the calculation with the Wenzel equation yields $\theta = 165°$, which is close to the experimentally observed values [42.49]. However, the simple Wenzel model may be not sufficient to explain the lotus effect, as the roughness structure forms a composite interface. Moreover, its structure has hierarchical roughness. So, a number of more sophisticated models have been developed to study the role of hierarchical roughness on contact angle [42.5,9–11,63–67,70]. A qualitative explanation for self-cleaning is that on a smooth surface contamination particles are mainly redistributed by a water droplet; on a rough surface they adhere to the droplet and are removed from the leaves when the droplet rolls off.

42.3.2 Characterization of Superhydrophobic and Hydrophilic Leaf Surfaces

In order to understand the mechanisms of hydrophobicity in plant leaves, a comprehensive comparative study of the superhydrophobic and hydrophilic leaf surfaces and their properties was carried out by *Bhushan* and *Jung* [42.26] and *Burton* and *Bhushan* [42.25]. Below is a discussion of the findings of the study.

Experimental Techniques

The static contact angles were measured using a Rame–Hart model 100 contact angle goniometer with droplets of deionized (DI) water [42.25, 26]. Droplets of $\approx 5\,\mu\mathrm{L}$ in volume (with diameter of a spherical droplet ≈ 2.1 mm) were gently deposited on the substrate using a microsyringe for the static contact angle determination. All measurements were made by five different points for each sample at $22 \pm 1\,°\mathrm{C}$ and $50 \pm 5\%$ relative humidity (RH). The measurement results were reproducible within $\pm 3°$.

An optical profiler (NT-3300, Wyko Corp., Tucson, AZ) was used to measure surface roughness for different surface structures [42.25, 26]. A greater z-range of the optical profiler of 2 mm is a distinct advantage over the surface roughness measurements with an AFM, which has a z-range on the order of 7 μm, but it has a maximum lateral resolution of only $\approx 0.6\,\mu\mathrm{m}$ [42.3, 4]. A commercial AFM (D3100, Nanoscope IIIa controller, Digital Instruments, Santa Barbara, CA) was used for additional surface roughness measurements with a high lateral resolution (sub nm) and for adhesion and friction measurements [42.26, 80]. The measurements were performed with a square-pyramidal Si(100) tip with a native oxide layer which had a nominal radius of 20 nm on a rectangular Si(100) cantilever with a spring constant of 3 N/m in tapping mode. Adhesion and friction force at various relative humidities

(RH) were measured using a 15 µm-radius borosilicate ball. A large tip radius was used to measure contributions from several microbumps and a large number of nanobumps. Friction force was measured under a constant load using a 90° scan angle at a velocity of 100 µm/s in 50 µm scans and at a velocity of 4 µm/s in 2 µm scans. The adhesion force was measured using the single point measurement of a force calibration plot.

SEM Micrographs

Figure 42.9 shows SEM micrographs of two superhydrophobic leaves – lotus (*Nelumbo nucifera*) and elephant ear or taro plant (*Colocasia esculenta*), referred to as lotus and colocasia, respectively – and two hydrophilic leaves – beech (*Fagus sylvatica*) and magnolia (*Magnolia grandiflora*), referred to as fagus and magnolia, respectively [42.26]. Lotus and colocasia are characterized by papillose epidermal cells responsible for the creation of papillae or microbumps on the surfaces, and an additional layer of 3-D epicuticular waxes which are a mixture of very long-chain fatty acids molecules (compounds with chains > 20 carbon atoms) and create nanostructure on the entire surface. Fagus and magnolia are characterized by rather flat tabular cells with a thin wax film with a 2-D structure [42.23]. The leaves are not self-cleaning, and contaminant particles from ambient are accumulated, which make them hydrophilic.

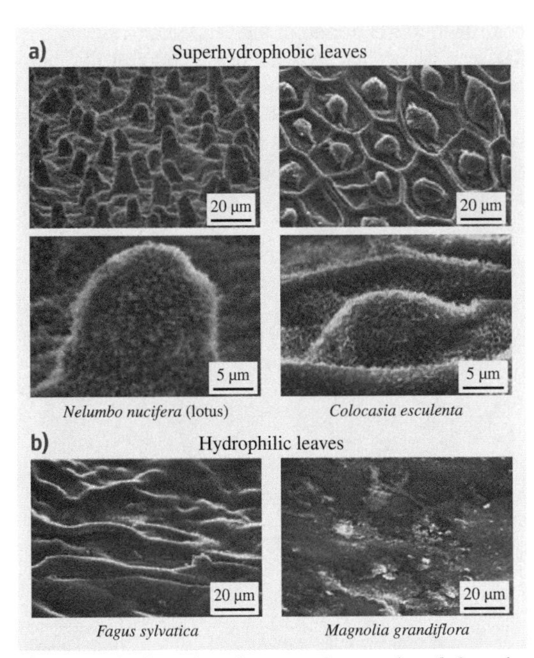

Fig. 42.9a,b Scanning electron micrographs of the relatively rough, water-repellent leaf surfaces of (**a**) *Nelumbo nucifera* (lotus) and *Colocasia esculenta*, and (**b**) the relatively smooth, wettable leaf surfaces of *Fagus sylvatica* and *Magnolia grandiflora* (after [42.26])

Contact Angle Measurements

Figure 42.10a shows the contact angles for the superhydrophobic and hydrophilic leaves before and after applying acetone. The acetone was applied in order to remove any wax present on the surface. As a result, for the superhydrophobic leaves, the contact angle dramatically reduced, whereas for the hydrophilic leaves, the contact angle was almost unchanged. It is known that there is a very thin, 2-D wax layer on the hydrophilic leaves, which introduces little roughness. In

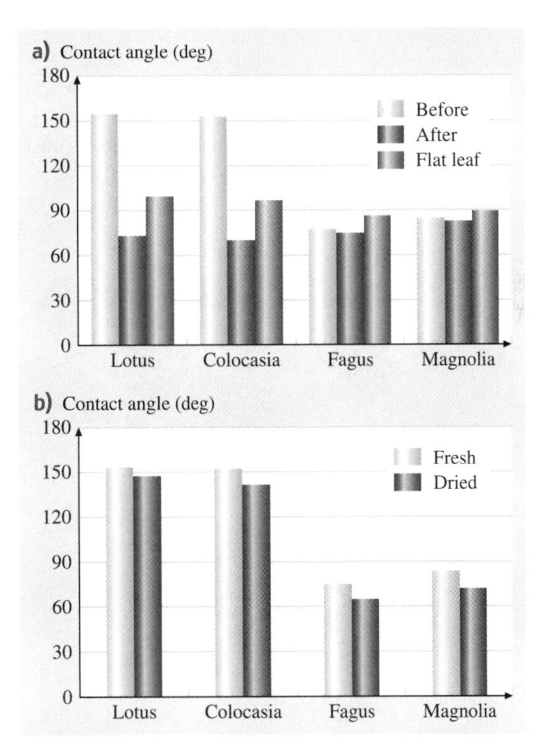

Fig. 42.10a,b Contact angle measurements and calculations for leaf surfaces: (**a**) before and after removing the surface layer as well as calculated values, and (**b**) for fresh and dried leaves. The contact angle on a smooth surface for the four leaves was obtained using the calculated roughness factor (after [42.26])

contrast, superhydrophobic leaves are known to have a thin 3-D wax layer on their surface, consisting of nanoscale roughness over microroughness created by the papillae, which results in a hierarchical roughness. The combination of this wax and the roughness of the leaf creates a superhydrophobic surface.

Bhushan and *Jung* [42.26] calculated the contact angles for leaves with smooth surfaces using the Wenzel equation and the calculated R_f and contact angle of the four leaves. The results are presented in Fig. 42.10a. The approximate values of R_f for lotus and colocasia are 5.6 and 8.4 and for fagus and magnolia are 3.4 and 3.8, respectively. Based on the calculations, the contact angles on smooth surfaces were $\approx 99°$ for lotus and 96° for colocasia. For both fagus and magnolia, the contact angles for the smooth surfaces were found to be $\approx 86°$ and 88°. A further discussion on the effect of R_f on the contact angle will be presented later.

Figure 42.10b shows the contact angles for both fresh and dried states for the four leaves. There is a decrease in the contact angle for all four leaves when they are dried. For lotus and colocasia, this decrease is present because it is found that a fresh leaf has taller bumps than a dried leaf (data to be presented later), which will give a larger contact angle, according to the Wenzel equation. When the surface area is at a maximum compared with the footprint area, as with a fresh leaf, the roughness factor will be at a maximum and will only reduce when shrinking has occurred after drying. To understand the reason for the decrease of contact angle after drying of hydrophilic leaves, dried magnolia leaves were also measured using an AFM. It was found that the dried leaf [peak–valley (P–V) height = 7 μm, mid-width = 15 μm, and peak radius = 18 μm] has taller bumps than a fresh leaf (P–V height = 3 μm, mid-width = 12 μm, and peak radius = 15 μm), which increases the roughness, and the contact angle decreases, leading to a more hydrophilic surface. The mid width is defined as the width of the bump at a height equal to half of peak value to mean line value.

Surface Characterization Using an Optical Profiler

The use of an optical profiler allows measurements to be made on fresh leaves which have a large P–V distance. Three different surface height maps for superhydrophobic and hydrophilic leaves are shown in Figs. 42.11 and 42.12 [42.26]. In each figure, a 3-D map and a flat map along with a 2-D profile in a given location of the flat 3-D map are shown. A scan size of $60 \times 50\,\mu m^2$ was used to obtain a sufficient amount of bumps to charac-

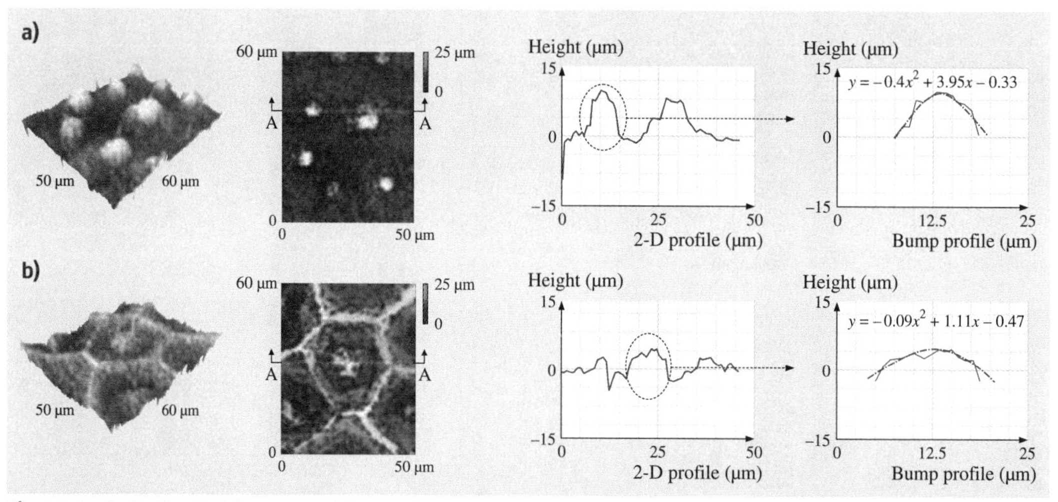

Fig. 42.11a,b Surface height maps and 2-D profiles of hydrophobic leaves obtained using an optical profiler. **(a)** For lotus leaf, a microbump is defined as a single, independent microstructure protruding from the surface. **(b)** For colocasia leaf, a microbump is defined as a single, independent protrusion from the leaf surface, whereas a ridge is defined as a structure that surrounds each bump and is completely interconnected on the leaf. A curve has been fitted to each profile to show exactly how the bump shape behaves. The radius of curvature is calculated from the parabolic curve fit to the bump (after [42.26])

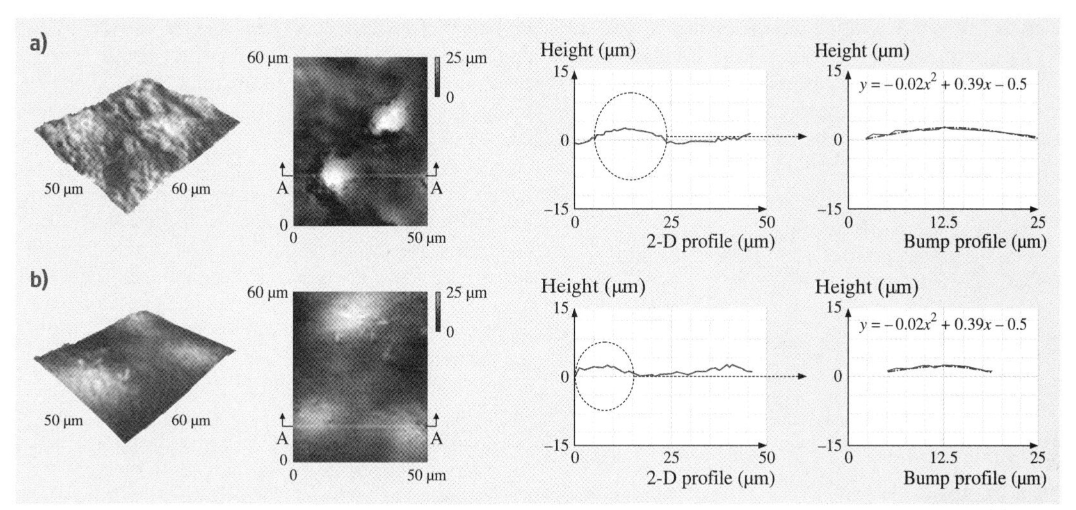

Fig. 42.12a,b Surface height maps and 2-D profiles of hydrophilic leaves obtained using an optical profiler. For (**a**) fagus and (**b**) magnolia leaves, a microbump is defined as a single, independent microstructure protruding from the surface. A curve has been fitted to each profile to show exactly how the bump shape behaves. The radius of curvature is calculated from the parabolic curve fit to the bump (after [42.26])

terize the surface but also to maintain enough resolution to obtain an accurate measurement.

The structures found with the optical profiler correlate well with the SEM images shown in Fig. 42.9. The bumps on the lotus leaf are distributed on the entire surface, but the colocasia leaf shows a very different structure to that of the lotus. The surface structure for colocasia not only has bumps similar to lotus, but also surrounding each bump is a ridge that keeps the bumps separated. With these ridges, the bumps have a hexag-

Table 42.3 Microbump and nanobump map statistics for hydrophobic and hydrophilic leaves, measured both fresh and dried, using an optical profiler and AFM (after [42.26])

Leaf		Microbump, scan size ($50 \times 50\,\mu m^2$)			Nanobump, scan size ($2 \times 2\,\mu m^2$)		
		P–V height (μm)	Mid-width (μm)	Peak radius (μm)	P–V height (μm)	Mid-width (μm)	Peak radius (μm)
Lotus							
Fresh		13[a]	10[a]	3[a]	0.78[b]	0.40[b]	0.15[b]
Dried		9[b]	10[b]	4[b]	0.67[b]	0.25[b]	0.10[b]
Colocasia							
Fresh	Bump	9[a]	15[a]	5[a]	0.53[b]	0.25[b]	0.07[b]
	Ridge	8[a]	7[a]	4[a]	0.68[b]	0.30[b]	0.12[b]
Dried	Bump	5[b]	15[b]	7[b]	0.48[b]	0.20[b]	0.06[b]
	Ridge	4[b]	8[b]	4[b]	0.57[b]	0.25[b]	0.11[b]
Fagus							
Fresh		5[a]	10[a]	15[a]	0.18[b]	0.04[b]	0.01[b]
		4[b]	5[b]	10[b]			
Magnolia							
Fresh		4[a]	13[a]	17[a]	0.07[b]	0.05[b]	0.04[b]
		3[b]	12[b]	15[b]			

[a] Data measured using optical profiler
[b] Data measured using AFM

onal (honeycomb) packing geometry that allows for the maximum number of bumps in a given area. The bumps of lotus and both bumps and ridges of colocasia contribute to the superhydrophobic nature since they both increase the R_f factor and result in air pockets between the droplet of water and the surface. In the fagus and magnolia height maps, short bumps can be seen on the surface. This means that, with decreased bump height, the probability of air pocket formation decreases, and bumps have a less beneficial effect on the contact angle.

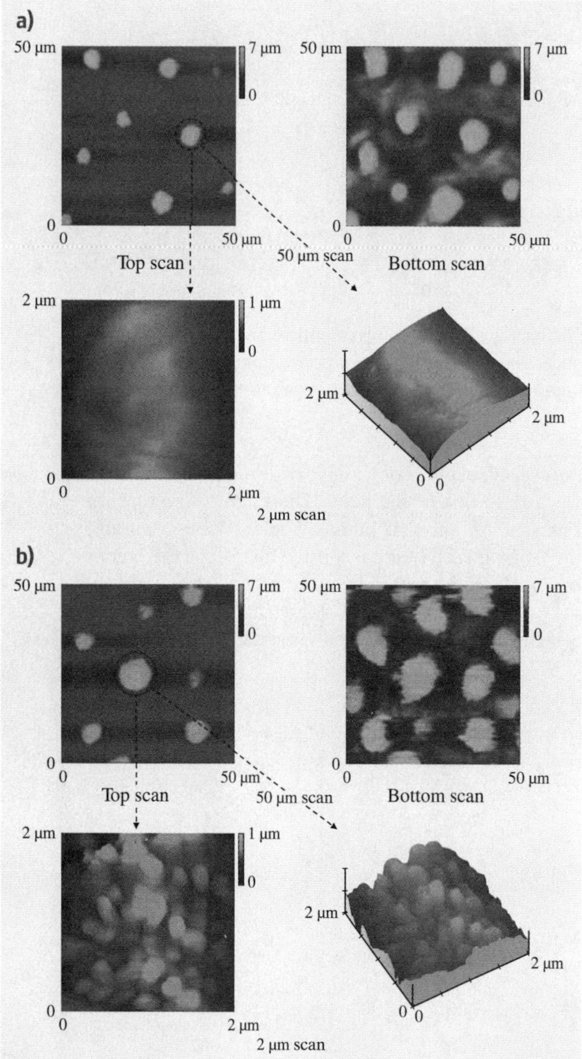

Fig. 42.13a,b Surface height maps showing the top scan and bottom scan in a 50 µm scan size and the bump peak scan selected in a 2 µm scan size for a lotus leaf in (a) contact mode and (b) tapping mode. Two methods were tested to obtain the high-resolution nanotopography of a lotus leaf (after [42.26])

As shown in the 2-D profiles of superhydrophobic and hydrophilic leaves shown in Figs. 42.11 and 42.12, a curve has been fitted to each profile to show exactly how the bump shape behaves. For each leaf a second-order curve fit has been applied to the profiles to show how closely the profile is followed. By using the second-order curve fitting of the profiles, the radius of curvature can be found [42.25, 26].

Using these optical surface height maps, different statistical parameters of bumps and ridges can be found to characterize the surface: P–V height, mid-width, and peak radius [42.3, 4]. Table 42.3 presents these quantities found in the optical height maps for the four leaves. Comparing the superhydrophobic and hydrophilic leaves, it can be seen that the P–V height for bumps of lotus and colocasia is much taller than that for the bumps of fagus and magnolia. The peak radius for the bumps of lotus and colocasia is also smaller than that for the bumps of fagus and magnolia. However, the values of mid-width for the bumps of the four leaves are similar.

Surface Characterization, Adhesion, and Friction Using an AFM

Comparison of Two AFM Measurement Techniques. To measure topographic images of the leaf surfaces, both the contact and tapping modes were first used [42.26]. Figure 42.13 shows surface height maps of dried lotus obtained using the two techniques. In the contact mode, local height variation for lotus leaf was observed in 50 µm scan size. However, little height variation was obtained in a 2 µm scan, even at loads as low as 2 nN. This could be due to the substantial frictional force generated as the probe scanned over the sample. The frictional force can damage the sample. The tapping-mode technique allows high-resolution topographic imaging of sample surfaces that are easily damaged, loosely held to their substrate or difficult to image by other AFM techniques [42.3, 4]. As shown in Fig. 42.13, with the tapping-mode technique, the soft and fragile leaves can be imaged successfully. Therefore tapping-mode technique was used to examine the surface roughness of the superhydrophobic and hydrophilic leaves using an AFM.

Surface Characterization. The AFM has a z-range on the order of 7 μm, and cannot be used for measurements in a conventional way because of the high P–V distances of a lotus leaf. *Burton* and *Bhushan* [42.25] developed a new method to fully determine the bump profiles. In order to compensate for the large P–V distance, two scans were made for each height: one measurement that scans the tops of the bumps and another measurement that scans the bottom or valleys of the bumps. The total height of the bumps is embedded within the two scans. Figure 42.14 shows the 50 μm surface height maps obtained using this method [42.26]. The 2-D profiles in the right-side column take the profiles from the top scan and the bottom scan for each scan size and splice them together to obtain the total profile of the leaf. The 2 μm surface height maps for both fresh and dried lotus can also be seen in Fig. 42.14. This scan area was selected on the top of a microbump obtained in the 50 μm surface height map. It can be seen that nanobumps are randomly and densely distributed over the entire surface of lotus.

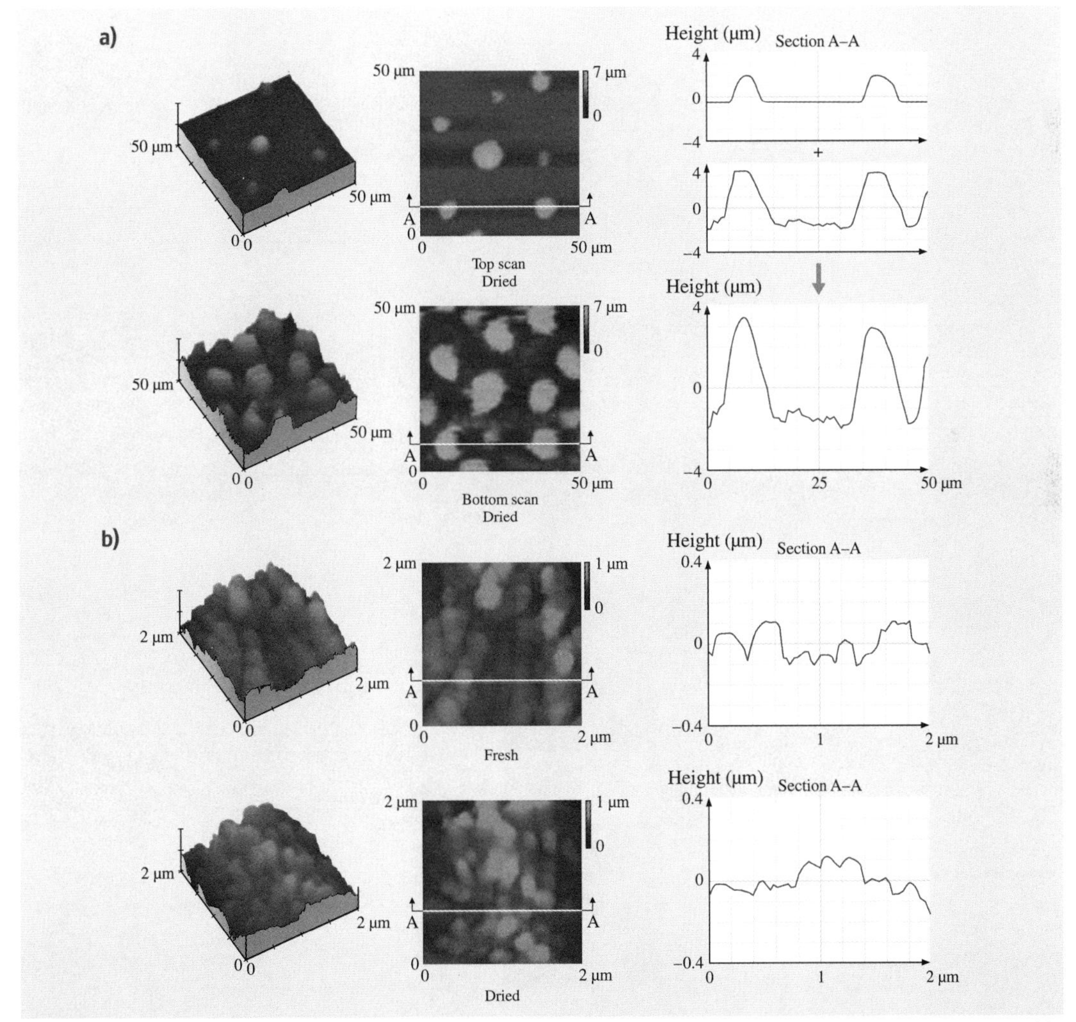

Fig. 42.14a,b Surface height maps and 2-D profiles showing the top scan and bottom scan of a dried lotus leaf in (**a**) 50 μm scan (because the P–V distance of a dried lotus leaf is greater than the z-range of an AFM), and (**b**) both fresh and dried lotus leaf in a 2 μm scan (after [42.26])

Bhushan and *Jung* [42.26] also measured surface height maps for hydrophilic leaves in both 50 and 2 μm scan sizes, as shown in Fig. 42.15. For fagus and magnolia, microbumps were found on the surface, and the P–V distance of these leaves is lower than that of lotus and colocasia. It can be seen in the 2 μm surface height maps that nanobumps selected on the peak of the microbump have an extremely low P–V distance.

Using the AFM surface height maps, different statistical parameters of bumps and ridges can be obtained: P–V height, mid-width, and peak radius. These quantities for the four leaves are listed in Table 42.3. It can be seen that the values correlate well with the values obtained from optical profiler scans except for the bump height, which decreases by more than half because of leaf shrinkage.

Adhesive Force and Friction. Adhesive force and coefficient of friction of superhydrophobic and hydrophilic leaves using AFM are presented in Fig. 42.16 [42.26]. For each type of leaf, adhesive force measurements were made for both fresh and dried leaves using

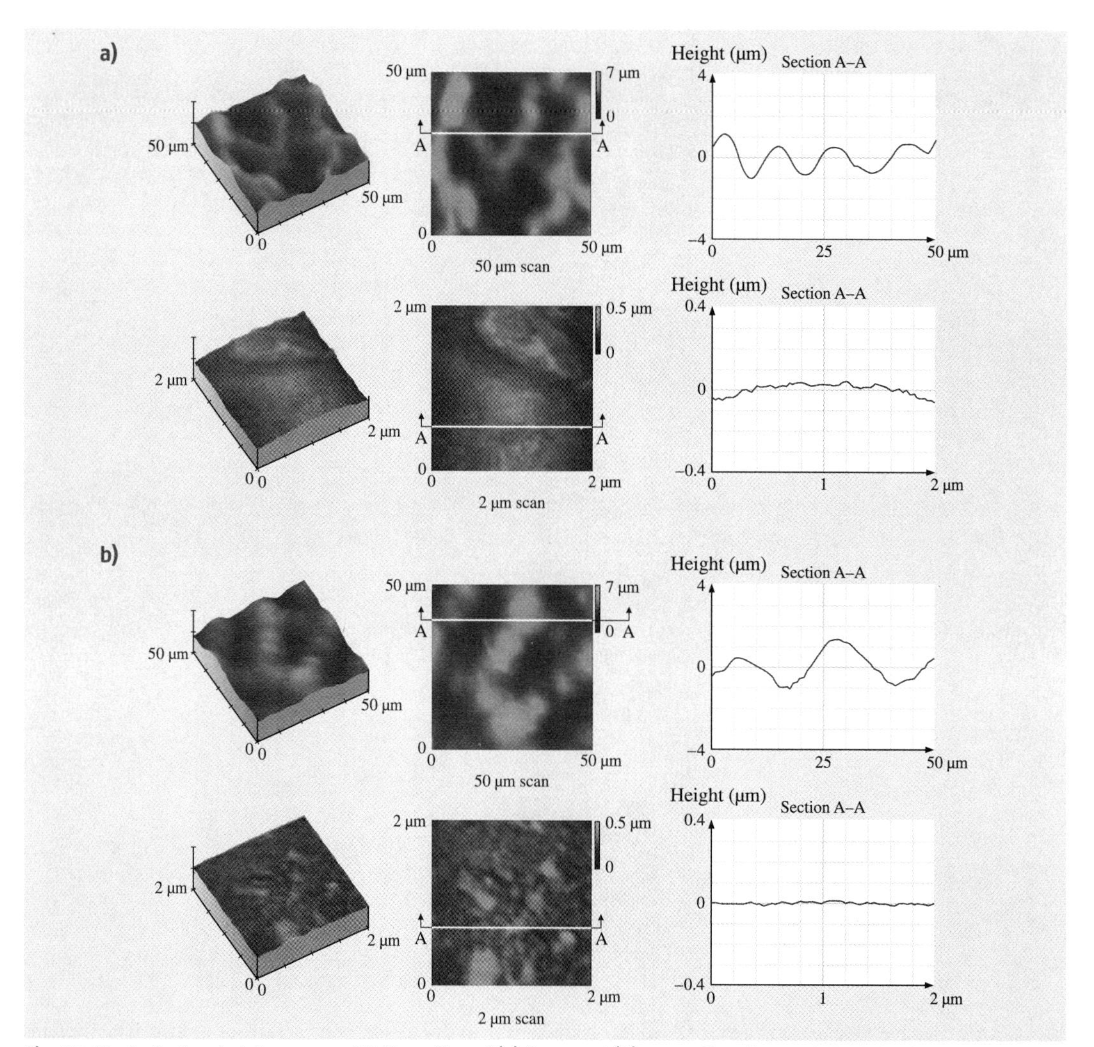

Fig. 42.15a,b Surface height maps and 2-D profiles of (a) fagus and (b) magnolia using an AFM in both 50 μm and 2 μm scans (after [42.26])

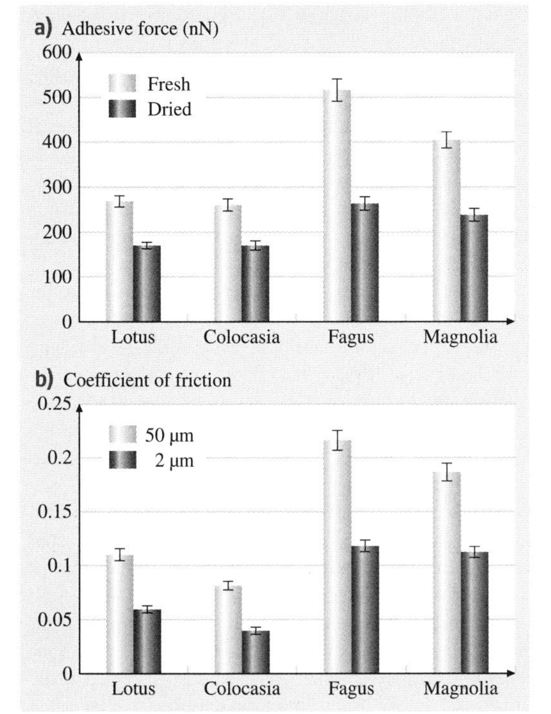

Fig. 42.16a,b Adhesive force for (**a**) fresh and dried leaves, and (**b**) the coefficient of friction for dried leaves for 50 and 2 μm scan sizes for hydrophobic and hydrophilic leaves. All measurements were made using a 15 μm-radius borosilicate tip. Reproducibility for both adhesive force and coefficient of friction is ±5% for all measurements (after [42.26])

a 15 μm-radius tip. It is found that the dried leaves had a lower adhesive force than the fresh leaves. Adhesive force arises from several sources in changing the presence of a thin liquid film, such as an adsorbed water layer that causes meniscus bridges to build up around the contacting and near-contacting bumps as a result of surface energy effects [42.3, 4]. When the leaves are fresh there is moisture within the plant material that causes the leaf to be soft, and when the tip comes into contact with the leaf sample, the sample will deform, and a larger real area of contact between the tip and sample will occur, and the adhesive force will increase. After the leaf has dried, the moisture that was in the plant material is gone, and there is not as much deformation of the leaf when the tip comes into contact with the leaf sample. Hence, the adhesive force is decreased because the real area of contact has decreased.

The adhesive force of fagus and magnolia is higher than that of lotus and colocasia. The reason is that the real area of contact between the tip and leaf surface is expected to be higher in hydrophilic leaves than in superhydrophobic leaves. In addition, fagus and magnolia are hydrophilic and have high affinity for water. The combination of high real area of contact and affinity for water are responsible for the higher meniscus forces [42.3, 4]. The coefficient of friction was only measured on a dried plant surface with the same sliding velocity (10 μm/s) in different scan sizes rather than including the fresh surface, because the P–V was too large to scan back and forth with the AFM to obtain the friction force. As expected, the coefficient of friction for superhydrophobic leaves is lower than that for hydrophilic leaves due to the real area of contact between the tip and leaf sample, similar to the adhesive force results. When the scan size decreases from the microscale to the nanoscale, the coefficient of friction also decreases for each leaf. The reason for this dependence is the scale-dependent nature of the roughness of the leaf surface. Figures 42.14 and 42.15 show AFM topography images and 2-D profiles of the surfaces for different scan sizes. The scan-size dependence of the coefficient of friction has been reported previously [42.119–121].

Part F | 42.3

Role of the Hierarchical Roughness

The approximation of the roughness factor for the leaves on the micro- and nanoscale was made using AFM scan data [42.26]. Roughness factors for various leaves are presented in Table 42.4. As mentioned earlier, the open space between asperities on a surface has the potential to collect air, and the probability of this occurring appears to be higher for nanobumps, as the distance between bumps on the nanoscale is smaller than for those on the microscale. Using roughness factor values, along with the contact angles (θ) from both superhydrophobic and hydrophilic surfaces, 153° and 152° in lotus and colocasia, and 76° and 84° in fagus and magnolia, respectively, the contact angles (θ_0) for the smooth surfaces can be calculated using the Wenzel equation (42.6) for microbumps and the Cassie–Baxter equation (42.9) for nanobumps. The contact angle ($\Delta\theta$) calculated using R_f on the smooth surface can be found in Table 42.4. It can be seen that the roughness factors and the differences ($\Delta\theta$) between θ and θ_0 on the nanoscale are higher than those on the microscale. This means that nanobumps on the top of a microbump increase the contact angle more effectively than do microbumps. In the case of hydrophilic leaves, the values of R_f and $\Delta\theta$ change very little on both scales.

Table 42.4 Roughness factor and contact angle ($\Delta\theta = \theta - \theta_0$) calculated using R_f on smooth surface for hydrophobic and hydrophilic leaves measured using an AFM, both microscale and nanoscale (after [42.26])

Leaf (contact angle)	Scan size	State	R_f	$\Delta\theta$ (deg)
Lotus (153°)	50 μm	Dried	5.6	54^a
	2 μm	Fresh	20	61^b
		Dried	16	60^b
Colocasia (152°)	50 μm	Dried	8.4	56^a
	2 μm bump	Fresh	18	60^b
		Dried	14	59^b
	2 μm ridge	Fresh	18	60^b
		Dried	15	59^b
Fagus (76°)	50 μm	Fresh	3.4	-10^a
	2 μm	Fresh	5.3	2^b
Magnolia (84°)	50 μm	Fresh	3.8	-4^a
	2 μm	Fresh	3.6	14^b

[a] Calculations made using Wenzel equation
[b] Calculations made using Cassie–Baxter equation. We assume that the contact area between the droplet and air is half of the whole area of the rough surface

Based on the data in Fig. 42.16, the coefficient of friction values on the nanoscale are much lower than those on the microscale. It is clearly observed that friction values are scale dependent. The height of a bump and the distance between bumps on the microscale is much larger than those on the nanoscale, which may be responsible for the larger values of friction force on the microscale.

One difference between micro- and nanobumps for surface enhancement of water repellency is the effect on contact angle hysteresis, in other words, the ease with which a droplet of water can roll on the surface. It has been stated earlier that contact angle hysteresis decreases and contact angle increases due to the decreased contact with the solid surface caused by air pockets beneath the droplet. The surface with nanobumps has a high roughness factor compared with that of microbumps. With large distances between microbumps, the probability of air pocket formation decreases and is responsible for high contact angle hysteresis. Therefore, on the surface with nanobumps, the contact angle is high and the contact angle hysteresis is low, and drops rebound easily and can be set into rolling motion with a small tilt angle [42.26].

Natural water-repellent and self-cleaning surfaces such as the lotus leaf [42.28, 29] or water-strider leg [42.31] have a hierarchical structure. However, the functionality of this hierarchical roughness remains a subject of discussion, and several explanations have been suggested. *Nosonovsky* and *Bhushan* [42.5, 9–11, 63–67, 70] showed that the mechanisms involved in superhydrophobicity are scale dependent, and thus the roughness must be hierarchical in order to respond to these mechanisms. The surface must be able to repel both macroscopic and microscopic droplets. *Fürstner* et al. [42.122] pointed out that artificial surfaces with one level of roughness can well repel large *artificial rain* droplets; however, they cannot repel small *artificial fog* droplets trapped in the valleys between the bumps, so the hierarchy may have to do with the ability to repel droplets of various size ranges. According to *Gao* and *McCarthy* [42.123], the large bumps allow the composite interface to be maintained, while the small ones enhance the contact angle in accordance with the Wenzel model. *Jung* and *Bhushan* [42.60] showed that a droplet with radius of ≈ 100–400 μm on a micropatterned surface goes through a transition from the composite interface to the solid–liquid interface as the pitch increases, and *Bhushan* et al. [42.54, 56, 57] showed that the hierarchical structure can prevent the gaps between the pillars from filling with liquid until the droplet evaporates completely.

42.4 How to Make a Superhydrophobic Surface

Fabrication of superhydrophobic surfaces has been an active area of research since the mid-1990s. In general, the same techniques that are used for micro- and nanostructure fabrication, such as lithography, etching, deposition, and self-assembly, have been utilized for producing superhydrophobic surfaces (Fig. 42.17, Table 42.5). Pros and cons of these techniques are summarized in Table 42.6. Among especially interesting developments is the creation of switchable surfaces that can be turned from hydrophobic to hydrophilic by surface energy modification through electrowetting, light and x-ray irradiation, dynamic effects, optical effects (e.g., the transparency, reflectivity or nonreflectivity) combined with the lotus effect, hydrophobic interactions, and so on [42.110, 124–127]. An important requirement for potential applications for optics and self-cleaning glasses is the creation of transparent superhydrophobic surfaces. In order for the surface to be transparent, roughness details should be smaller than the wavelength of visible light ($\approx$ 400–700 nm) [42.128].

Two main requirements for a superhydrophobic surface are that the surface should be rough and that it should be hydrophobic (low surface energy). These two requirements lead to two methods of producing a superhydrophobic surface: first, it is possible to make a rough surface from an initially hydrophobic material and, second, to modify a rough hydrophilic surface by modifying its surface chemistry or applying a hydrophobic material upon it. Note that roughness is usually a more critical property than the low surface energy, since both moderately hydrophobic and very hydrophobic materials can exhibit similar wetting behavior when roughened.

42.4.1 Roughening to Create One-Level Structure

Lithography is a well-established technique, applied for creating a large area of periodic micro-/nanopatterns. It includes photo, e-beam, x-ray, and soft lithography. *Bhushan* and *Jung* [42.52] produced patterned Si using photolithography. To obtain a sample that is hydrophobic, a self-assembled monolayer (SAM) of 1,1,-2,2,-tetrahydroperfluorodecyltrichlorosilane (PF_3) was deposited on the sample surfaces using a vapor-phase deposition technique. They obtained a superhydrophobic surface with a contact angle up to 170°. *Martines* et al. [42.129] fabricated ordered arrays of nanopits and nanopillars by using electron-beam lithography. They obtained a superhydrophobic surface with a static contact angle of 164° and contact angle hysteresis of 1° for a surface consisting of tall pillars with cusped tops after hydrophobization with octadecyltrichlorosilane (OTS). *Fürstner* et al. [42.122] created silicon wafers with regular patterns of spikes by x-ray lithography. The wafer was hydrophobized by sputtering a layer of gold and subsequent immersion in a hexadecanethiol solution. AFM can be used in nanolithography to produce a nanostructure with the aid of solvent [42.130] or electric field [42.131] on polystyrene (PS) and polymethylmethacrylate (PMMA), respectively. *Jung* and *Bhushan* [42.81] created low-aspect-ratio asperities (LAR, 1 : 1 height-to-diameter ratio), high-aspect-ratio asperities (HAR, 3 : 1 height-to-diameter ratio), and a lotus pattern (replica from the lotus leaf), all on a PMMA surface using soft lithography. A self-assembled monolayer (SAM) of perfluorodecyltriethoxysilane (PFDTES) was deposited on the patterned surfaces using a vapor-phase deposition technique.

One well-known and effective way to make rough surfaces is etching using either plasma, laser, chemical or electrochemical techniques [42.132]. *Jansen* et al. [42.133] etched a silicon wafer using a fluorine-based plasma by utilizing the black silicon method to obtain isotropic, positively, and negatively tapered as well as vertical walls with smooth surfaces. *Coulson* et al. [42.134] described an approach in plasma chemical roughening of poly(tetrafluoroethylene) (PTFE) substrates followed by the deposition of low-surface-energy plasma polymer layers, which give rise to high repellency towards polar and nonpolar probe liquids. A different

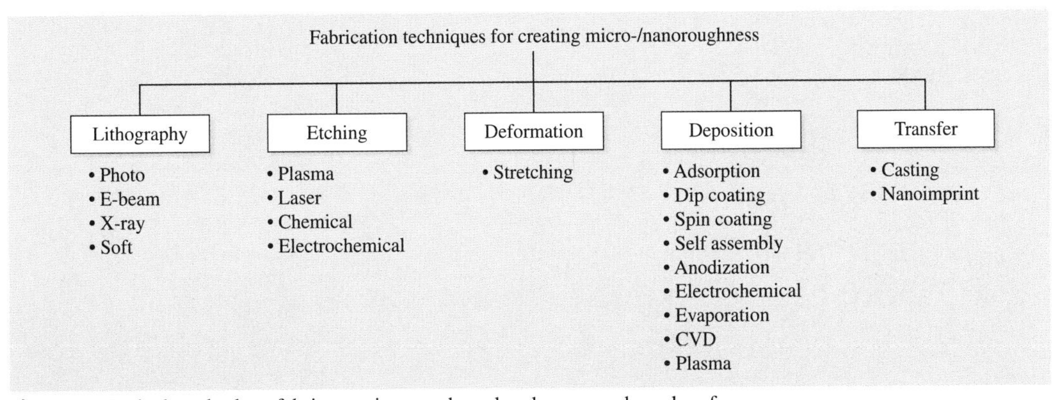

Fig. 42.17 Typical methods to fabricate microroughened and nanoroughened surface

Table 42.5 Typical materials and corresponding techniques to produce micro-/nanoroughness

Material	Technique	Contact angle (deg)	Notes	Source
Teflon	Plasma	168		*Zhang* et al. [42.139]; *Shiu* et al. [42.135]
Fluorinated block polymer solution	Casting under humid environment	160	Transparent	*Yabu* and *Shimomura* [42.140]
PFOS	Electro- and chemical polymerization	152	Reversible (electric potential)	*Xu* et al. [42.124]
PDMS	Laser treatment	166		*Khorasani* et al. [42.137]
PS-PDMS block copolymer	Electrospining	> 150		*Ma* et al. [42.141]
PS, PC, PMMA	Evaporation	> 150		*Bormashenko* et al. [42.142]
PS nanofiber	Nanoimprint	156		*Lee* et al. [42.143]
Polyaniline nanofiber	Chemical polymerization	175		*Chiou* et al. [42.144]
PET	Oxygen plasma etching	> 150		*Teshima* et al. [42.136]
Organo-triethoxysilanes	Sol–gel	155	Reversible (temperature)	*Shirtcliffe* et al. [42.125]
Al	Chemical etching	> 150		*Qian* and *Shen* [42.138]
Copper	Electrodeposition	160	Hierarchical	*Shirtcliffe* et al. [42.145]
Si	Photolithography	170		*Bhushan* and *Jung* [42.52]
Si	E-beam lithography	164		*Martines* et al. [42.129]
Si	X-ray lithography	> 166		*Fürstner* et al. [42.122]
PS, PMMA	AFM nanolithography			*Martin* et al. [42.131]; *Cappella* and *Bonaccurso* [42.130]
Si	Casting	158	Plant leaf replica	*Sun* et al. [42.47]; *Fürstner* et al. [42.122]
Si (black Si)	Plasma etching	> 150	For liquid flow	*Jansen* et al. [42.133]
Silica	Sol–gel	150		*Hikita* et al. [42.146]; *Shang* et al. [42.147]
Silica	Layer-by-layer assembly	160	Hierarchical	*Zhao* et al. [42.148]
Polyelectrolyte multilayer surface overcoated with silica nanoparticles	Self-assembly	168		*Zhai* et al. [42.149]
Epoxy resin with synthetic and plant waxes	Replication and self-assembly	173	Hierarchical	*Bhushan* et al. [42.54–57]; *Koch* et al. [42.30]
Nanosilica spheres	Dip coating	105		*Klein* et al. [42.150]
Silica colloidal particles in PDMS	Spin coated	165	Hierarchical	*Ming* et al. [42.151]
Au clusters	Electrochemical deposition	> 150		*Zhang* et al. [42.152]
Carbon nanotubes	Chemical vapor deposition	> 165		*Lau* et al. [42.153]
Carbon nanotubes	Chemical vapor deposition	159	Hierarchical	*Huang* et al. [42.154]
ZnO, TiO_2 nanorods	Sol–gel	> 150	Reversible (UV irradiation)	*Feng* et al. [42.110]

Table 42.6 Pros and cons of various fabrication techniques

Techniques	Pros	Cons
Lithography	Accuracy, large area	Slow process, high cost
Etching	Fast	Chemical contamination, less control
Deposition	Flexibility, cheap	Can be high temperature, less control
Self-assembly	Flexibility, cheap	Requires suitable precursor

approach was taken by *Shiu* et al. [42.135], who treated a Teflon film with oxygen plasma and obtained a superhydrophobic surface with a contact angle of 168°. Fluorinated materials have limited solubility, which makes it difficult to roughen them. However, they may be linked or blended with other materials, which are often easier to roughen, in order to make superhydrophobic surfaces. *Teshima* et al. [42.136] obtained a transparent superhydrophobic surface from a poly(ethylene terephthalate) (PET) substrate via selective oxygen plasma etching followed by plasma-enhanced chemical vapor deposition using tetramethylsilane (TMS) as the precursor. *Khorasani* et al. [42.137] produced porous polydimethylsiloxane (PDMS) surfaces with contact angle of 175° using a CO_2 pulsed-laser etching method as an excitation source for the surface. *Qian* and *Shen* [42.138] described a simple surface roughening method by dislocation selective chemical etching on polycrystalline metals such as aluminum. After treatment with fluoroalkylsilane, the etched metallic surfaces exhibited superhydrophobicity. *Xu* et al. [42.124] fabricated a reversible superhydrophobic surface with a double-roughened perfluorooctanesulfonate (PFOS)-doped conducting polypyrrole (PPy) film by a combination of electropolymerization and chemical polymerization. Reversibility was achieved by switching between superhydrophobic doped or oxidized states and superhydrophilicity dedoped or neutral states with changing the applied electrochemical potential.

A stretching method can be used to produce a superhydrophobic surface. *Zhang* et al. [42.139] stretched a Teflon film and converted it into fibrous crystals with a large fraction of void space in the surface, leading to high roughness and superhydrophobicity.

Deposition methods also can be used to make a substrate rough. There are several ways to make a rough surface including adsorption, dip coating, electrospinning, anodization, electrochemical, evaporation, chemical vapor deposition (CVD), and plasma. Solidification of wax can be used to produce a superhydrophobic surface. *Shibuichi* et al. [42.48] used alkylketene dimer (AKD) wax on a glass plate to spontaneously form a fractal structure on its surfaces. They obtained a surface with a contact angle > 170° without any fluorination treatments. *Klein* et al. [42.150] obtained superhydrophobic surfaces by simply dip-coating of a substrate with a slurry containing nanosilica spheres, which adhered to the substrate after a low-temperature heat treatment. After reaction of the surface with a fluoroalkyltrichlorosilane, the hydrophobicity increased with a decreasing area fraction of spheres. *Ma* et al. [42.141] produced block copolymer poly(styrene-b-dimethylsiloxane) fibers with submicrometer diameters in the range 150–400 nm by electrospinning from a solution in tetrahydrofuran and dimethylformamide. They obtained superhydrophobic nonwoven fibrous mats with a contact angle of 163°. *Shiu* et al. [42.135] produced self-organized close-packed superhydrophobic surfaces by spin-coating the monodispersed polystyrene beads solution on a substrate surface. *Abdelsalam* et al. [42.155] studied the wetting of structured gold surfaces formed by electrodeposition through a template of submicrometer spheres and discussed the role of the pore size and shape in controlling wetting. *Bormashenko* et al. [42.142] used evaporated polymer solutions of polystyrene (PS), polycarbonate (PC), and polymethylmethacrylate (PMMA) dissolved in chlorinated solvents, dichloromethane (CH_2Cl_2), and chloroform ($CHCl_3$) to obtain self-assembled structure with hydrophobic properties. Chemical/physical vapor deposition (CVD/PVD) has been used for the modification of surface chemistry as well. *Lau* et al. [42.153] created superhydrophobic carbon nanotube forests by modifying the surface of vertically aligned nanotubes with plasma-enhanced chemical vapor deposition (PECVD). Superhydrophobicity was achieved down to the microscopic level, where essentially spherical, micrometer-sized water droplets can be suspended on top of the nanotube forest. *Zhu* et al. [42.156] and *Huang* et al. [42.154] prepared surfaces with two-scale roughness by controlled growth of carbon nanotube (CNT) arrays by CVD. *Zhao* et al. [42.157] also synthesized vertically aligned multi-walled carbon nanotube (MWCNT) arrays by CVD on

Si substrates using a thin film of iron (Fe) as a catalyst layer and aluminum (Al) film.

Attempts to create superhydrophobic surfaces by casting and nanoimprint methods have been successful. *Yabu* and *Shimomura* [42.140] prepared a porous superhydrophobic transparent membrane by casting a fluorinated block-polymer solution under humid environment. Transparency was achieved because the honeycomb-patterned films had a subwavelength pore size. *Sun* et al. [42.47] reported a nanocasting method to make a superhydrophobic PDMS surface. They first made a negative PDMS template using a lotus leaf as an original template and then used the negative template to make a positive PDMS template – a replica of the original lotus leaf. *Zhao* et al. [42.158] prepared a superhydrophobic surface by casting a micellar solution of a copolymer poly(styrene-b-dimethylsiloxane) (PS-PDMS) in humid air based on the cooperation of vapor-induced phase separation and surface enrichment of PDMS block. *Lee* et al. [42.143] produced vertically aligned PS nanofibers by using nanoporous anodic aluminum oxide as a replication template in a heat- and pressure-driven nanoimprint pattern-transfer process. As the aspect ratio of the polystyrene (PS) nanofibers increased, the nanofibers could not stand upright but formed twisted bundles, resulting in a three-dimensionally rough surface with a contact angle of $\approx 155°$.

42.4.2 Coating to Create One-Level Hydrophobic Structures

Modifying the surface chemistry with a hydrophobic coating widens the potential applications of superhydrophobic surfaces. There are several ways to modify the chemistry of a surface including sol–gel, dip-coating, self-assembly, electrochemical, and chemical/physical vapor deposition. *Shirtcliffe* et al. [42.125] prepared porous sol–gel foams from organotriethoxysilanes which exhibited switching between superhydrophobicity and superhydrophilicity when exposed to different temperatures. *Hikita* et al. [42.146] used colloidal silica particles and fluoroalkylsilane as the starting materials and prepared a sol–gel film with superliquid-repellency by hydrolysis and condensation of alkoxysilane compounds. *Feng* et al. [42.110] produced superhydrophobic surfaces using ZnO nanorods by sol–gel method. They showed that superhydrophobic surfaces can be switched into hydrophilic surfaces by alternation of ultraviolet (UV) irradiation. *Shang* et al. [42.147] did not blend low-surface-energy materials in the sols, but described a procedure to make transparent superhydrophobic surfaces by modifying silica-based gel films with a fluorinated silane. In a similar way, *Wu* et al. [42.159] made a microstructured ZnO-based surface via a wet chemical process and obtained superhydrophobicity after coating the surface with long-chain alkanoic acids. *Chiou* et al. [42.144] fabricated polyaniline nanofibers using chemical oxidative polymerization to produce uniform aligned nanofibers and treated with CF_4 plasma treatment to create superhydrophobic surfaces with a contact angle of 175°.

Zhai et al. [42.149] used a layer-by-layer (LBL) self-assembly technique to create a poly(allylamine hydrochloride)/poly(acrylic acid) (PAH/PAA) multilayer which formed a honeycomb-like structure on the surface after an appropriate combination of acidic treatments. After cross-linking the structure, they deposited silica nanoparticles on the surface via alternate dipping of the substrates into an aqueous suspension of the negatively charged nanoparticles and an aqueous PAH solution, followed by a final dipping into the nanoparticle suspension. Superhydrophobicity was obtained after the surface was modified by a chemical vapor deposition of (tridecafluoro-1,1,2,2-tetrahydrooctyl)-1-trichlorosilane followed by thermal annealing.

Zhang et al. [42.152] showed that the surface covered with dendritic gold clusters, which was formed by electrochemical deposition onto an indium tin oxide (ITO) electrode modified with a polyelectrolyte multilayer, showed superhydrophobic properties after further deposition of a *n*-dodecanethiol monolayer. *Han* et al. [42.160] described the fabrication of lotus-leaf-like superhydrophobic metal surfaces by using electrochemical reaction of Cu or Cu-Sn alloy plated on steel sheets with sulfur gas, and subsequent perfluorosilane treatment. Chemical bath deposition (CBD) has also been used to make nanostructured surfaces, thus *Hosono* et al. [42.161] fabricated a nanopin film of brucite-type cobalt hydroxide (BCH) and achieved a contact angle of 178° after further modification with lauric acid (LA). *Shi* et al. [42.162] described the use of galvanic cell reaction as a facile method to chemically deposit Ag nanostructures on *p*-silicon wafer on a large scale. When the Ag-covered silicon wafer was further modified with a self-assembled monolayer of *n*-dodecanethiol, a superhydrophobic surface was obtained with a contact angle of $\approx 154°$ and a tilt angle $< 5°$.

42.4.3 Methods to Create Two-Level (Hierarchical) Structures

Two-level (hierarchical) roughness structures are typical for superhydrophobic surfaces in nature, as was discussed above. Recently, many efforts have been devoted to fabricating these hierarchical structures in various ways. *Shirtcliffe* et al. [42.145] prepared a hierarchical (double-roughened) copper surface by electrodeposition from acidic copper sulfate solution onto flat copper and a patterning technique of coating with a fluorocarbon hydrophobic layer. Another way to obtain a rough surface for superhydrophobicity is assembly from colloidal systems. *Ming* et al. [42.151] prepared a hierarchical (double-roughened) surface consisting of silica-based raspberry-like particles. First is the attachment of epoxy and amino groups onto the silica microparticles of ≈ 700 nm and nanoparticles of ≈ 70 nm, respectively, using established synthetic procedures. Two suspensions in ethanol are created, one with the microparticles and another one with nanoparticles. In the next step, the suspension with silica microparticles is added dropwise to the suspension with the nanoparticles. The nanoparticles attach to the microparticles due to the reaction between the epoxy and amino groups present on the surface of the particles. Then, the suspension is centrifuged to separate any unreacted particles. A next step involves depositing these micro-/nanostructured particles into an epoxy film (on silicon). Finally, since the resulting micro-/nanoparticle surface is initially hydrophilic, it is made hydrophobic by deposition of monoepoxy-end-capped poly(dimethylsiloxane) (PDMS). *Northen* and *Turner* [42.163] fabricated arrays of flexible silicon dioxide platforms supported by single high-aspect-ratio silicon pillars down to 1 μm in diameter and with heights up to ≈ 50 μm. When these platforms were coated with polymeric organorods ≈ 2 μm tall and 50–200 nm in diameter, the surface was highly hydrophobic with a water contact angle of 145°. *Chong* et al. [42.164] fabricated hierarchically ordered nanowire arrays with periodic voids at the microscale and hexagonally packed nanowires at the nanoscale. This hierarchical surface was created by selective electrodeposition using nanoporous anodic alumina as a template and a porous gold film as a working electrode that is patterned by microsphere monolayers. *Wang* et al. [42.165] also developed a novel precursor hydrothermal redox method with $Ni(OH)_2$ as the precursor to fabricate a hierarchical structure consisting of nickel hollow microspheres with nickel nanoparticles in situ. The created hierarchical hollow structure exhibited enhanced coercivity and remnant magnetization as compared with hollow nickel submicrometer spheres, hollow nickel nanospheres, bulk nickel, and free Ni nanoparticles.

Kim et al. [42.166] fabricated a hierarchical structure which looks like the structures of the lotus leaf. First, nanoscale porosity was generated by anodic aluminum oxidation, and then the anodized porous alumina surface was replicated by polytetrafluoroethylene. The polymer sticking phenomenon during the replication created the submicrostructures on the negative polytetrafluoroethylene nanostructure replica. The contact angle of the created hierarchical structure was $\approx 160°$ and the tilting angle was $< 1°$. *Del Campo* and *Greiner* [42.167] reported that SU-8 hierarchical patterns comprising features with lateral dimensions ranging from 5 to 2 mm and heights from 10 to 500 μm were obtained by photolithography, which comprises a step of layer-by-layer exposure in soft contact-printed shadow masks which are embedded into the SU-8 multilayer. *Bhushan* et al. [42.54–57] and *Koch* et al. [42.30] produced hierarchical structures by replication of a micropatterned silicon surface and lotus leaf microstructure using an epoxy resin and by self-assembly of synthetic and plant waxes as thin hydrophobic three-dimensional crystals to create hydrophobic nanostructures. The fabrication technique used is a low-cost two-step process, which provides flexibility in the fabrication of a variety of hierarchical structures. They showed that a hierarchical structure has a high propensity for air pocket formation and leads to a static contact angle of 173° and contact angle hysteresis, and a tilt angle of $\approx 2°$. *Zhao* et al. [42.148] fabricated a hierarchical structure by using layer-by-layer assembly of silica nanoparticles on a microsphere-patterned polyimide precursor substrate combined with fluoroalkylsilane treatment. The microstructures were created by replica molding of polyamide using two-dimensional PS microsphere arrays. They obtained a superhydrophobic surface with a static contact angle of 160° and sliding angle of $< 10°$. *Cortese* et al. [42.168] applied plasma CF_4 treatment on micropattern PDMS and obtained contact angle of 170°. *Kuan* et al. [42.169] produced a hierarchical structure by imprinting ZnO precursor films using gratings with 830 nm and 50 μm dimensions. They achieved a contact angle of 141° by using nanostructures deposited on saw-tooth patterns without modifying the surface chemistry.

42.5 Fabrication and Characterization of Micro-, Nano-, and Hierarchical Patterned Surfaces

In this section, we will discuss experimental measurements of the wetting properties of micro-, nano-, and hierarchical patterned surfaces.

42.5.1 Experimental Techniques

Contact Angle, Surface Roughness, and Adhesion

The static and dynamic (advancing and receding) contact angles were measured using a Rame–Hart model 100 contact angle goniometer and droplets of DI water [42.8, 30, 52, 54–57, 59, 80, 81]. For measurement of the static contact angle, the droplet size should be smaller than the capillary length, but larger than the dimension of the structures present on the surfaces. Droplets of $\approx 5\,\mu l$ in volume (with the diameter of a spherical droplet ≈ 2.1 mm) were gently deposited on the substrate using a microsyringe for measurement of the static contact angle. The advancing and receding contact angles were measured by the addition and removal of water from a sessile DI water droplet using a microsyringe. The contact angle hysteresis was calculated as the difference between the measured advancing and receding contact angles, and the tilt angle was measured by using a simple tilting stage [42.8,30,52,54–57]. All measurements were made at five different points for each sample at $22 \pm 1\,°C$ and $50 \pm 5\%$ RH. The measurements were reproducible to within $\pm 3°$.

For surface roughness measurement, an optical profiler (NT-3300, Wyko Corp., Tucson, AZ) was used for different surface structures [42.8, 25, 26, 52]. The optical profiler has one advantage due to its greater z-range (2 mm) compared with the AFM (z-range $\approx 7\,\mu m$), but it has a maximum lateral resolution of only $\approx 0.6\,\mu m$ [42.3, 4]. Experiments were performed using tips of three different radii to study the effect of scale dependence. Large-radius atomic force microscope (AFM) tips were primarily used in this study. A borosilicate ball with $15\,\mu m$ radius and a silica ball with $3.8\,\mu m$ radius were mounted on a gold-coated triangular Si_3N_4 cantilever with a nominal spring constant of 0.58 N/m. A square-pyramidal Si_3N_4 tip with a nominal radius of 30–50 nm on a triangular Si_3N_4 cantilever with a nominal spring constant of 0.58 N/m was used for the smaller radius tip. Adhesive force was measured using the single point measurement of a force calibration plot [42.3, 4, 17].

Droplet Evaporation Studies

Droplet evaporation was observed and recorded by a digital camcorder (Sony, DCRSR100) with a 10× optical and 120× digital zoom for every run of the experiment. Then the decrease in the diameter of the droplets with time was determined [42.54, 57, 59, 60]. The frame speed of the camcorder was 0.03 s per frame. An objective lens placed in front of the camcorder during recording gave a total magnification of 10–20 times. Droplet diameter as small as a few hundred microns could be measured with this method. Droplets were gently deposited on the substrate using a microsyringe, and the whole process of evaporation was recorded. Images obtained were analyzed using Imagetool software (University of Texas Health Science Center) for the contact angle. To find the dust trace remaining after droplet evaporation, an optical microscope with a charge-coupled device (CCD) camera (Nikon, Optihot-2) was used. All measurements were made in a controlled environment at $22 \pm 1\,°C$ and $45 \pm 5\%$ RH [42.54,56,57,59,60].

Bouncing Droplet Studies

The process of dynamic impact was recorded by a high-speed camera (Kodak Ektapro HS Motion Analyzer, model 4540) operated at 500 frames/s for each experimental run and then measuring the dynamic impact behavior of the droplet as a function of time. The impact velocity was calculated by varying the droplet release height. The size of the droplet was the same as that of a droplet for the static contact angle. All measurements were made in a controlled environment at $22 \pm 1\,°C$ and $45 \pm 5\%$ RH [42.61].

Vibrating Droplet Studies

The process of dynamic behavior was obtained by a system producing vertical vibrations [42.62]. The system consists of an electrodynamic shaker (Labworks Inc., model ET-126A/B) connected to a signal generator and power amplifier, a digital camcorder (Sony, DCRSR100, Tokyo) with an objective lens, and a lamp (Digital Instruments Inc., model F0-50) as a light source. The specimen was placed on top of the shaker, and a droplet was gently deposited using a microsyringe. The size of the droplet was the same as that for the static contact angle as reported earlier. The vibration frequency was controlled between 0 and 300 Hz at 0.4 mm amplitude for measurement of the resonance frequency of a droplet. For wetting behavior of a droplet on the surface, a frequency of 30 Hz, which was less than the resonance frequency, was chosen, and the vibration amplitude was controlled between 0 and 3 mm. The vibration time applied to the droplet was 1 min for each experiment. All measurements were made in a controlled environment at $22 \pm 1\,°C$ and $45 \pm 5\%$ RH [42.62].

Microdroplet Condensation and Evaporation Studies Using ESEM

A Philips XL30 ESEM equipped with a Peltier cooling stage was used to study smaller droplets [42.60]. ESEM uses a gaseous secondary-electron detector (GSED) for imaging. The ESEM column is equipped with a multi-stage differential pressure-pumping unit. The pressure in the upper part is $\approx 10^{-6}$–10^{-7} Torr, but a pressure of ≈ 1–15 Torr can be maintained in the observation chamber. When the electron beam (primary electrons) ejects secondary electrons from the surface of the sample, the secondary electrons collide with gas molecules in the ESEM chamber, which in turn acts as a cascade amplifier, delivering the secondary-electron signal to the positively biased GSED. The positively charged ions are attracted toward the specimen to neutralize the negative charge produced by the electron beam. Therefore, the ESEM can be used to examine electrically isolated specimens in their natural state. In ESEM, adjusting the pressure of the water vapor in the specimen chamber and the temperature of the cooling stage allows water to condense on the sample in the chamber. For the measurement of the static and dynamic contact angles on patterned surfaces, video images were recorded. The voltage of the electron beam was 15 kV, and the distance of the specimen from the final aperture was ≈ 8 mm. If the angle of observation is not parallel to the surface, the electron beam is not parallel to the surface but inclined at an angle, which will produce a distortion in the projection of the droplet profile. A mathematical model to calculate the real contact angle from the ESEM images was used to correct for tilting of the surfaces during imaging [42.60, 171].

Generation of Submicron Droplets

In order to generate submicron droplets, *Jung* and *Bhushan* [42.170] developed an AFM-based technique using a modified nanoscale dispensing (NADIS) probe, as shown in Fig. 42.18. The NADIS probe was fabricated by modifying commercially available silicon nitride (Si_3N_4) cantilevers (Olympus OMCL-RC800) with lengths of 100 and 200 μm, spring constants of 0.8 and 0.1 N/m, and resonance frequencies of 68.94 and 122.02 kHz, respectively (Swiss Center for Electronics and Microtechnology). The probe consisted of a loading area (30 μm diameter) for the liquid on the upper side of the cantilever. The loading area was produced by removing the material locally in and around the tip with a reflective gold layer using focused ion-beam milling. The remaining gold was made hydrophobic using hexadecanethiol (in liquid phase), whereas the bare silicon nitride in the milling area remained hydrophilic. The hydrophilic–hydrophobic transition prevents spreading of the loaded liquid over the entire cantilever.

A droplet of a certain volume V is deposited on the surface. Figure 42.19 shows an idealized spherical capped droplet. Based on the thickness of the droplet h and contact diameter d, the contact angle is obtained by the following equation for a simple spherical capped geometry of droplet [42.170]

$$\theta = \sin^{-1}\left\{\frac{3d}{2[3V/(\pi h^2)+h]}\right\} . \tag{42.23}$$

For calculation of the contact angle of the droplet, we use the following three steps:

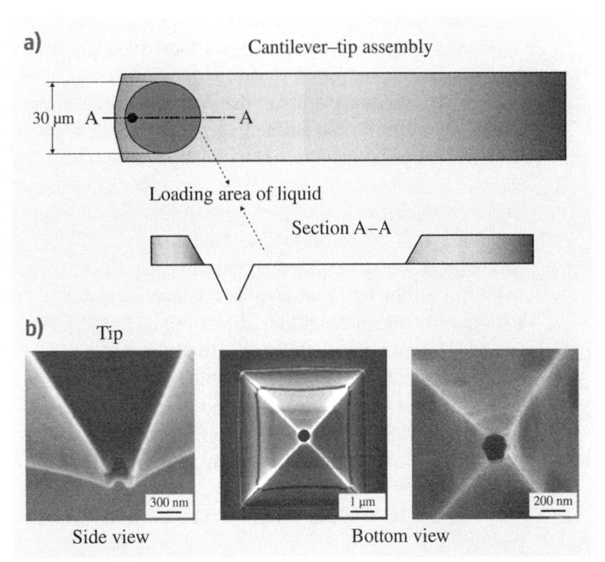

Fig. 42.18 **(a)** Schematic of the modified nanoscale dispensing (NADIS) probe used for the generation of submicron-size droplets. The loaded liquid is limited to the loading area (30 μm-diameter circle), **(b)** scanning electron micrograph of the tip in side view and bottom view with different aperture sizes (500 and 200 nm) at its apex (after [42.170])

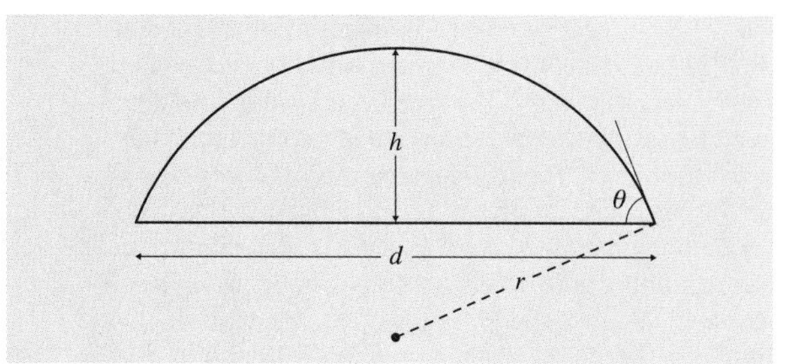

Fig. 42.19 Droplet of liquid in contact with a surface at a contact angle of θ. The thickness of the droplet is h. The contact diameter between the droplet and surface is d. The radius of curvature of the droplet is r (after [42.170])

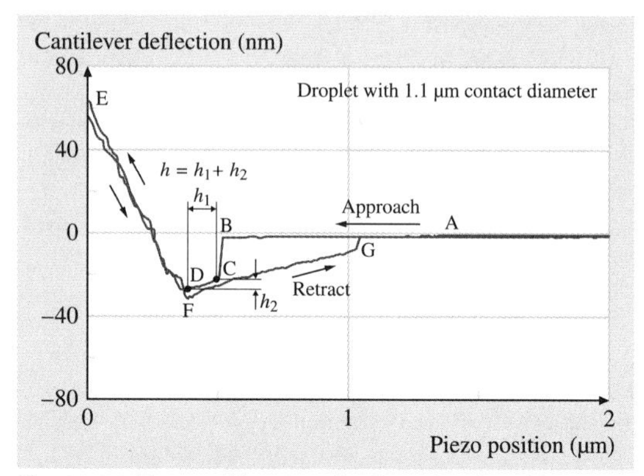

Fig. 42.20 Force calibration plot for a droplet with 1.1 μm contact diameter on a Si surface. The h is a measure of the droplet thickness on the surface (after [42.170])

1. For the measurement of the volume of a droplet deposited on the surface, the change in resonance frequency of the cantilever before and after depositing the droplet on the surface is measured. The resonance frequency of the cantilever is measured by performing a frequency sweep of the voltage-driven oscillations by the thermal tuning method [42.172].
2. For the measurement of the thickness of a droplet deposited on the surface, the distance between the tip snap-in and the position where the tip is in contact with the surface is measured in the force calibration mode.
3. For the measurement of the contact diameter between a droplet and surface, the image of the droplet after evaporation is measured using a Si tip.

For the thickness of a droplet deposited on the surface, the force calibration plot was used [42.3, 4, 17]. The droplet was deposited in the first approach. The force calibration plot was obtained at a second approach to measure the thickness of the droplet [42.173–175]. Figure 42.20 shows a typical force calibration plot curve for the Si surface. In particular, the liquid film thickness h is the sum of the travel distance of the piezo (described as h_1) and the deflection of cantilever (described as h_2), as shown in Fig. 42.20. The h overestimates the thickness of the droplet because of the van der Waals attraction between the tip and the droplet and the liquid pick-up on the tip.

The resolution of volume (mass), thickness, and contact diameter was $\approx 1 \times 10^{-4}\,\mu m^3$ (0.12 fg), 0.1 nm, and < 1 nm, respectively. The resolution of volume was calculated from the measured data of the shift in the resonance frequency of the cantilever from 122.01 to 122.02 kHz during an evaporation time of 10 min. The resolutions of the thickness and contact diameter measurements were calculated from the calibration data of the z piezo and x–y piezo by the AFM vendor (Veeco), respectively. The accuracy of volume, thickness, and contact diameter measurement was about $\pm 10\%$, 10%, and < 1 nm, respectively.

42.5.2 Micro- and Nanopatterned Polymers

Jung and *Bhushan* [42.81] studied two types of polymers: poly(methyl methacrylate) (PMMA) and polystyrene (PS). PMMA and PS were chosen because they are widely used in MEMS/NEMS devices. Both hydrophilic and hydrophobic surfaces can be produced by using these two polymers, as PMMA has polar (hydrophilic) groups with high surface energy while PS has electrically neutral and nonpolar(hydrophobic) groups with low surface energy. Furthermore, a PMMA structure can be made hydrophobic by treating it appropriately, for example, by coating with a hydrophobic self-assembled monolayer (SAM).

Four types of surface patterns were fabricated from PMMA: a flat film, low-aspect-ratio asperities (LAR, 1 : 1 height-to-diameter ratio), high-aspect-ratio asperities (HAR, 3 : 1 height-to-diameter ratio), and a replica of the lotus leaf (the lotus pattern). Two types of surface patterns were fabricated from PS: a flat film and the lotus pattern. Figure 42.21 shows SEM images of the two types of nanopatterned structures, LAR and HAR, and the one type of micropatterned structure, the lotus pattern, all on a PMMA surface [42.80, 81]. Both micro- and nanopatterned structures were manufactured using soft lithography. For nanopatterned structures, PMMA film was spin-coated on the silicon wafer. A UV-cured mold of polyurethane acrylate (PUA) resin with nanopatterns of interest was made, which enables one to create sub-100 nm patterns with high aspect ratio [42.176]. The mold was placed on the PMMA film, a slight pressure of $\approx 10\,g/cm^2$ (≈ 1 kPa) was applied, and annealing was carried out at 120 °C. Finally, the PUA mold was removed from the PMMA film. For micropatterned structures, a polydimethylsiloxane (PDMS) mold was first made by casting PDMS against a lotus leaf, followed by heating. Then, the mold was placed on the PMMA and PS film to create a positive replica of lotus leaf. As shown in Fig. 42.21, only microstructures exist on the surface of the lotus pattern [42.81].

Since PMMA itself is hydrophilic, in order to obtain a hydrophobic sample, a SAM of perfluorodecyltriethoxysilane (PFDTES) was deposited on the sample surfaces using a vapor-phase deposition technique. PFDTES was chosen because of its hydrophobic nature. The deposition conditions for PFDTES were 100 °C temperature, 400 Torr pressure, 20 min deposition time, and 20 min annealing time. The polymer surface was exposed to an oxygen plasma treatment (40 W, O_2 187 Torr, 10 s) prior to coating [42.177]. The oxygen plasma treatment is necessary to oxidize any organic contaminants on the polymer surface and also to alter the surface chemistry to allow enhanced bonding between the SAM and the polymer surface.

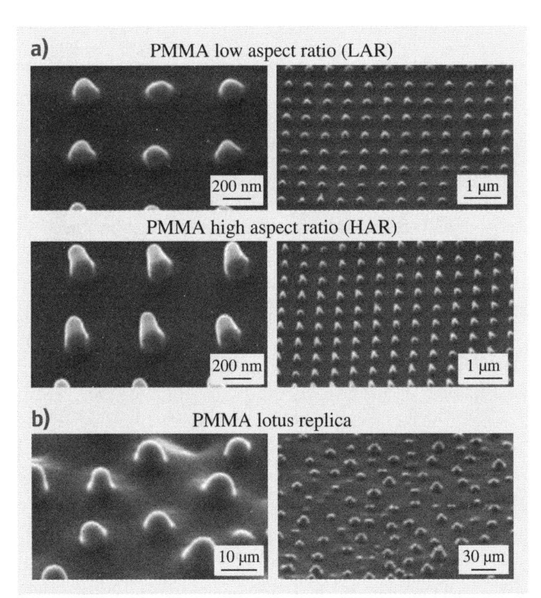

Fig. 42.21a,b Scanning electron micrographs of (a) the two nanopatterned polymer surfaces (shown using two magnifications to see both the asperity shape and the asperity pattern on the surface) and (b) the micropatterned polymer surface (lotus pattern, which has only microstructures on the surface) (after [42.80, 81])

Contact Angle Measurements

Jung and *Bhushan* [42.81] measured the static contact angle of water with the patterned PMMA and PS structures (Fig. 42.22). Since the Wenzel roughness factor is the parameter that often determines wetting behavior, the roughness factor was calculated, and is presented in Table 42.7 for various samples. The data show that the contact angle of the hydrophilic ma-

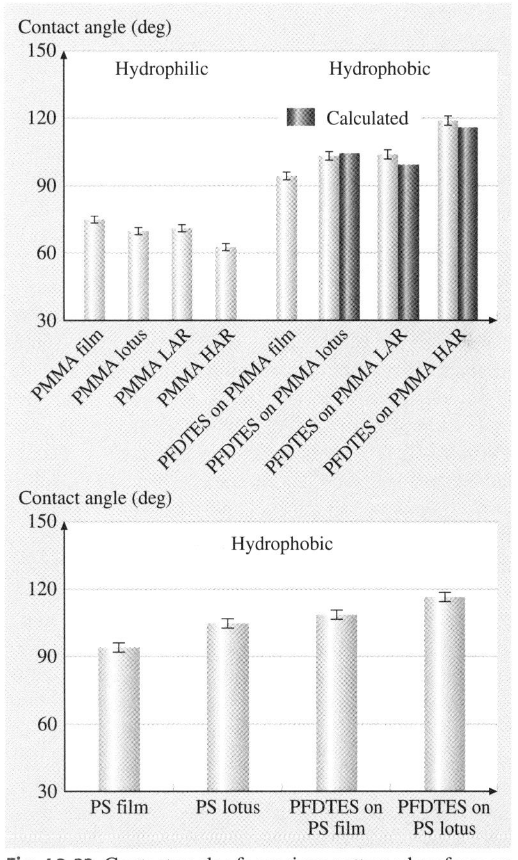

Fig. 42.22 Contact angles for various patterned surfaces on PMMA and PS polymers and values calculated using the Wenzel equation (after [42.81])

Table 42.7 Roughness factor for micro- and nanopatterned polymers (after [42.81])

	LAR	HAR	Lotus
R_f	2.1	5.6	3.2

terials decreases with increasing roughness factor, as predicted by the Wenzel model. When the polymers were coated with PFDTES, the film surface became hydrophobic. Figure 42.22 also shows the contact angle for various PMMA samples coated with PFDTES. For a hydrophobic surface, the standard Wenzel model predicts an increase of contact angle with roughness factor, which is what happens in the case of patterned samples. The calculated values of the contact angle for various patterned samples based on the contact angle of the smooth film and Wenzel equation are also presented. The measured contact angle values for the lotus pattern were comparable to the calculated values, whereas for the LAR and HAR patterns they are higher. This suggests that nanopatterns benefit from air pocket formation. For the PS material, the contact angle of the lotus pattern also increased with increasing roughness factor.

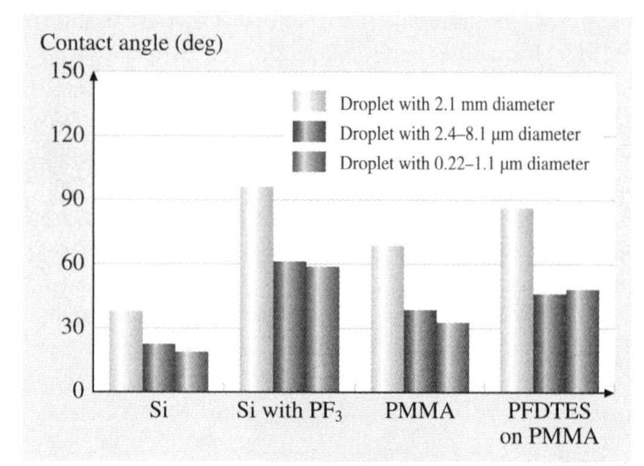

Fig. 42.23 Contact angle measurements for different droplet sizes on various surfaces (after [42.170])

Effect of Submicron Droplet

Wetting phenomena have been well studied and understood at the macroscale; however, micro- and nanoscale wetting mechanisms require further investigation. The actual contact angle under which the liquid–vapor interface comes into contact with the solid surface at the micro- and nanoscale is known to decrease with decreasing droplet size [42.93, 178]. *Jung* and *Bhushan* [42.170] measured the contact angle of micro- and nanodroplets on various surfaces based on the contact diameter, thickness, and volume of droplets measured using the AFM-based technique. The contact angle for different droplet sizes on various surfaces is summarized in Fig. 42.23. The data for the microdroplets with 2.4–8.1 μm diameter and nanodroplets with 0.22–1.1 μm diameter were compared with conventional contact angle measurements obtained with a droplet with 2.1 mm diameter (5 μl volume). The measured values of micro- and nanodroplets using an AFM were found to be lower than those of the macrodroplet [42.93, 178]. There are several reasons for this scale dependence, such as the effect of contact line tension of a three-phase system (solid–liquid–vapor), which is the excess free energy of a solid–liquid–vapor system per unit length of contact line [42.64, 93, 178, 179]. Another reason may be surface heterogeneity [42.93]. For a thin fluid film present on a surface, the disjoining pressure of a film is repulsive, analogous to the repulsive van der Waals force across a film, and causes a film to spread on surfaces. It decreases with the liquid layer thickness [42.2]. This pressure may lead to a smaller contact angle at the nanoscale.

Scale Dependence of Adhesive Force

Jung and *Bhushan* [42.81] found that scale dependence of adhesion and friction forces is important for this study because the tip–surface interface area changes with size. The meniscus force will change due to either changing tip radius, the hydrophobicity of the sample or the number of contacting and near-contacting points. Figure 42.24 shows the dependence of tip radius and hydrophobicity on the adhesive force for PMMA and PFDTES coated on PMMA [42.81]. When the radius of the tip is changed, the contact angle of the sample is changed and asperities are added to the sample surface; the adhesive force will change due to the changes in the meniscus force and the real area of contact.

The two plots in Fig. 42.24 show the adhesive force on a linear scale for the different surfaces with varying tip radius. The top bar chart in Fig. 42.24 is for hydrophilic PMMA film, lotus pattern, LAR, and HAR, and shows the effect of tip radius and hydrophobicity on adhesive force. For increasing radius, the adhesive force increases for each material. With a larger radius, the real area of contact and the meniscus contribution increase, resulting in increased adhesion. The bottom bar chart in Fig. 42.24 shows the results for PFDTES coated on each material. These

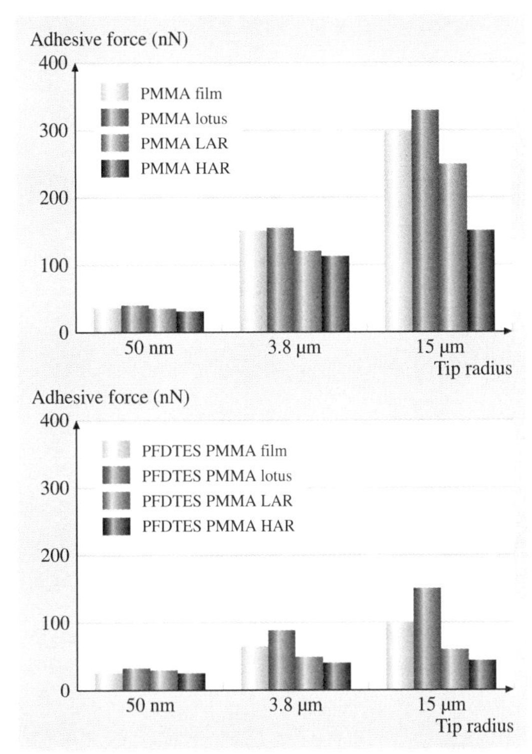

Fig. 42.24 Scale-dependent adhesive force for various patterned surfaces measured using AFM tips of various radii (after [42.81])

samples show the same trends as the film samples, but the increase in adhesion is not as dramatic. The hydrophobicity of PFDTES on material reduces the meniscus forces, which in turn reduces adhesion to the surface. The dominant mechanism for the hydrophobic material is real area of contact and not meniscus force, whereas with the hydrophilic material there is a combination of real area of contact and meniscus forces [42.81].

42.5.3 Micropatterned Si Surfaces

Micropatterned surfaces produced from single-crystal silicon (Si) by photolithography and coated with a SAM were used by *Jung* and *Bhushan* [42.59–61] in their study. Silicon has traditionally been the most commonly used structural material for micro-/nanocomponents. A Si surface can be made hydrophobic by coating it with a SAM. One of the purposes of this investigation was to study the transition from the Cassie–Baxter to Wenzel regime by changing the distance between the pillars. To create patterned Si, two series of nine samples each were fabricated using photolithography [42.100]. Series 1 had 5 µm diameter and 10 µm-high, flat-top, cylindrical pillars with different pitch values (7, 7.5, 10, 12.5, 25, 37.5, 45, 60, and 75 µm), and series 2 had 14 µm-diameter, 30 µm-high, flat-top, cylindrical pillars with different pitch values (21, 23, 26, 35, 70, 105, 126, 168, and 210 µm). The pitch is the spacing between the centers of two adjacent pillars. A SAM of PF_3 was deposited on the Si sample surfaces using a vapor-phase deposition technique [42.100]. PF_3 was chosen because of the hydrophobic nature of the surface. The thickness and root-mean-square (RMS) roughness of the SAM of PF_3 were 1.8 and 0.14 nm, respectively [42.180].

An optical profiler was used to measure the surface topography of the patterned surfaces [42.8, 52, 60, 61]. One sample each from the two series was chosen to characterize the surfaces. Two different surface height maps can be seen for the patterned Si in Fig. 42.25. In each case, a 3-D map and a flat map along with a 2-D profile at a given location of the flat 3-D map are shown. A scan size of $100\times 90\,\mu m^2$ was used to obtain a sufficient number of pillars to characterize the surface but also to maintain enough resolution to obtain an accurate measurement.

Let us consider the geometry of flat-top, cylindrical pillars of diameter D, height H, and pitch P, distributed in a regular square array as shown in Fig. 42.25. For the special case of droplet size much larger than P (of interest in this study), a droplet contacts the flat-top of the pillars forming the composite interface, and the cavities are filled with air. For this case $f_{LA} = 1 - \pi D^2/(4P^2) = 1 - f_{SL}$. Further, assume that the flat tops are smooth with $R_f = 1$. The contact angles for the Wenzel and Cassie–Baxter regimes are given by (42.6) and (42.9) as [42.52]

$$\text{Wenzel:} \qquad \cos\theta = \left(1 + \frac{\pi DH}{P^2}\right)\cos\theta_0\,, \tag{42.24}$$

$$\text{Cassie–Baxter:} \qquad \cos\theta = \frac{\pi D^2}{4P^2}(\cos\theta_0 + 1) - 1\,. \tag{42.25}$$

Geometrical parameters of the flat-top, cylindrical pillars in series 1 and 2 are used for calculating the contact angle for the two above-mentioned cases. Figure 42.26 shows plots of the predicted values of the contact angle as a function of pitch between the pillars for the two cases. The Wenzel and Cassie–Baxter

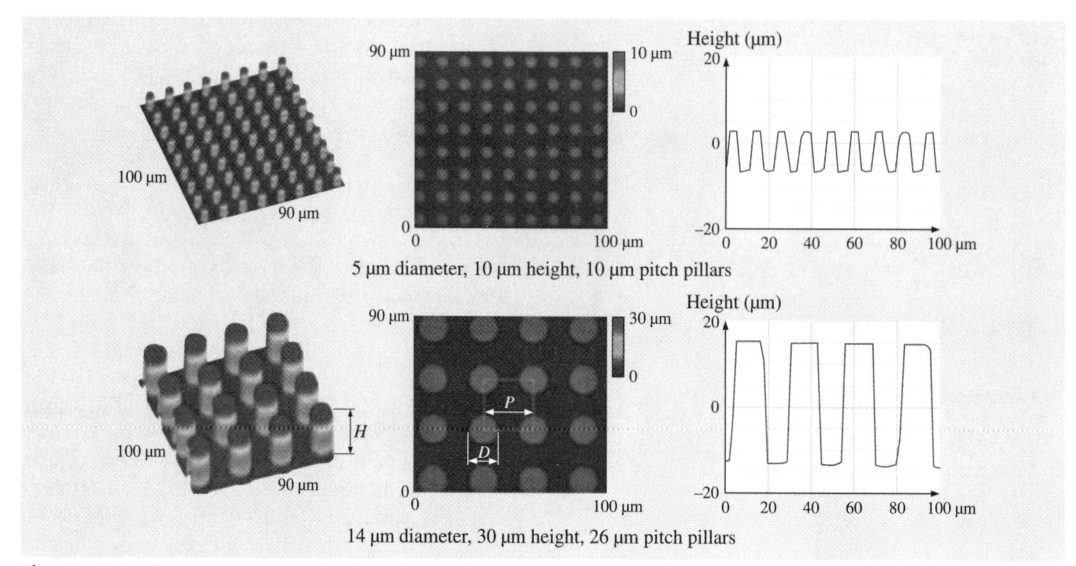

Fig. 42.25 Surface height maps and 2-D profiles of the patterned surfaces obtained using an optical profiler (after [42.52])

equations present two possible equilibrium states for a water droplet on the surface. This indicates that there is a critical pitch below which the composite interface dominates and above which the homogeneous interface dominates the wetting behavior. Therefore, one needs to find this critical point, which can be used to design superhydrophobic surfaces. Furthermore, even in cases where the liquid droplet does not contact the bottom of the cavities, the water droplet can be in a metastable state and can become unstable, with transition from the Cassie–Baxter to Wenzel regime occurring if the pitch is large.

Cassie–Baxter and Wenzel Transition Criteria

A stable composite interface is essential for successful design of superhydrophobic surfaces. However, the composite interface is fragile, and it may transform into the homogeneous interface. What triggers the transition between the regimes remains a subject of argument, although a number of explanations have been suggested. *Nosonovsky* and *Bhushan* [42.63] have studied destabilizing factors for the composite interface and found that a convex surface (with bumps) leads to a stable interface and high contact angle. Also, they have suggested the effects of a droplet's

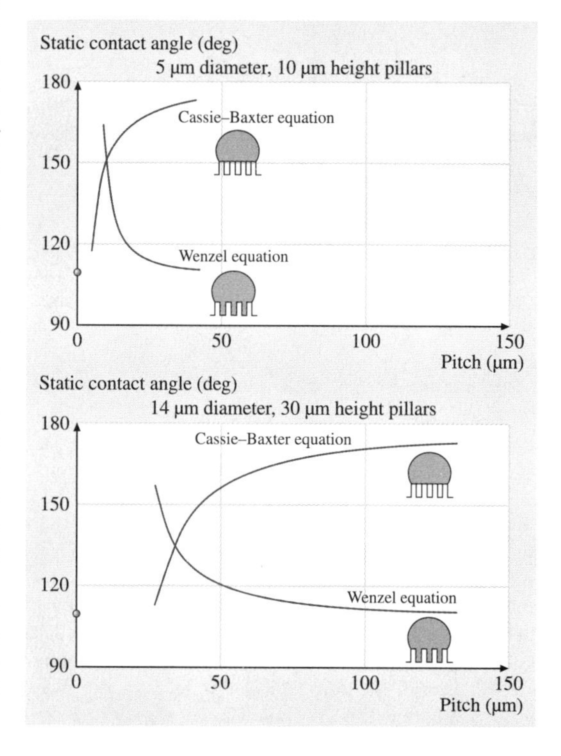

Fig. 42.26 Static contact angle as a function of geometric parameters for a given value of θ_0 calculated using the Wenzel and Cassie–Baxter equations for two series of patterned surfaces with different pitch values (after [42.52]) ▶

weight and curvature among the factors which affect the transition.

Bhushan and *Jung* [42.8, 52] and *Jung* and *Bhushan* [42.59–61] investigated the effect of droplet curvature on the Cassie–Baxter and Wenzel regime transition. First, they considered a small water droplet suspended on a superhydrophobic surface consisting of a regular array of circular pillars with diameter D, height H, and pitch P, as shown in Fig. 42.27. The local deformation for small droplets is governed by surface effects rather than gravity. The curvature of a droplet is governed by the Laplace equation, which relates the pressure inside the droplet to its curvature [42.1]. Therefore, the curvature is the same at the top and at the bottom of the droplet [42.64]. For the patterned surface considered here, the maximum droop of the droplet occurs in the center of the square formed by the four pillars, as shown in Fig. 42.27a. Therefore, the maximum droop of the droplet (δ) in the recessed region can be found in the middle of two diagonal pillars, as shown in Fig. 42.27b, which is $(\sqrt{2}P - D)^2/(8R)$. If the droop is greater than the depth of the cavity, then the droplet will just contact the bottom of the cavities between pillars. If it is much greater, transition from the Cassie–Baxter to Wenzel regime occurs

$$\frac{\left(\sqrt{2}P - D\right)^2}{R} \geq H \,. \tag{42.26}$$

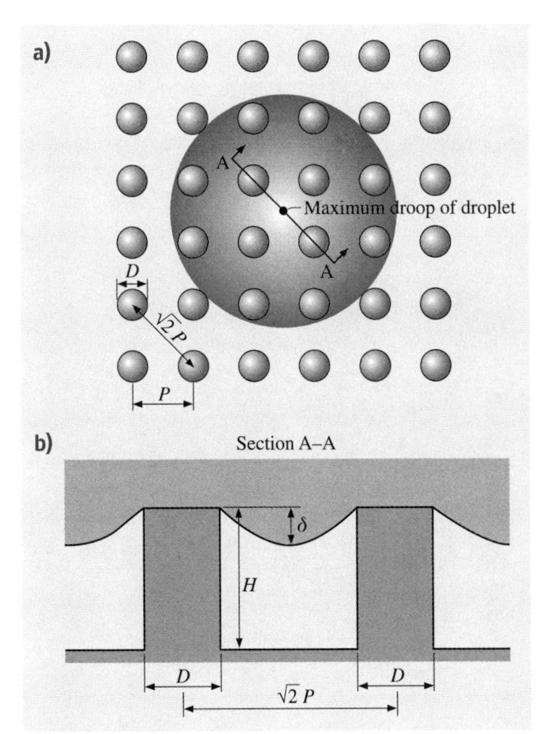

Fig. 42.27a,b A liquid droplet suspended on a superhydrophobic surface consisting of a regular array of circular pillars. **(a)** Plan view: the maximum droop of the droplet occurs at the center of the square formed by four pillars. **(b)** Side view in section A–A. The maximum droop of the droplet (δ) can be found equidistant between two diagonal pillars (after [42.59])

To investigate the dynamic effect of a bouncing water droplet on the Cassie–Baxter and Wenzel regime transition, *Jung* and *Bhushan* [42.61] considered a water droplet hitting a superhydrophobic surface, as shown in Fig. 42.27. As the droplet hits the surface at velocity v, a liquid–air interface below the droplet is formed when the dynamic pressure is less than the Laplace pressure. The Laplace pressure can be written as

$$p_\mathrm{L} = \frac{2\gamma}{R} = \frac{16\gamma\delta}{(\sqrt{2}P - D)^2} \,, \tag{42.27}$$

where γ is the surface tension of the liquid–air interface, and the dynamic pressure of the droplet is equal to

$$p_\mathrm{d} = \tfrac{1}{2}\rho v^2 \,, \tag{42.28}$$

where ρ is the mass density of the liquid droplet. If the maximum droop of the droplet (δ) is larger than the height of the pillars (H), the droplet contacts the bottom of the cavities between pillars. Determination of the critical velocity at which the droplet touches the bottom is obtained by equating the Laplace pressure to the dynamic pressure. To develop a composite interface, velocity should be smaller than the critical velocity given by

$$v < \sqrt{\frac{32\gamma H}{\rho\left(\sqrt{2}P - D\right)^2}} \,. \tag{42.29}$$

Furthermore, in the case of large distances between the pillars, the liquid–air interface can easily be destabilized due to dynamic effects. This leads to the formation of a homogeneous solid–liquid interface [42.63].

Nosonovsky and *Bhushan* [42.68] used the energy-barrier approach to study the Cassie–Baxter and Wenzel transition. The energy barrier is given by the product of the height of the pillars H, the pillar perimeter πD, the pillar density $1/P^2$, and the area A_0 required to initiate the transition; the corresponding change in the surface energy is

$$\Delta E = A_0 \frac{\pi HD}{P^2}(\gamma_{SL} - \gamma_{SA}) = -A_0 \frac{\pi HD}{P^2}\gamma_{LA}\cos\theta_0 , \quad (42.30)$$

where A_0 is $\pi(R\sin\theta)^2$.

For a short pitch, the net energy of the Cassie–Baxter state is lower than that of the Wenzel state, whereas for larger pitch values, the energy of the Wenzel state is lower (Fig. 42.8c). However, due to the energy barriers, a metastable Cassie–Baxter state with a higher energy than the Wenzel state may be found.

The energy barrier of the Cassie–Baxter and Wenzel transition can be estimated as the kinetic energy of the droplet. The kinetic energy of a droplet of radius R_0, mass m, and density ρ with velocity v is given by

$$E_{kin} = \frac{(4/3)\,\pi\rho R^3 v^2}{2} . \quad (42.31)$$

Contact Angle Measurements

In order to study the effect of pitch value on the transition from Cassie–Baxter to Wenzel regime, the static contact angles were measured on patterned Si coated with PF_3; the data are plotted as a function of pitch between the pillars in Fig. 42.28a [42.8, 52, 59–61]. A dot-dashed line represents the transition criteria range obtained using (42.26). The flat Si coated with PF_3 showed a static contact angle of 109°. The contact angle of selected patterned surfaces is much higher than that of the flat surfaces. It first increases with an increase in the pitch values, then drops rapidly to a value slightly higher than that of the flat surfaces. In the first portion, it jumps to a high value of 152° corresponding to a superhydrophobic surface and continues to increase to 170° at a pitch of 45 μm in series 1 and 126 μm in series 2 because open air space increases with increasing pitch, responsible for the propensity for air pocket formation. The sudden drop at a pitch value of about 50 μm in series 1 and 150 μm in series 2 corresponds to the transition from the Cassie–Baxter to the Wenzel regime. In series 1, the value predicted from the curvature transition criteria (42.26) is slightly higher than the experimental observations. However, in series 2, there is good agreement between the experimental data and the values theoretically predicted by

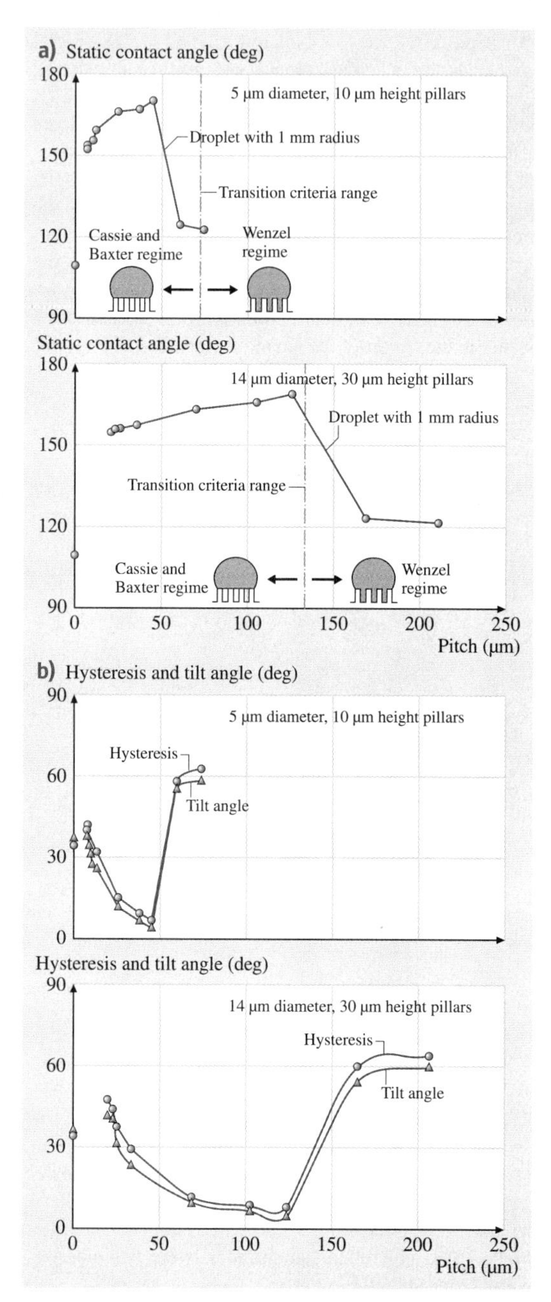

Fig. 42.28 (a) Static contact angle (a *dot-dashed line* represents the transition criteria range obtained using (42.26)). (b) Contact angle hysteresis and tilt angle as a function of geometric parameters for two series of patterned surfaces with different pitch values for a droplet 1 mm in radius (5 μl volume). Data at zero pitch correspond to a flat sample (after [42.52, 59]) ▶

Jung and *Bhushan* [42.59–61] for the Cassie–Baxter and Wenzel transition.

Figure 42.28b shows contact angle hysteresis and tilt angle as a function of pitch between the pillars [42.8, 52]. Both angles are comparable. The flat Si coated with PF_3 showed a contact angle hysteresis of 34° and tilt angle of 37°. The angle first increases with increasing pitch value, which has to do with pinning of the droplet at the sharp edges of the micropillars. Figure 42.29 shows droplets on patterned Si with 5 μm-diameter and 10 μm-height pillars with different pitch values. The asymmetrical shape of the droplet signifies pinning. The pinning on the patterned surfaces can be observed as compared with the flat surface. The patterned surface with low pitch (7 μm) has more pinning than the patterned surface with high pitch (37.5 μm), because the former has more sharp edges contacting with a droplet. As the pitch increases, there is a greater propensity for air pocket formation and fewer sharp edges per unit area, which is responsible for the sudden drop in the angle. The lowest contact angle hysteresis and tilt angle are 5° and 3°, respectively, which were observed on the patterned Si with 45 μm of series 1 and 126 μm of series 2. Above a pitch value of 50 μm in series 1 and 150 μm in series 2, the angle increases very rapidly because of the transition to the Wenzel regime.

These results suggest that air pocket formation and the reduction of pinning in the patterned surface play an important role for a surface with both low contact angle hysteresis and tilt angle [42.8, 52]. Hence, to create superhydrophobic surfaces, it is important that they are able to form a stable composite interface with air pockets between solid and liquid.

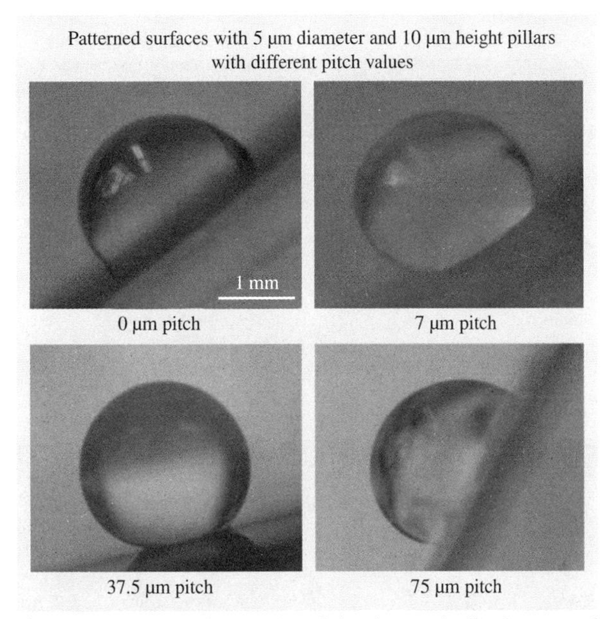

Fig. 42.29 Optical micrographs of droplets on inclined patterned surfaces with different pitch values. The images were taken when the droplet started to move down. Data at zero pitch correspond to a flat sample (after [42.52])

Observation of Transition During Droplet Evaporation

In order to study the effect of droplet size on the transition from a composite state to a wetted state, *Jung* and *Bhushan* [42.59, 60] performed droplet evaporation experiments to observe the Cassie–Baxter and Wenzel transition on two different patterned Si surfaces coated with PF_3. The series of four images in Fig. 42.30 shows successive photos of a droplet evaporating on the two patterned surfaces. The initial radius of the droplet was ≈ 700 μm, and the time interval between the first two photos was 180 s and between the latter sets was 60 s. In the first three photos, the droplet is shown in a Cassie–Baxter state, and its size gradually decreases with time. However, as the radius of the droplet reached 360 μm on the surface with 5 μm-diameter, 10 μm-high, 37.5 μm-pitch pillars, and 423 μm on the surface with 14 μm-diameter, 30 μm-high, 105 μm-pitch pillars, transition from the Cassie–Baxter to Wenzel regime occurred, as indicated by the arrow. The light passes below the first droplet, indicating that air pockets exist, so that the droplet is in the Cassie–Baxter state. However, an air pocket is not visible below the last droplet, so it is in the Wenzel state. This could result from an impalement of the droplet in the patterned surface, characterized by a smaller contact angle.

To find the contact angle before and after the transition, the values of the contact angle are plotted against the theoretically predicted value, based on the Wenzel (calculated using (42.6)) and Cassie–Baxter (calculated using (42.9)) models. Figure 42.31 shows the static contact angle as a function of geometric parameters for the experimental contact angles before (circles) and after (triangles) the transition compared with the Wenzel and Cassie–Baxter equations (solid lines) with a given value of θ_0 for two series of patterned Si with different pitch values coated with PF_3 [42.60]. The fit between the experimental data and the theoretically predicted values is good for the contact angles before and after transition.

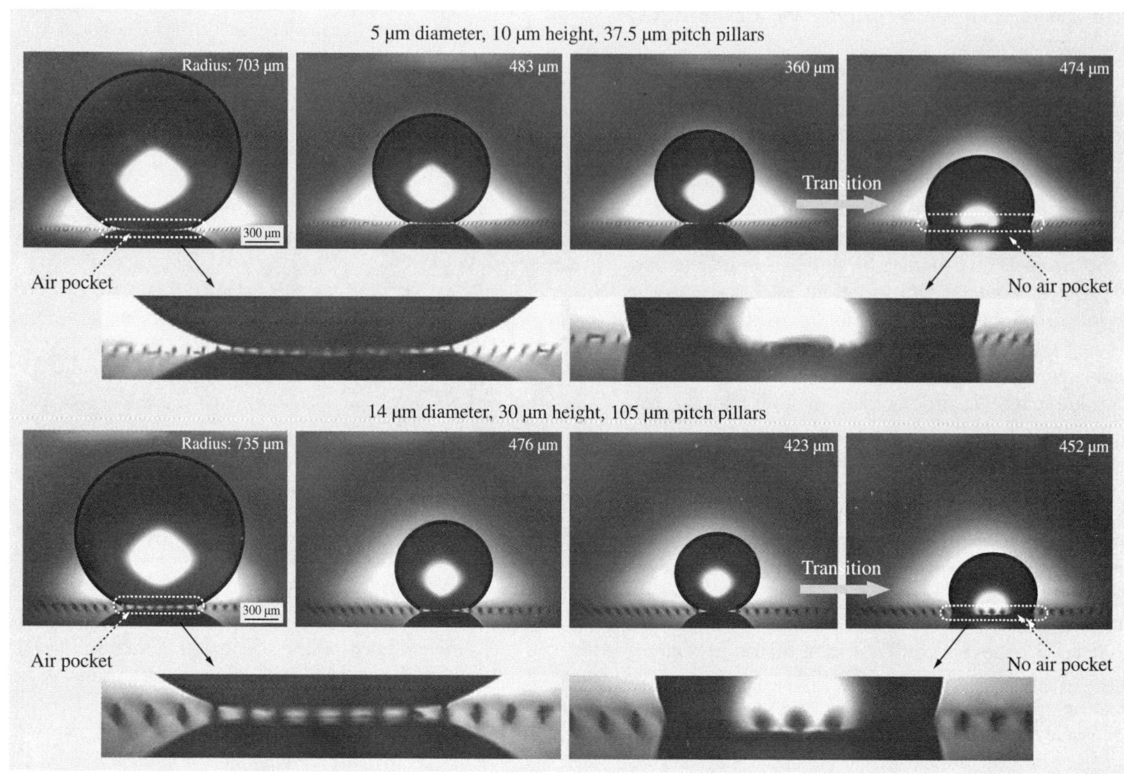

Fig. 42.30 Evaporation of a droplet on two different patterned surfaces. The initial radius of the droplet is $\approx 700\,\mu m$; the time interval between the first two photos was 180 s and between the latter two was 60 s. As the radius of the droplet reaches $360\,\mu m$ on the surface with $5\,\mu m$-diameter, $10\,\mu m$-height, $37.5\,\mu m$-pitch pillars, and $420\,\mu m$ on the surface with $14\,\mu m$-diameter, $30\,\mu m$-height, $105\,\mu m$-pitch pillars, the transition from the Cassie–Baxter regime to the Wenzel regime occurs, as indicated by the *arrow*. Before the transition, an air pocket is clearly visible at the bottom area of the droplet, but after the transition, the air pocket is not found at the bottom area of the droplet (after [42.60])

To prove the validity of the transition criteria in terms of droplet size, the critical radius of a droplet deposited on the patterned Si with different pitch values coated with PF_3 is measured during the evaporation experiment [42.59, 60]. Figure 42.32 shows the radius of a droplet as a function of geometric parameters for the experimental results (circles) compared with the transition criterion (42.26) from the Cassie–Baxter regime to the Wenzel regime (solid lines) for two series of patterned Si with different pitch values coated with PF_3. It is found that the critical radius of impalement is in good quantitative agreement with our predictions. The critical radius of the droplet increases linearly with the geometric parameter (pitch). For the surface with small pitch, the critical droplet radius can become quite small.

To verify the transition, *Jung* and *Bhushan* [42.59, 60] used another approach using dust mixed in water. Figure 42.33 presents the dust trace remaining after a droplet with 1 mm radius ($5\,\mu l$ volume) evaporated from the two patterned Si surfaces. As shown in the top image, after the transition from the Cassie–Baxter regime to the Wenzel regime, the dust particles remained not only at the top of the pillars but also at the bottom, with a footprint size of $\approx 450\,\mu m$. However, as shown in the bottom image, the dust particles remained on only a few pillars with a footprint size of $\approx 25\,\mu m$ until the end of the evaporation process. From Fig. 42.32, it is observed that the transition occurs at a droplet radius $\approx 300\,\mu m$ on the $5\,\mu m$-diameter, $10\,\mu m$-high pillars with $37.5\,\mu m$ pitch, but the transition does not occur on the patterned Si surface with

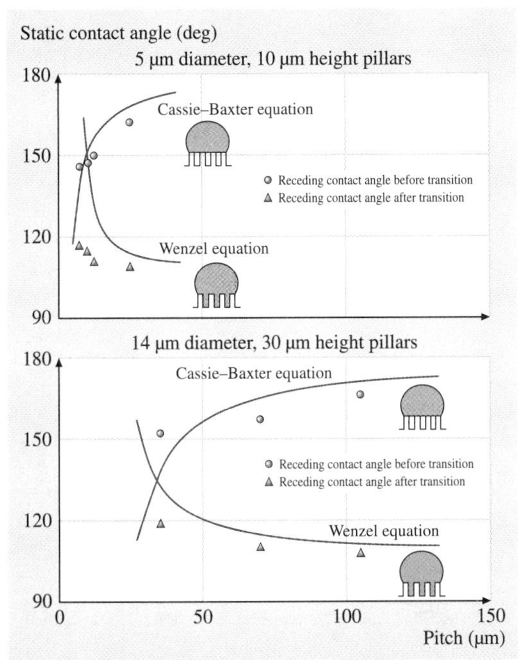

Fig. 42.31 Receding contact angle as a function of geometric parameters before (*circles*) and after (*triangles*) transition compared with predicted static contact angle values calculated using the Wenzel and Cassie–Baxter equations (*solid lines*) with a given value of θ_0 for two series of patterned surfaces with different pitch values (after [42.60])

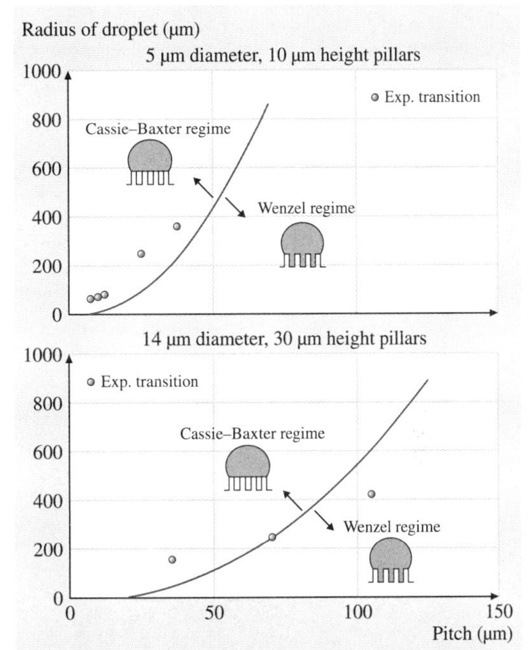

Fig. 42.32 Droplet radius as a function of geometric parameters for the experimental results (*circles*) compared with the transition criteria from the Cassie–Baxter regime to Wenzel regime (*solid lines*) for two series of patterned surfaces with different pitch values (after [42.60])

pitch $<\approx 5\,\mu m$. These experimental observations are consistent with model predictions. In the literature, it has been shown that, on superhydrophobic natural lotus, the droplet remains in the Cassie–Baxter regime during the evaporation process [42.181]. This indicates that the distance between the pillars should be minimized to improve the ability of the droplet to resist sinking.

Scaling of the Cassie–Baxter to Wenzel Transition for Different Series

Nosonovsky and *Bhushan* [42.9, 10, 64, 65, 70] studied the data for the Cassie–Baxter and Wenzel transition with the two series of surfaces using the nondimensional spacing factor

$$S_f = \frac{D}{P} \,. \tag{42.32}$$

The values of the droplet radius at which the transition occurs during the evaporation plotted against the spacing factor scale well for the two series of experimental results, yielding virtually the same straight line. Thus the two series of patterned surfaces scale well with each other, and the transition occurs at the same value of the spacing factor multiplied by the droplet radius (Fig. 42.34a). The physical mechanism leading to this observation remains to be determined; however, it is noted that this mechanism is different from that suggested by (42.26). The observation suggests that the transition is a linear one-dimensional (1-D) phenomenon and that neither droplet droop (which would involve P^2/H) nor droplet weight (which would involve R^3) are responsible for the transition, but rather linear geometric relations are involved. Note that the experimental values approximately correspond to the values of the ratio $RD/P = 50\,\mu m$, or the total area of the pillar tops under the droplet $(\pi D^2/4)\pi R^2/P^2 = 6200\,\mu m^2$.

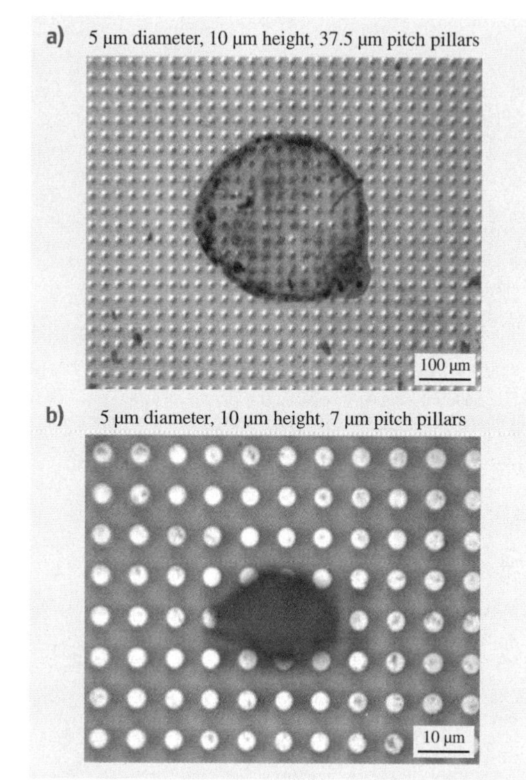

Fig. 42.33a,b Dust traces remaining after droplet evaporation for the patterned surface. In (**a**) the transition occurred at 360 μm droplet radius, and in (**b**) the transition occurred at ≈ 20 μm droplet radius during the process of droplet evaporation. The footprint size is ≈ 450 and 25 μm for (**a**) and (**b**), respectively (after [42.60])

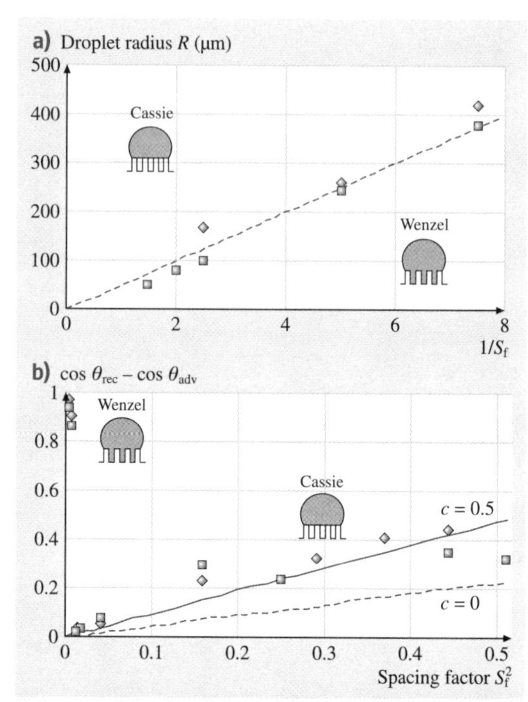

Fig. 42.34 (**a**) Droplet radius R for the Cassie–Baxter to Wenzel transition as a function of $P/D = 1/S_f$. It is observed that the transition takes place at a constant value of $RD/P \approx 50\,\mu m$ (*dashed line*). This shows that the transition is a linear phenomenon. (**b**) Contact angle hysteresis as a function of S_f for the first (*squares*) and second (*diamonds*) series of experiments, compared with the theoretically predicted values of $\cos\theta_{adv} - \cos\theta_{rec} = (D/P)^2(\pi/4)(\cos\theta_{adv0} - \cos\theta_{rec0}) + c(D/P)^2$, where c is a proportionality constant. It is observed that, when only the adhesion hysteresis/interface energy term is considered ($c = 0$), the theoretical values are underestimated by ≈ 0.5, whereas $c = 0.5$ provides a good fit. Therefore, the contribution of the adhesion hysteresis is of the same order of magnitude as the contribution kinetic effects (after [42.70])

Contact Angle Hysteresis and Wetting/Dewetting Asymmetry

Contact angle hysteresis can be viewed as a result of two factors that act simultaneously. First, the changing contact area affects the contact angle hysteresis, since a certain value of contact angle hysteresis is inherent for even a nominally flat surface. Decreasing the contact area by increasing the pitch between the pillars leads to a proportional decrease of the contact angle hysteresis. This effect is clearly proportional to the contact area between the solid surface and the liquid droplet. Second, the edges of the pillar tops prevent the motion of the triple line. This roughness effect is proportional to the contact line density, and its contribution was, in the experiment, comparable to the contact area effect. Interestingly, the effect of the edges is much more significant for the advancing than for the receding contact angle.

Nosonovsky and *Bhushan* [42.9, 10, 64, 65, 70] studied the wetting of two series of patterned Si surfaces with different pitch values coated with PF_3 based on the spacing factor (42.32). They found that the contact angle hysteresis involves two terms (Fig. 42.34b): the term $S_f^2(\pi/4)(\cos\theta_{adv0} - \cos\theta_{rec0})$ corresponding to the adhesion hysteresis (which is found even at a nominally flat surface and is a result of molecular-scale imperfectness) and the term $H_r \propto D/P^2$ corresponding to microscale roughness and proportional to the edge line density. Thus the contact angle hysteresis is given, based on (42.18) and (42.32), by using $R_f = 1 + \pi DH/P^2$ and $f_{LA} = 1 - \pi D^2/(4P^2) = 1 - f_{SL}$, as [42.53, 63]

$$\cos\theta_{adv} - \cos\theta_{rec} = \frac{\pi}{4} S_f^2 (\cos\theta_{adv0} - \cos\theta_{rec0}) + H_r \,. \quad (42.33)$$

Besides the contact angle hysteresis, the asymmetry of the Wenzel and Cassie–Baxter states is the result of the wetting/dewetting asymmetry. While the fragile metastable Cassie–Baxter state is often observed, as well as its transition to the Wenzel state, the opposite transition never happens. Using (42.24) and (42.25), the contact angle with patterned surfaces is given by [42.53, 63]

$$\cos\theta = \left(1 + 2\pi S_f^2\right)\cos\theta_0 \,, \quad \text{(Wenzel state)} \,, \quad (42.34)$$

$$\cos\theta = \frac{\pi}{4} S_f^2 (\cos\theta_0 + 1) - 1 \,, \quad \text{(Cassie–Baxter state)} \,. \quad (42.35)$$

For a perfect macroscale system, the transition between the Wenzel and Cassie–Baxter states should occur only at the intersection of the two regimes (the point at which the contact angle and net energies of the two regimes are equal, corresponding to $S_f = 0.51$). It is observed, however, that the transition from the metastable Cassie–Baxter to stable Wenzel occurs at much lower values of the spacing factor $0.083 < S_f < 0.111$. As shown in Fig. 42.35a, the stable Wenzel state (i) can transform into the stable Cassie–Baxter state with increasing S_f (ii). The metastable Cassie–Baxter state (iii) can abruptly transform (iv) into the stable Wenzel state (i). The transition point (i/ii) corresponds to equal free energies in the Wenzel and Cassie–Baxter states, whereas the transition (iv) corresponds to a Wenzel energy much lower than the Cassie–Baxter energy and thus involves significant energy dissipation and is irreversible [42.70]. The solid and dashed straight lines correspond to the values of the contact angle, calculated from (42.34) and (42.35) using the contact angle for a nominally flat surface $\theta_0 = 109°$. The two series of the experimental data are shown with squares and diamonds.

Figure 42.35b shows the values of the advancing contact angle plotted against the spacing factor (42.32). The solid and dashed straight lines correspond to the values of the contact angle for the Wenzel and Cassie–Baxter states, calculated from (42.34) and (42.35) using the advancing contact angle for a nominally flat surface,

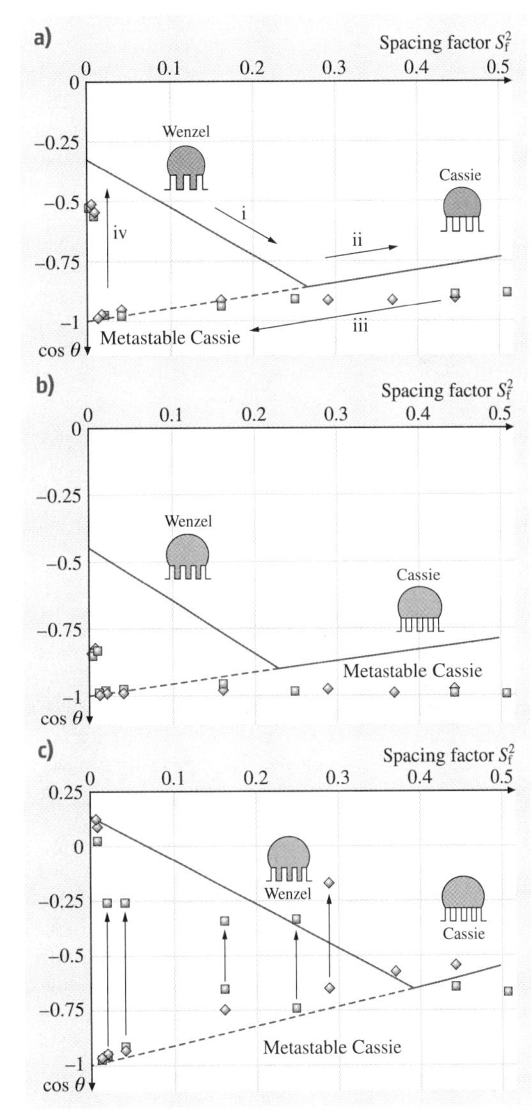

Fig. 42.35a–c Theoretical (*solid* and *dashed lines*) and experimental (*squares* for the first series, *diamonds* for the second series) results for: (a) the contact angle as a function of the spacing factor, (b) the advancing contact angle, and (c) the receding contact angle, and values of the contact angle observed after the transition during evaporation (after [42.70])

$\theta_{advO} = 116°$ [42.70]. It is observed that the calculated values underestimate the advancing contact angle, especially for large S_f (small distance between the pillars or pitch P). This is understandable, because the calculation only takes into account the effect of the contact area and ignores the effect of roughness and edge line density (it corresponds to $H_r = 0$ in (42.33)), while this effect is more pronounced for high pillar density (large S_f). In a similar manner, the contact angle is underestimated for the Wenzel state, since the pillars constitute a barrier for the advancing droplet.

Figure 42.35c shows the values of the contact angle after the transition took place (squares and diamonds), as was observed during evaporation [42.70]. For both series, the values almost coincide. For comparison, the values of the receding contact angle measured for millimeter-sized water droplets are also shown (squares and diamonds), since evaporation constitutes removing liquid, and thus the contact angle during evaporation should be compared with the receding contact angle. The solid and dashed straight lines correspond to the values of the contact angle, calculated from (42.34) and (42.35) using the receding contact angle for a nominally flat surface $\theta_{rec0} = 82°$. Figure 42.35c demonstrates agood agreement between the experimental data and (42.34) and (42.35).

In the analysis of the evaporation data of micropatterned surfaces, *Nosonovsky* and *Bhushan* [42.10] found several effects specific to the multiscale character of this process. First, they discussed applicability of the Wenzel and Cassie–Baxter equations for average surface roughness and heterogeneity. These equations relate the local contact angle with the apparent contact angle of a rough/heterogeneous surface. However, it is not obvious what should be the size for the roughness/heterogeneity averaging, since the triple line at which the contact angle is defined has two very different length scales: its width is of molecular size scale while its length is of the order of the size of the droplet (that is, μm or mm). They presented an argument that, in order for the averaging to be valid, the roughness details should be small compared with the size of the droplet (and not the molecular size). They showed that, while for uniform roughness/heterogeneity the Wenzel and Cassie–Baxter equations can be applied, for a more complicated case of nonuniform heterogeneity, the generalized equations should be used. The proposed generalized Cassie–Baxter and Wenzel equations are consistent with a broad range of available experimental data. The generalized equations are valid both in the cases when the classical Wenzel and Cassie–Baxter equations can be applied as well as in the cases when the latter fail.

The macroscale contact angle hysteresis and Cassie–Baxter and Wenzel transition cannot be determined from the macroscale equations and are governed by micro- and nanoscale effects, so wetting is a multiscale phenomenon [42.9, 10, 64, 65, 70]. The kinetic effects associated with contact angle hysteresis should be studied at the microscale, whereas the effects of adhesion hysteresis and the Cassie–Baxter and Wenzel transition involve processes at the nanoscale. Their theoretical arguments are supported by the experimental data on micropatterned surfaces. Experimental study of contact angle hysteresis demonstrates that two different processes are involved: the changing solid–liquid area of contact, and pinning of the triple line. The latter effect is more significant for the advancing than for the receding contact angle. The transition between wetting states was observed for evaporating microdroplets, and the droplet radius scales well with the geometric parameters of the micropattern.

Observation of Transition for a Bouncing Droplet

Jung and *Bhushan* [42.61] performed bouncing-droplet experiments to observe how impact velocity influences the Cassie–Baxter and Wenzel transition during the droplet hitting the surface on two different micropatterned Si surfaces with PF_3. Figure 42.36 shows snapshots of a droplet with 1 mm radius hitting the surfaces. The impact velocity was obtained just prior to the droplet hitting the surface. As shown in the images in the first row for the two sets of surfaces, the droplet hitting the surface under an impact velocity of 0.44 m/s first deformed and then retracted, and bounced off the surface. Finally, the droplet sat on the surface and had a high contact angle, which suggests the formation of a solid–air–liquid interface. Next, they repeated the impact experiment with an increased impact velocity. The bounce off does not occur, and the wetting of the surface (and possibly the pinning of the droplet) occurred at an impact velocity of 0.88 and 0.76 m/s, respectively, referred to as the critical velocity (described earlier). The second row of the two sets of images shows the droplet at the critical velocity. After the droplet hit the surface, it wetted the surface (possibly the droplet was also pinned) after the deformation of the droplet. This is because air pockets do not exist below the droplet as a result of droplet impalement by the pillars, characterized by a smaller contact angle. These observations indicate the transition from a Cassie–Baxter to a Wenzel regime.

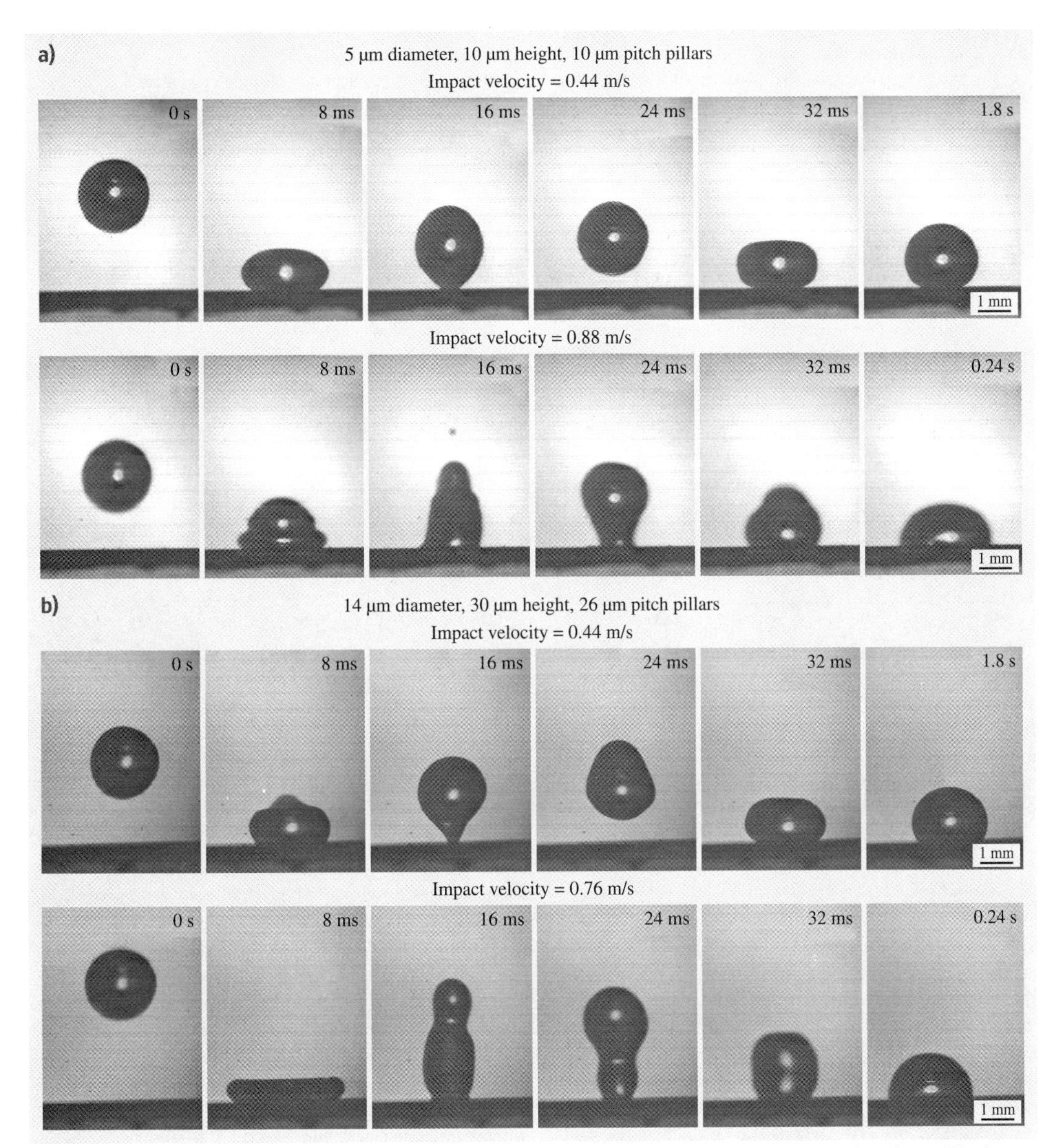

Fig. 42.36a,b Snapshots of a droplet with 1 mm radius impacting on two different micropatterned surfaces. The impact velocity was obtained just prior to the droplet hitting the surface. The pinning of the droplet on the surface with (**a**) 5 μm-diameter, 10 μm-height, 10 μm-pitch pillars, and on (**b**) the surface with 14 μm-diameter, 30 μm-height, 26 μm-pitch pillars, occurred at impact velocities of 0.88 and 0.76 m/s, respectively (after [42.61])

To identify whether one is in the Wenzel or Cassie–Baxter regime, the contact angle data in the static condition and after bounce off were plotted [42.61]. Figure 42.37 shows the measured static contact angle as a function of geometric parameters for the droplet with 1 mm radius gently deposited on the surface (circles) and for the droplet with 1 mm radius after hitting the surface at 0.44 m/s (triangles). The data are compared

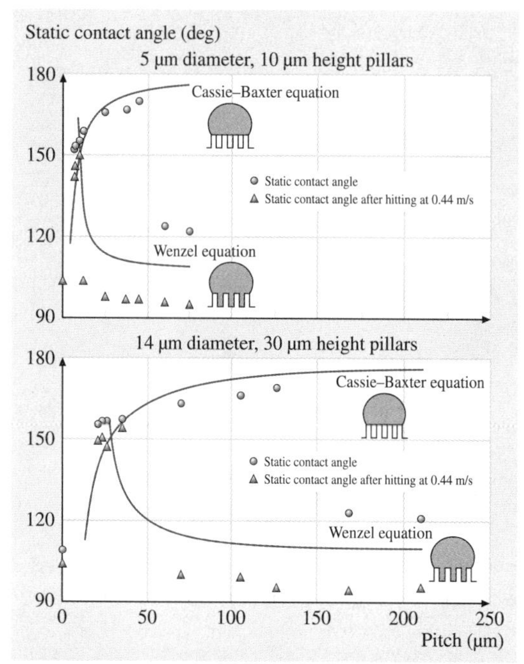

Fig. 42.37 Measured static contact angle as a function of geometric parameters for a droplet with 1 mm radius gently deposited on the surface (*circles*) and for the same sized droplet hitting the surface at 0.44 m/s (*triangles*). The data are compared with predicted static contact angle values obtained using the Wenzel and Cassie–Baxter equations (*solid lines*) with a given value of θ_0 (109°) for a smooth surface for two series of patterned Si with different pitch values (after [42.61])

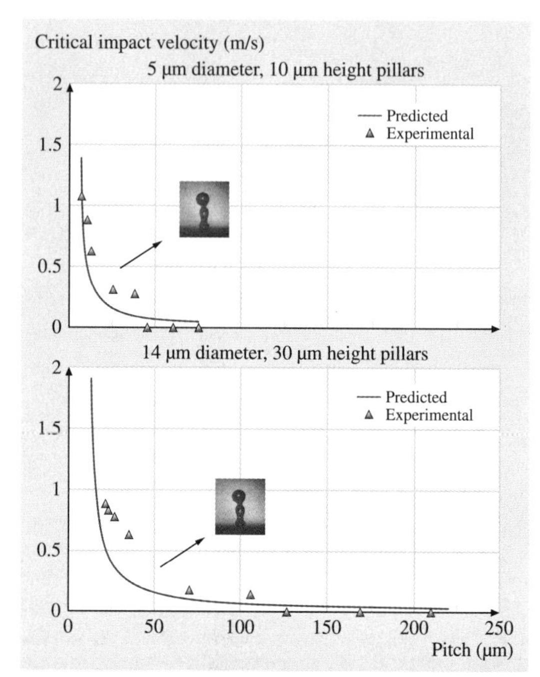

Fig. 42.38 Measured critical impact velocity for a droplet with 1 mm radius as a function of geometric parameters (*triangles*). The data are compared with the criterion of impact velocity for droplet pinning (*solid lines*) for two series of patterned Si with different pitch values (after [42.61])

with predicted static contact angle values obtained using the Wenzel and the Cassie–Baxter equations (solid lines) with a given value of θ_0 (109°) for a smooth surface for two series of the micropatterned surfaces. In the case of the droplet gently deposited on the surface, as the pitch increases up to 45 μm of series 1 and 126 μm of series 2, the static contact angle first increases gradually from 152° to 170°. Then, the contact angle starts decreasing sharply. The increase in the contact angle occurs because of an increase in the roughness factor and the formation of composite surface [42.52]. The decrease in the contact angle at pitch values > 60 μm for series 1 and > 168 μm for series 2 occurs due to the transition from composite interface to solid–liquid interface. In the case of the droplet hitting the surface at 0.44 m/s, it is shown that the liquid–air interface can easily be destabilized due to the dynamic impact on the surface with a pitch value > 12.5 μm for series 1 and > 70 μm for series 2, although the droplet is in the Cassie–Baxter regime when it is gently deposited on the surface. The static contact angle of the droplet after hitting at 0.44 m/s is lower than that of the gently deposited droplet. It can be interpreted that, after hitting, the droplet contacts with the bottom of the cavities between pillars and pushes out the entrapped air under the droplet, resulting in an abrupt increase of the solid–liquid surface area by dynamic impact. It will be shown in the following paragraph that the critical velocity at which wetting occurs for series 1 and series 2 samples is equal to ≈ 0.44 m/s at pitch values > 12.5 μm and > 70 μm, respectively. Thus, wetting at the velocity used here is expected.

To study the validity of the transition criterion (42.29), the critical impact velocity at which wetting of the surface (and possibly pinning of droplet) occurs was measured [42.61]. For calculations, the surface tension of the water–air interface (γ) was taken as 0.073 N/m, the mass density (ρ) was 1000 kg/m^3 (for water), and

$1\,\mathrm{kg\,m/s^2} = 1\,\mathrm{N}$ [42.1]. Figure 42.38 shows the measured critical impact velocity of a droplet with 1 mm radius as a function of geometric parameters (triangles). The trends are compared with the predicted curve (solid lines). It is found that the critical impact velocity at which wetting occurs is in good quantitative agreement with our predictions. The critical impact velocity of the droplet decreases with the geometric parameter (pitch). For the surface with small pitch, the critical impact velocity of the droplet can be large.

The energy barrier of the Cassie–Baxter and Wenzel transition can be estimated from the kinetic energy of the droplets [42.68]. Figure 42.39 shows the dependence of the kinetic energy corresponding to the transition E_{kin} on $\Delta E/(A_0 \cos\theta_0)$ calculated from (42.30). It is observed that the dependence is close to linear; however, the series of smaller pillars has larger energies of transition. The value of A_0 is in the range $0.11\,\mathrm{mm}^2 < A_0 < 0.18\,\mathrm{mm}^2$ for series 1 and $0.05\,\mathrm{mm}^2 < A_0 < 0.11\,\mathrm{mm}^2$ for series 2, which is of the same order as the actual area under the droplet.

These results suggest that the energy barrier for the Cassie–Baxter and Wenzel transition is given by (42.30) and is proportional to the area under the droplet. For droplets sitting on the surface or evaporating, the transition takes place when the size of the barrier decreases to the value of the vibrational energy U. The vibrational energy of the droplet is the energy associated with the vibration of the droplet due to surface waves, thermal vibration, etc. Assuming that $U = \mathrm{const.}$, the proportionality of P/D and R suggests that the energy barrier is proportional to RD/P. This is indeed true, since the area under the droplet is $A_0 = \pi(R\sin\theta)^2$. Substituting $\sin^2\theta = 0.1$, $\cos\theta_0 = \cos 109° = -0.33$, and $\gamma_{\mathrm{LA}} = 0.072\,\mathrm{J/m^2}$ into (42.30) and taking the observed value $RD/P = 50\,\mu\mathrm{m}$ yields an estimated value of the vibrational energy of $U = 1.2\times10^{-10}\,\mathrm{J}$. The transition happens because the size of the droplet is decreased or because the pitch between the pillars that cover the surface is increased. A different way to overcome the barrier is to hit the surface with a droplet with a certain kinetic energy.

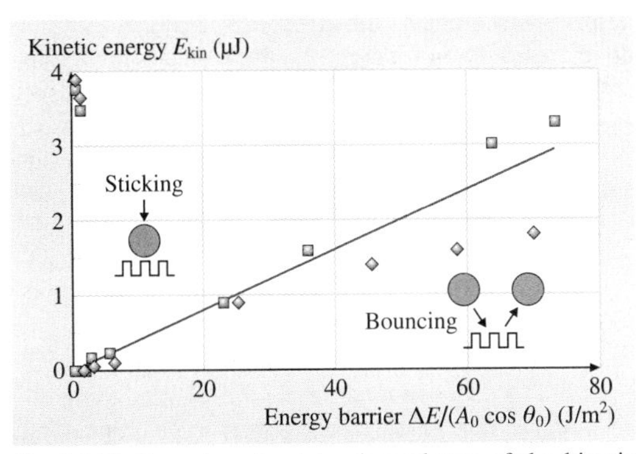

Fig. 42.39 Bouncing droplets: dependency of the kinetic energy of a droplet corresponding to the regime transition upon the energy barrier calculated from (42.30). The fit (*solid line*) is shown for $A_0 = 0.12\,\mathrm{mm}^2$ (after [42.68])

The vibrational energy U also plays a role in overcoming energy barriers that lead to contact angle hysteresis during liquid flow [42.38]. To estimate the effect of the energy barriers on contact angle hysteresis, we assume, based on (42.5), that the difference between the advancing and receding contact angle is given by

$$\cos\theta_{\mathrm{rec}} - \cos\theta_{\mathrm{adv}} = \frac{\Delta W}{\gamma_{\mathrm{LA}}}\,, \tag{42.36}$$

where ΔW corresponds to the energy barrier associated with the wetting–dewetting cycle. Assuming that this energy barrier is of the same order as the vibrational energy per contact area $\Delta W = U/A_0$, and taking $A_0 = 0.1\,\mathrm{mm}^2$, we end up with $\Delta W = 10^{-3}\,\mathrm{J/m^2}$. For water ($\gamma_{\mathrm{LA}} = 0.072\,\mathrm{J/m^2}$), (42.36) leads to a realistic value of hysteresis on a superhydrophobic surface of $\cos\theta_{\mathrm{rec}} - \cos\theta_{\mathrm{adv}} = 0.014$. This number provides an estimate for contact angle hysteresis in the limit of small energy barriers comparable to U. The actual values for a micropatterned surface are dependent upon the solid–liquid contact area (which provides energy barriers due to so-called adhesion hysteresis) and the density of the solid–air–liquid contact line (which provides additional pinning) and were found to be between 0.0144 and 0.440 [42.53], thus showing good agreement with the value calculated based on U as the lower limit. This indicates that the value of U is relevant for both the Cassie–Baxter and Wenzel regime transition and contact angle hysteresis.

Observation and Measurement of Contact Angle Using ESEM

Figure 42.40 shows how water droplets grow and merge in an ESEM [42.60] that was used as a contact angle analysis tool. Microdroplets (with diameter < 1 mm) were distributed on a patterned surface coated with PF_3 using condensation by decreasing the temperature. At the beginning, some small water droplets appeared, i. e., water droplets at locations 1, 2, and 3 in the left image. During further condensation with decreasing temperature, droplets at locations 1 and 3 gradually grew while droplets at location 2 merged. With further condensa-

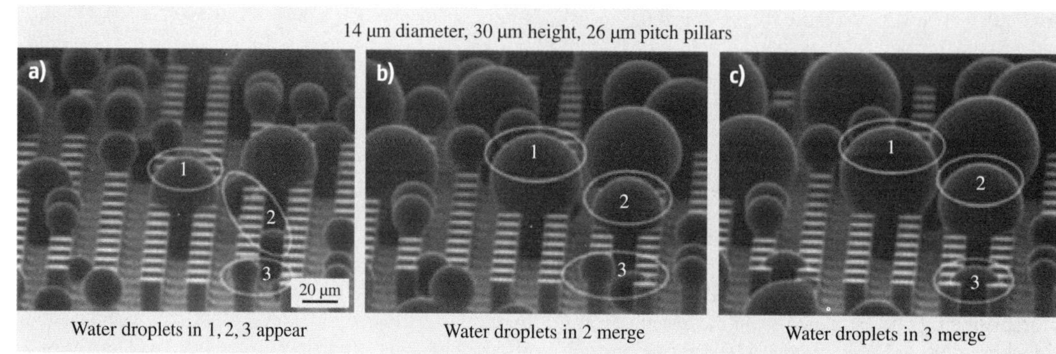

Fig. 42.40a–c Microdroplets (< 1 mm diameter) growing and merging process under ESEM during increasing condensation due to decreasing temperature. **(a)** Some small water droplets appear at the beginning, i. e., water droplets 1, 2, and 3. **(b)** Water droplets at locations 1 and 3 increase in size, and water droplets at location 2 merge to form one large droplet. **(c)** Water droplets at locations 1 and 2 increase in size, and water droplets at location 3 merge to form one large droplet (after [42.60])

tion, droplets at locations 1 and 2 gradually grew while droplets at location 3 merged into one big droplet in the rightmost image. In all cases, condensation was initiated at the bottom; therefore, the droplets were in the Wenzel regime.

Compared with the conventional contact angle measurement, ESEM is able to provide detailed information about the contact angle of microdroplets on patterned surfaces. The diameter of the water droplets used for the contact angle measurement was 10 μm, so that the size limit pointed out by *Stelmashenko* et al. [42.87] was avoided. For droplet size < 1 μm, substrate backscattering can distort the intensity profile such that the images are inaccurate.

As shown in Fig. 42.41, the static contact angle and contact angle hysteresis of the microdroplets on flat and patterned surfaces were obtained from the images using the methodology described earlier. Once the microdroplet's condensation and evaporation has reached a dynamic equilibrium, static contact angles were determined. The flat Si coated with PF_3 showed a static contact angle of 98°. The patterned surfaces coated with PF_3 had increased static contact angle compared with the flat surface coated with PF_3 due to the effect of roughness. Advancing and receding contact angles were measured during condensation/evaporation by decreasing/increasing the temperature of the cooling stage, and the contact angle hysteresis was then calculated [42.60].

Figure 42.42 shows the contact angle hysteresis as a function of geometric parameters for microdroplets formed in the ESEM (triangles) for two series of patterned Si with different pitch values coated with PF_3. Data at zero pitch correspond to a flat Si sample. The droplets with ≈ 20 μm radii, which are larger than the pitch, were selected in order to look at the effect of pillars in contact with the droplet. These data were compared with conventional contact angle measurements with the droplet with 1 mm radius (5 μl volume) (circle and solid lines) [42.52]. When the distance between pillars increases above a certain value, the contact area between the patterned surface and the droplet decreases, resulting in a decrease of the contact angle hysteresis. Both droplets with 1 mm and 20 μm radii showed the same trend. The contact angle hysteresis for the patterned surfaces with low pitch are higher compared with the flat surface due to the effect of sharp edges on the pillars, resulting in pinning [42.49]. Contact angle hysteresis for a flat surface can arise from roughness and surface heterogeneity. For a droplet advancing forward on the patterned surfaces, the line of contact of the solid, liquid, and air will be pinned at the edge point until it is able to move, resulting in increased contact angle hysteresis. The contact angle hysteresis for the microdroplet from ESEM is lower compared with that for the droplet with 1 mm radius. The difference of contact angle hysteresis between a microdroplet and a droplet with 1 mm radius could result from the different pinning effects, because the latter has more sharp edges contacting with a droplet compared with the former. These results show how droplet size can affect the wetting properties of patterned Si surfaces [42.60].

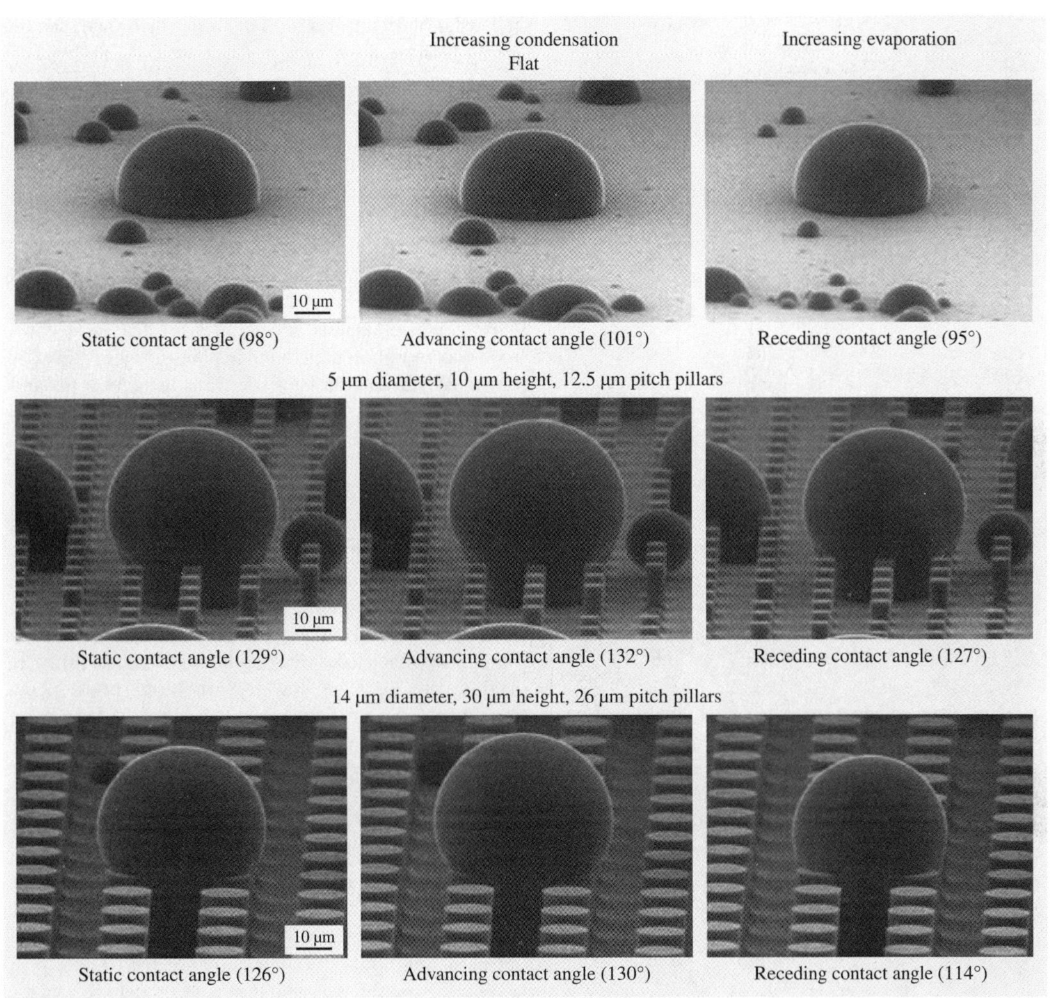

Fig. 42.41 Microdroplets on flat and two patterned surfaces using ESEM. The second column of images were taken during increasing condensation, and the third column of images were taken during increasing evaporation. The static contact angle was measured when the droplet was stable. The advancing contact angle was measured after increasing condensation by decreasing the temperature of the cooling stage. The receding contact angle was measured after decreasing evaporation by increasing the temperature of the cooling stage (after [42.60])

42.5.4 Ideal Surfaces with Hierarchical Structure

It has been reported earlier that a hierarchical surface is needed to develop a composite interface with high stability. The structure of an ideal hierarchical surface is shown in Fig. 42.43. The asperities should be high enough that the droplet does not touch the valleys. As an example, for a structure with circular pillars, the following relationship should hold for a composite interface: $(\sqrt{2}P - D)^2/R < H$ (42.26). As an example, for a droplet with a radius on the order of 1 mm or larger, a value of H on the order of 30 μm, D on the order of 15 μm, and P on the order of 130 μm (Fig. 42.28) is optimum. Nanoasperities can pin the liquid–air interface and thus prevent liquid from filling the valleys between asperities. They are also required to support nanodroplets, which may condense in the valleys be-

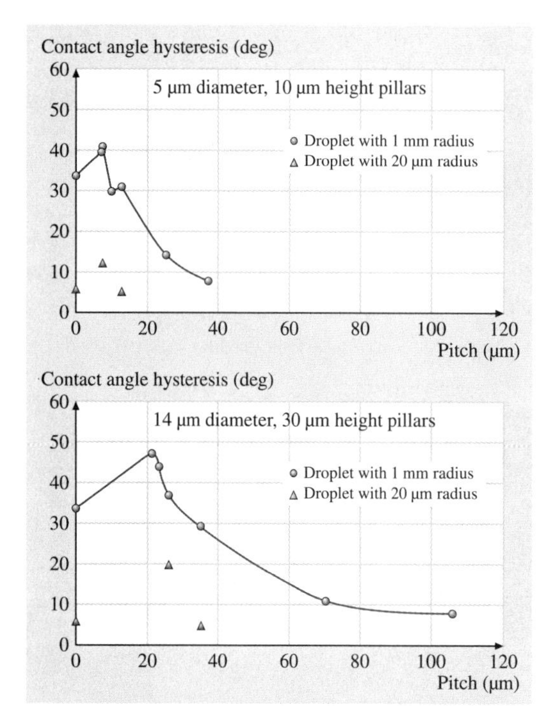

Fig. 42.42 Contact angle hysteresis as a function of geometric parameters for a microdroplet with $\approx 20\,\mu m$ radius obtained by ESEM (*triangles*) compared with a droplet with 1 mm radius (5 μl volume) (*circles* and *solid lines*) for two series of patterned surfaces with different pitch values. Data at zero pitch correspond to a flat sample (after [42.60])

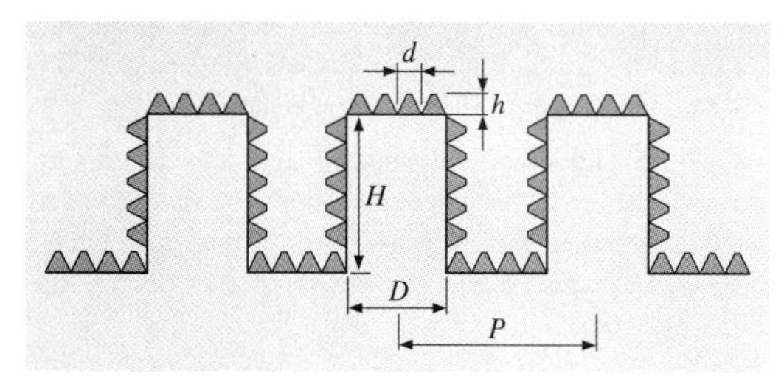

Fig. 42.43 Schematic of the structure of an ideal hierarchical surface. Microasperities consist of circular pillars with diameter D, height H, and pitch P. Nanoasperities consist of pyramidal nanoasperities of height h and diameter d with rounded tops

tween large asperities. Therefore, nanoasperities should have a small pitch to handle nanodroplets, < 1 mm down to a few nanometers radius. The values of h on the order of 10 nm and d on the order of 100 nm can be easily fabricated.

42.5.5 Hierarchical Structured Surfaces

A hierarchical structure is composed of at least two levels of structuring on different length scales. *Bhushan* et al. [42.54–57] and *Koch* et al. [42.30] fabricated surfaces with a hierarchical structure with micropatterned epoxy replicas, and lotus leaf microstructure, and created a second level of structuring with wax platelets and wax tubules. Platelets and tubules are the most common wax morphologies found in plant surfaces and exist on superhydrophobic leaves such as those of lotus and *Colocasia esculenta*. The developed structures mimic the hierarchical structures of superhydrophobic leaves. Two steps of the fabrication process include the production of microstructured surfaces by soft lithography and the subsequent development of nanostructures on top by self-assembly of plant waxes and artificial wax components.

A two-step molding process was used to fabricate several structurally identical copies of micropatterned Si surface and lotus leaves. The technique used is a fast, precise, and low-cost molding process for biological and artificial surfaces [42.182, 183]. The technique was used to mold a microstructured Si surface with pillars of 14 μm diameter and 30 μm height with 23 μm pitch [42.30, 54–57], fabricated by photolithography [42.100]. Before replication of the lotus leaf, the epicuticular wax tubules were removed in areas of $\approx 6\,cm^2$. For this purpose, a two-component fast-hardening glue was applied on the upper side of the leaves and carefully pressed onto the leaf. After hardening, the glue with the embedded waxes was removed from the leaf, and the same procedure was repeated [42.30]. The replication is a two-step molding process, in which first a negative replica of a template and then a positive replica are generated, as shown in Fig. 42.44a [42.56]. A polyvinylsiloxane dental wax was applied onto the surface and immediately pressed down with a glass plate. After complete hardening of the molding mass (at room temperature for ≈ 5 min), the silicon master surface and the mold (negative) were separated. After a relaxation time of 30 min for the molding material, the negative replicas were filled with a liquid epoxy resin. Specimens were immediately transferred to a vacuum chamber at 750 mTorr (100 Pa) pressure for 10 s to remove trapped air and to increase resin infiltration through the structures. After hardening at room temperature (24 h at 22 °C), the positive replica

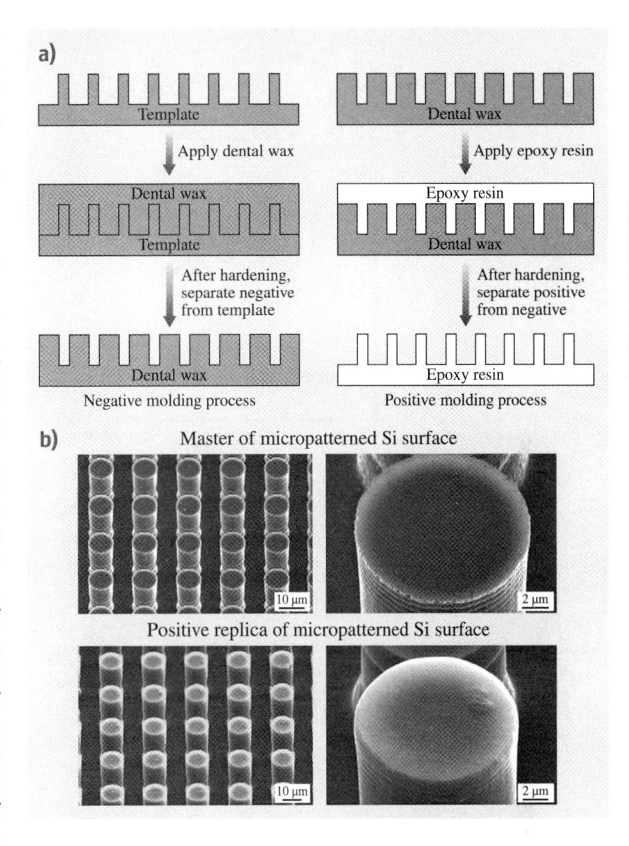

Fig. 42.44 (a) Schematic of the two-step molding process used to fabricate microstructure, in which first a negative is generated and then a positive, and (b) SEM micrographs of the master of the micropatterned Si surface and the positive replica fabricated from the master surface, measured at 45° tilt angle (shown using two magnifications) (after [42.56]) ▶

was separated from the negative replica. The second step can be repeated to generate a number of replicas. The pillars of the master surface have been replicated without any morphological changes, as shown in Fig. 42.44b [42.56]. The nanogrooves of a couple of hundred nanometers in lateral dimension present on the pillars of the master surface are shown to be reproduced faithfully in the replica.

Bhushan et al. [42.54–57] and *Koch* et al. [42.30] created a nanostructure by self-assembly of synthetic and plant waxes deposited by thermal evaporation. The alkane *n*-hexatriacontane ($C_{36}H_{74}$) has been used for the development of platelet nanostructures. Tubule-forming waxes which were isolated from leaves of *Tropaeolum majus* (L.) and *Nelumbo nucifera*, in the following referred to as *T. majus* and lotus, were used to create tubule structures. The chemical structure of the major components of the wax-forming tubule and alkane *n*-hexatriacontane are shown in Table 42.8. The complete chemistry of the plant waxes used is presented in *Koch* et al. [42.118]. For homogenous deposition of the waxes and alkane, a thermal evaporation system, as shown in Fig. 42.45, has been used [42.56]. Specimens of smooth surfaces (flat silicon replicas) and microstructured replicas were placed in a vacuum chamber at 30 mTorr (4 kPa), 20 mm above a heating plate loaded with waxes of *n*-hexatriacontane (300, 500 or 1000 μg), *T. majus* wax (500, 1000, 1500 or 2000 μg), and lotus

Table 42.8 Chemical structure of the major components of *n*-hexatriacontane, and *T. majus* and lotus waxes. The major component is shown first (after [42.30, 54, 57])

n-Hexatriacontane		$C_{36}H_{74}$
Tropaeolum majus	Nonacosan-10-ol	$CH_3-(CH_2)_8-CH(OH)-(CH_2)_{18}-CH_3$
	Nonacosane-4,10-diol	$CH_3-(CH_2)_2-CH(OH)-(CH_2)_5-CH(OH)-(CH_2)_{18}-CH_3$
Lotus	Nonacosane-10,15-diol	$CH_3-(CH_2)_8-CH(OH)-(CH_2)_4-CH(OH)-(CH_2)_{13}-CH_3$
	Nonacosan-10-ol	$CH_3-(CH_2)_8-CH(OH)-(CH_2)_{18}-CH_3$

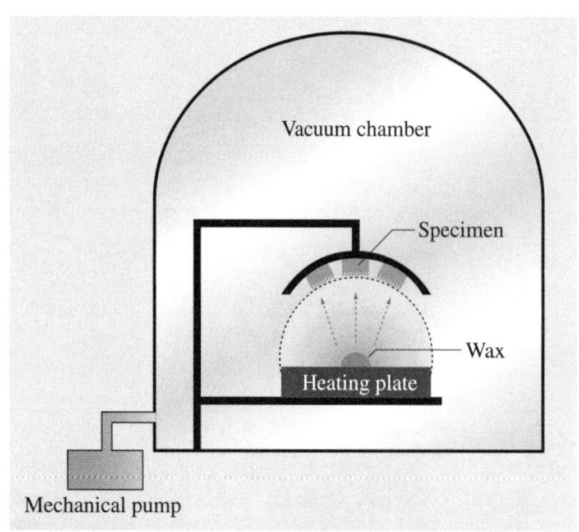

Fig. 42.45 Schematic of the thermal evaporation system used for self-assembly of wax. Evaporation from the point source to the substrate occurs over a hemispherical region (*dotted line*) (after [42.56])

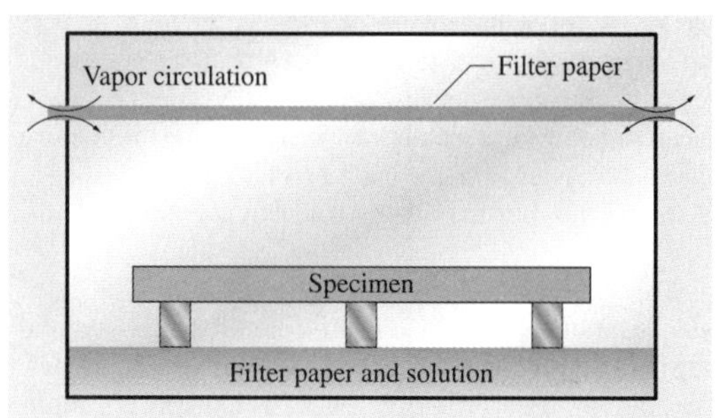

Fig. 42.46 Schematic of the glass recrystallization chamber used for tubule formation. The filter paper placed at the bottom of the chamber was wetted with 20 ml of the solvent, and slow evaporation of the solvent was provided by placing a thin filter paper between the glass body and the cap placed above. The total volume of the chamber is about 200 cm^3 (after [42.57])

wax (2000 μg) [42.30, 54–57]. The wax was evaporated by heating it to 120 °C. In a vacuum chamber, evaporation from the point source to the substrate occurs in a straight line; thus, the amount of sublimated material is equal in a hemispherical region over the point source [42.184]. In order to estimate the amount of sublimated mass, the surface area of the half-sphere was calculated by using the formula $2\pi r^2$, whereby the radius (r) represents the distance between the specimen to be covered and the heating plate with the substance to be evaporated. The amounts of wax deposited on the specimen surfaces were 0.12, 0.2, and 0.4 $\mu g/mm^2$ for n-hexatriacontane, 0.2, 0.4, 0.6, and 0.8 $\mu g/mm^2$ for *T. majus*, and 0.8 $\mu g/mm^2$ for lotus waxes, respectively.

After coating, the specimens with n-hexatriacontane were placed in a desiccator at room temperature for 3 days for crystallization of the alkanes. A stable stage was indicated by no further increase of crystal sizes.

For the plant waxes, which are a mixture of aliphatic components, different crystallization conditions have been chosen. It has been reported by *Niemietz* et al. [42.185] that an increase of temperature from 21 °C (room temperature) to 50 °C had a positive effect on the mobilization and diffusion of wax molecules, required for separation of the tubule-forming molecules. It is also known that chemical ambient has an influence on the propensity for wax crystallization, thus specimens with evaporated plant waxes (*T. majus* and lotus) were stored for 3 days at 50 °C in a crystallization chamber, where they were exposed to a solvent (ethanol) in vapor phase (Fig. 42.46). Specimens were placed on metal posts and a filter paper wetted with 20 mL of the solvent was placed below the specimens. Slow diffusive loss of the solvent in the chamber was provided by placing a thin filter paper between the glass body and the lid. After evaporation of the solvent, specimens were left in the oven at 50 °C in total for 7 days. Figure 42.47 shows the nanostructures formed by lotus wax, 7 days after wax deposition on flat surfaces. Figure 42.47a shows the nanostructure after storage at 21 °C; in these no tubules were grown. Figure 42.47b shows that lotus waxes exposed to ethanol vapor for 3 days at 50 °C formed wax tubules. A detailed description of the nanostructure sizes and nanoroughness is given in the following.

Flat films of n-hexatriacontane and wax tubules were made by heating the substances above their melting point and rapidly cooling. This procedure interrupts the crystallization and leads to smooth films [42.30, 54–57].

Nanostructures with Various Platelet Crystal Densities

Figure 42.48a shows scanning electron microscope (SEM) micrographs of a flat surface and nanostructures fabricated with various masses of n-hexatriacontane [42.54]. The nanostructure is formed by three-dimensional platelets of n-hexatriacontane, as shown in detail in Fig. 42.48b. Platelets are flat crystals, grown perpendicular to the substrate surface.

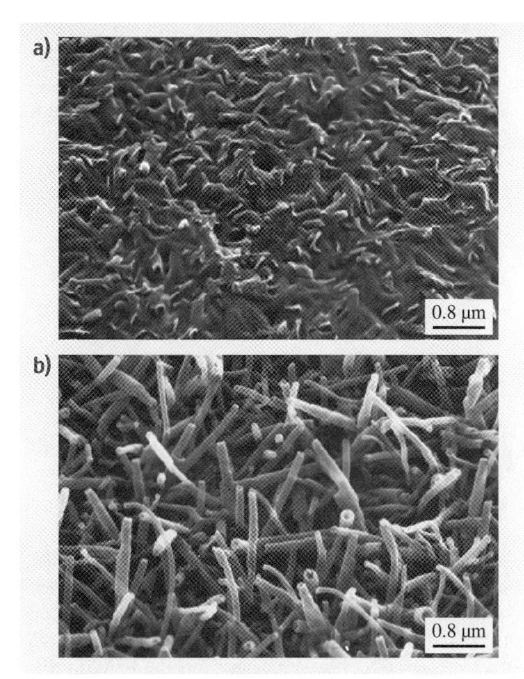

Fig. 42.47a,b SEM micrographs showing the morphology of lotus wax deposited on a flat epoxy replica surface after two treatments of specimens, measured at 45° tilt angle: (a) after 7 days at 21 °C, nanostructure on the flat epoxy replica was found with no tubules, and (b) after 7 days at 50 °C with ethanol vapor; tubular nanostructures with random orientation were found on the surface (after [42.30])

They are randomly distributed on the surface, and their shapes and sizes show some variations. Some of the single platelets are connected to their neighboring crystals at their lateral ends. This arrangement leads to a kind of cross-linking of the single platelets. As shown in Fig. 42.48b, and based on additional specimens, the platelet thickness varied between 50 and 100 nm, and their length varied between 500 and 1000 nm. Self-assembly of n-hexatriacontane used here, and of most long-chain hydrocarbons, leads to layered structures with lamellar order. In these structures, the molecular axis is orientated parallel to the substrate surface. The growth of these layers results in an ordered, crystalline 3-D structure [42.186]. The created nanostructures are comparable to the wax crys-

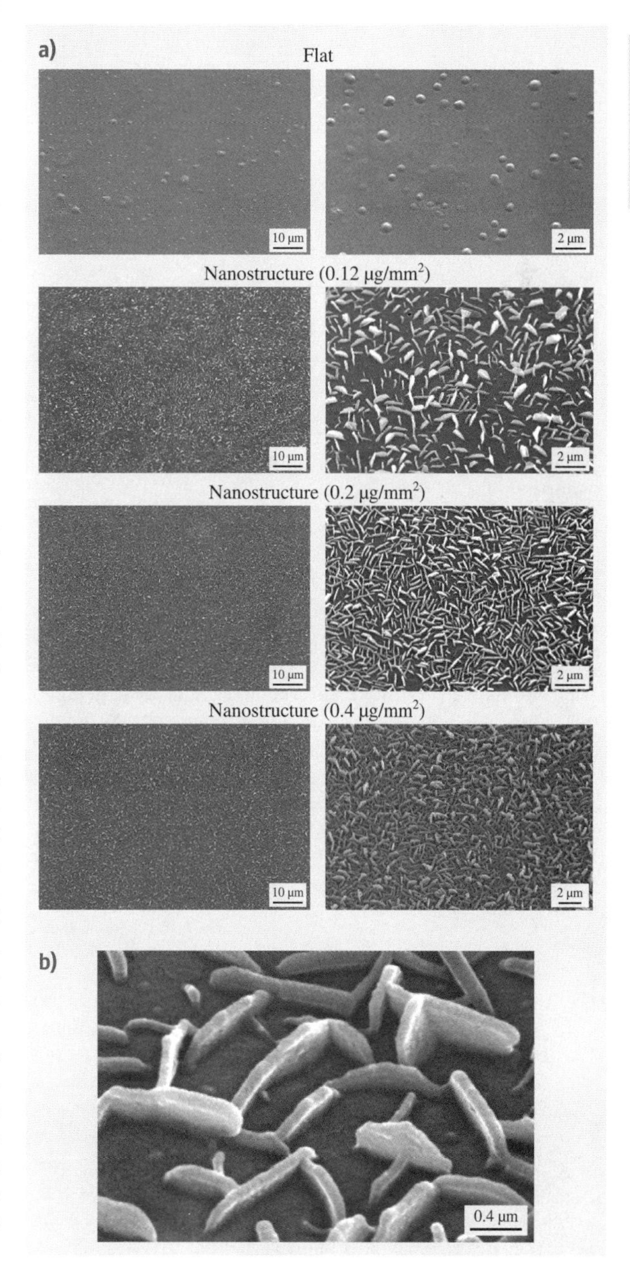

Fig. 42.48a,b SEM micrographs taken at 45° tilt angle (shown using two magnifications) of (a) the flat surface and nanostructures fabricated with various masses of n-hexatriacontane and (b) three-dimensional platelets forming nanostructures on the surface fabricated with 0.2 μg/mm^2 n-hexatriacontane. All samples are fabricated with epoxy resin coated with n-hexatriacontane (after [42.54]) ▶

Table 42.9 Roughness statistics for nanostructured surface measured using an AFM (scan size $10 \times 10\,\mu m^2$). Nanostructures were fabricated with n-hexatriacontane, and *T. majus* and lotus waxes (after [42.30, 54, 57])

	RMS height (nm)	Peak-to-valley height (nm)	η (μm^{-2})
***n*-Hexatriacontane**			
Nanostructure ($0.12\,\mu g/mm^2$)	46	522	0.78
Nanostructure ($0.2\,\mu g/mm^2$)	65	663	1.39
Nanostructure ($0.4\,\mu g/mm^2$)	82	856	1.73
***Tropaeolum majus* wax**			
Nanostructure ($0.8\,\mu g/mm^2$)	180	1570	0.57
Lotus wax			
Nanostructure ($0.8\,\mu g/mm^2$)	187	1550	1.47

η – summit density

tal morphology found on superhydrophobic leaves, e.g., *Colocasia esculenta* [42.22] and *Triticum aestivum* (wheat) [42.187]. SEM micrographs of the nanostructures fabricated with three different masses of n-hexatriacontane show different densities of crystals. An atomic force microscope (AFM) was used to characterize the nanostructures. Statistical parameters of nanostructures [root-mean-square (RMS) height, peak-to-valley height, and summit density (η)] were calculated and are presented in Table 42.9 [42.3, 4]. A summit is defined as a point whose height is greater than its four nearest-neighboring points, above a threshold value of 10% of RMS height to avoid measurement errors. The measurement results were reproducible within $\pm 5\%$.

To study the effect of nanostructures with different crystal density on superhydrophobicity, static contact angle, contact angle hysteresis and tilt angle, and adhesive forces were measured [42.54]. For contact angle hysteresis, the advancing and receding contact angles were measured at the front and back of the droplet moving along the tilted surface, respectively. The data are shown in Fig. 42.49. The static contact angle of a flat surface coated with a film of n-hexatriacontane was $91°$.

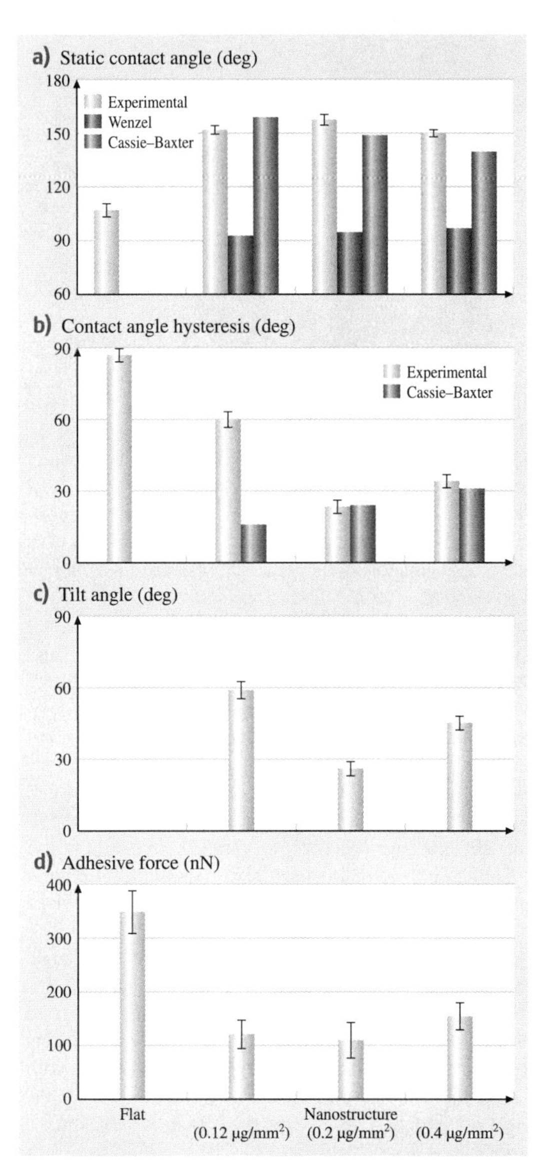

Fig. 42.49a–d Bar chart showing the measured (**a**) static contact angle, (**b**) contact angle hysteresis, and (**c**) tilt angle; also shown are calculated static contact angles obtained using the Wenzel and Cassie–Baxter equations with a given value of θ_0, plus contact angle hysteresis calculated using the Cassie–Baxter equation on flat surface and nanostructures fabricated with various masses of n-hexatriacontane. The droplet on the flat surface does not move along the surface even at a tilt angle of $90°$. (**d**) Adhesive forces for various structures, measured using a $15\,\mu m$-radius borosilicate tip (after [42.54]) ▶

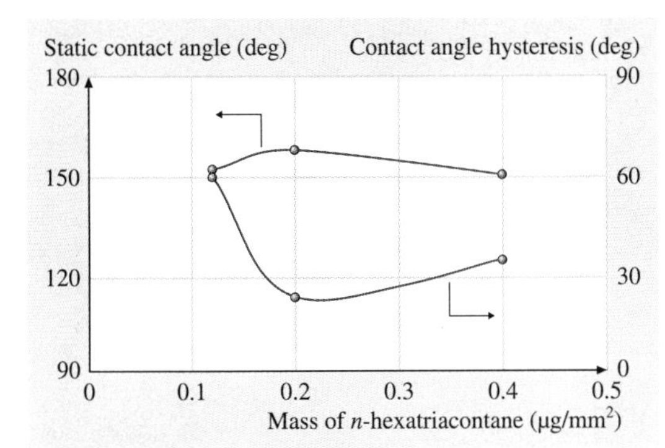

Fig. 42.50 Static contact angle and contact angle hysteresis as a function of mass of n-hexatriacontane deposited on nanostructures (after [42.54])

It showed a contact angle hysteresis of 87°, and the droplet still adhered at a tilt angle of 90°. Nanostructuring of flat surfaces with n-hexatriacontane platelets creates superhydrophobic surfaces with high static contact angle and a reduction of contact angle hysteresis and tilt angle. The values are a function of crystal density. Figure 42.50 shows a plot of static contact angle and contact angle hysteresis as a function of the mass of n-hexatriacontane deposited. As the mass of n-hexatriacontane increased, the static contact angle first increased, and the contact angle hysteresis decreased. Then, above a mass of $0.2\,\mu g/mm^2$, static contact angle and contact angle hysteresis gradually decreased and increased, respectively, with increasing mass. The highest static contact angle and lowest contact angle hysteresis were 158° and 23° at a mass of $0.2\,\mu g/mm^2$. As shown in Fig. 42.49, the adhesive force measured using a $15\,\mu m$-radius borosilicate tip in an AFM also shows a similar trend as the wetting properties. Adhesive forces of the nanostructured surfaces were lower than for the flat surface because the contact between the tip and surface was lower than on the flat surface, because the contact between the tip and surface was reduced by surface structuring [42.3, 4].

In order to identify wetting regimes (Wenzel or Cassie–Baxter) as well as to understand the effect of crystal density on the propensity for air pocket formation for the nanostructured surfaces, roughness factor (R_f) and fractional liquid–air interface (f_{LA}) are needed. The R_f for the nanostructures was calculated using the AFM map [42.25, 26]. The calculated results were reproducible within ±5%. The R_f for the nanostructured surfaces with masses of 0.12, 0.2, and $0.4\,\mu g/mm^2$ were found to be 3.4, 4.9, and 6.8, respectively. For calculation of f_{LA} of the nanostructures, only the higher crystals are assumed to come into contact with a water droplet. The fractional geometrical area of the top surface for the nanostructures was calculated from SEM micrographs with top view (0° tilt angle). The SEM images were converted to high-contrast black-and-white images using Adobe Photoshop. The increase of contrast in the SEM image eliminates the smaller platelet structures that were visible in the original SEM image. The higher crystals led to white signals in the SEM figure. The fractional geometrical area of the top nanostructured surfaces with masses of 0.12, 0.2, and $0.4\,\mu g/mm^2$ was found to be 0.07, 0.15, and 0.24, leading to f_{LA} of 0.93, 0.85, and 0.76, respectively. The calculated results were reproducible within ±5%. The values of static contact angle in the Wenzel and Cassie–Baxter regimes for the nanostructured surfaces were calculated using the values of R_f and f_{LA} and are presented in Fig. 42.49. The values of contact angle hysteresis in the Cassie–Baxter regime for various surfaces were calculated using (42.20); the data are presented in Fig. 42.49.

As shown in Fig. 42.49, the experimental static contact angle and contact angle hysteresis values for the two nanostructured surfaces with 0.2 and $0.4\,\mu g/mm^2$ were comparable to the calculated values in Cassie–Baxter regime. The results suggest that a droplet on two nanostructured surfaces should exist in the Cassie–Baxter regime. However, the experimental static contact angle and contact angle hysteresis values for the nanostructured surface with $0.12\,\mu g/mm^2$ were lower and higher, respectively, than the calculated values in the Cassie–Baxter regime. It is believed that neighboring crystals are separated at lower crystal density and any trapped air can be squeezed out, whereas neighboring crystals are interconnected at higher densities and air remains trapped. At the highest crystal density with a mass of $0.4\,\mu g/mm^2$, there is less open volume compared with that at $0.2\,\mu g/mm^2$, which explains the droplet static contact angle going from 158° to 150° [42.54].

The Influence of Hierarchical Structure with Wax Platelets on Superhydrophobicity

Bhushan et al. [42.55, 56] created surfaces with n-hexatriacontane at $0.2\,\mu g/mm^2$ to study the influence of hierarchical structure on superhydrophobicity. Figure 42.51 shows scanning electron microscope (SEM) micrographs of a flat surface and nano-, micro-, and hierarchical structures. To study the effect of structure on superhydrophobicity, static contact angle, contact

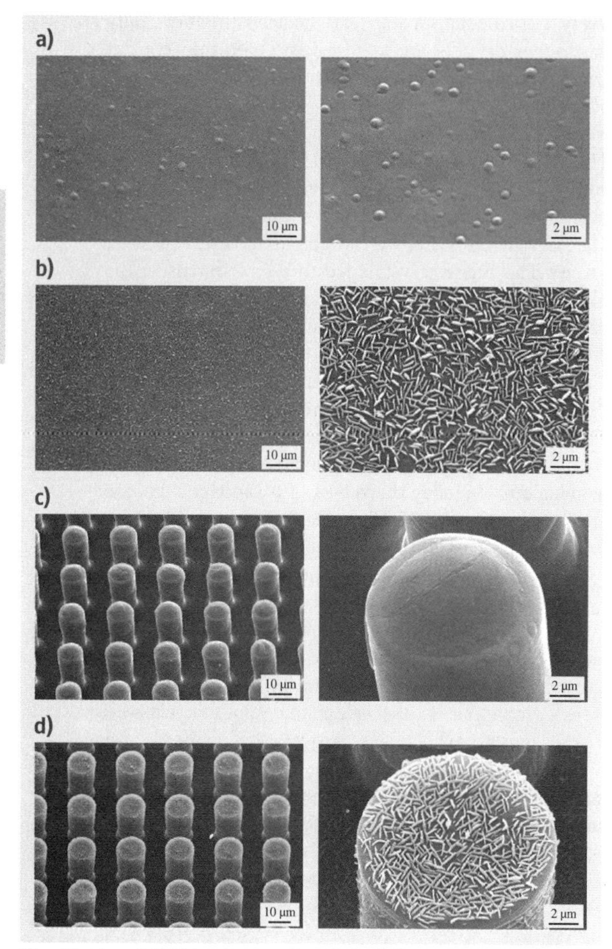

Fig. 42.51a–d SEM micrographs of the (a) flat surface, (b) nanostructure (0.2 μg/mm^2), (c) microstructure, and (d) hierarchical structure (0.2 μg/mm^2) measured at 45° tilt angle (shown using two magnifications). All samples are fabricated with epoxy resin coated with n-hexatriacontane (after [42.55])

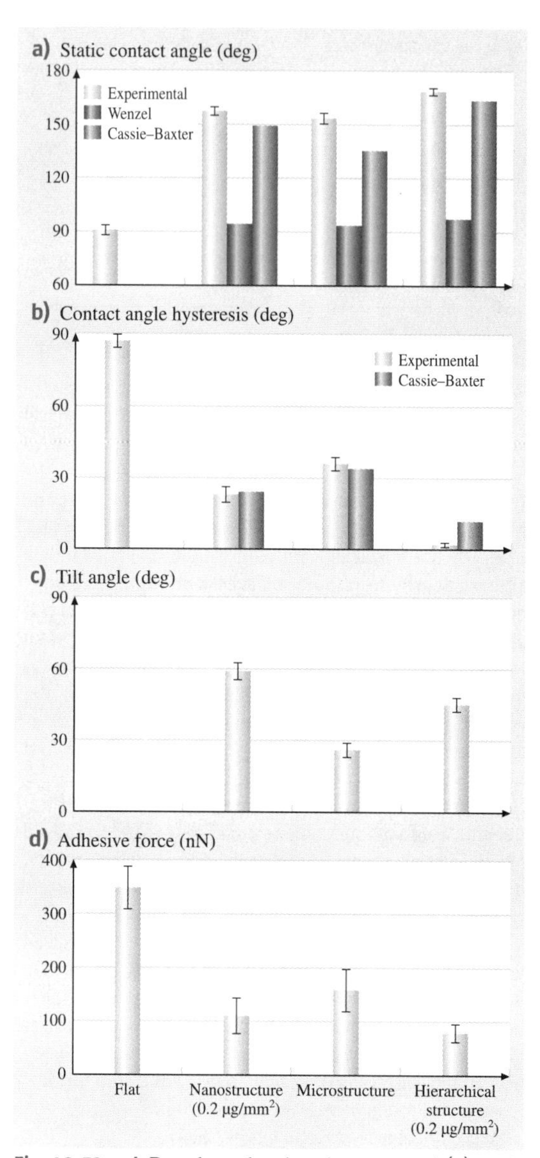

Fig. 42.52a–d Bar chart showing the measured (a) static contact angle, (b) contact angle hysteresis, and (c) tilt angle, plus static contact angles calculated using the Wenzel and Cassie–Baxter equations with a given value of θ_0, and calculated contact angle hysteresis using the Cassie–Baxter equation for various structures. The droplet on the flat surface does not move along the surface even at a tilt angle of 90°. (d) Adhesive forces for various structures, measured using a 15 μm-radius borosilicate tip (after [42.55])

angle hysteresis and tilt angle, and adhesive forces of four structures were measured. The data are shown in Fig. 42.52. The static contact angle of a flat surface coated with a film of n-hexatriacontane was 91°, and increased to 158° when n-hexatriacontane formed a nanostructure of platelets on it. The static contact angle on a flat specimen with a microstructure was 154°, but it increased to 169° for the hierarchical surface structure. Contact angle hysteresis and tilt angle for flat, micro-, and nanostructured surfaces show similar

Table 42.10 Summary of static contact angles and contact angle hysteresis measured and calculated for droplets in the Wenzel and Cassie–Baxter regimes on various surfaces with n-hexatriacontane using the calculated values of R_f and f_{LA} (after [42.55, 56])

	R_f	f_{LA}	Static contact angle (deg)			Contact angle hysteresis (deg)	
			Measured	Calculated in Wenzel regime	Calculated in Cassie–Baxter regime	Measured	Calculated in Cassie–Baxter regime
Flat	–	–	91	–	–	87[a]	–
Nanostructure	4.9	0.85	158	95	149	23	24
Microstructure	3.5	0.71	154	94	136	36	34
Hierarchical structure	8.4	0.96	169	98	164	2	12

[a] Advancing and receding contact angles are 141° and 54°, respectively.

trends. The flat surface showed a contact angle hysteresis of 87°, and the droplet still adhered at a tilt angle of 90°. The superhydrophobic micro- and nanostructured surfaces showed a reduction of contact angle hysteresis and tilt angle, but a water droplet still needs a tilt angle of 26° and 51°, respectively, before sliding. Only the hierarchical surface structure with static contact angle of 169° and low contact angle hysteresis of 2° exceeds the basic criteria for superhydrophobic and self-cleaning surfaces [42.8]. Adhesive force measured using a 15 μm-radius borosilicate tip in an AFM also shows a similar trend as the wetting properties. The adhesion force of the hierarchical surface structure was lower than that of both micro- and nanostructured surfaces because the contact between the tip and surface was lower as a result of the contact area being reduced [42.3, 4].

In order to identify wetting regimes (Wenzel or Cassie–Baxter) for the various surfaces, roughness factor (R_f) and fractional liquid–air interface (f_{LA}) are needed. The R_f for the nanostructure was described earlier. The R_f for microstructure was calculated for the geometry of flat-top, cylindrical pillars of diameter D, height H, and pitch P, distributed in a regular square array. For this case, the roughness factor for the microstructure is $(R_f)_{micro} = (1 + \pi DH/P^2)$. The roughness factor for the hierarchical structure is the sum of $(R_f)_{micro}$ and $(R_f)_{nano}$. The values calculated for various surfaces are summarized in Table 42.10.

For the calculation of f_{LA}, we make the following assumptions. For the microstructure, we consider that a droplet much larger in size than the pitch P contacts only the flat top of the pillars in the composite interface, and the cavities are filled with air. For the microstructure, the fractional flat geometrical area of the liquid–air interface under the droplet is $(f_{LA})_{micro} = (1 - \pi D^2/(4P^2))$ [42.8]. The fractional geometrical area of the top surface for the nanostructure was described earlier. For the hierarchical structure, the fractional flat geometrical area of the liquid–air interface, is $(f_{LA})_{hierarchical} = 1 - [\pi D^2/(4P^2)][1 - (f_{LA})_{nano}]$. The values of contact angle hysteresis in the Cassie–Baxter regime for various surfaces were calculated using (42.20). The values are summarized in Table 42.10.

The values of static contact angle in the Wenzel and Cassie–Baxter regimes for various surfaces were calculated using the values of R_f and f_{LA} (Table 42.10). As shown in Fig. 42.52, the experimental static contact angle values for the three structured surfaces were larger than the values calculated for the Cassie–Baxter regime. The results suggest that the droplets on the three structured surfaces were in the Cassie–Baxter regime. This indicates that the microstructure and nanostructure surface induce air pocket formation. For the contact angle hysteresis, there is good agreement between the experimental data and the theoretically predicted values for the Cassie–Baxter regime. As shown in Fig. 42.53,

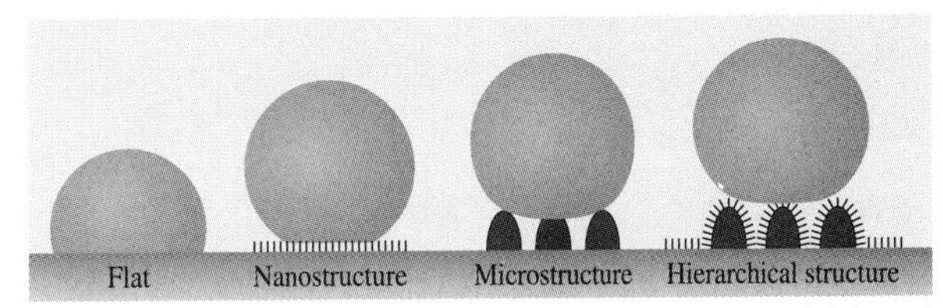

Fig. 42.53 Schematic and wetting of the four different fabricated surfaces. The largest contact area between the droplet and the surface is given by the flat and microstructured surfaces, but is reduced in the nanostructured surfaces and minimized in the hierarchical structured surfaces

these results show that air pocket formation in the micro- and nanostructure decreases the solid–liquid contact [42.30]. In hierarchical structured surfaces, air pocket formation further decreases the solid–liquid contact and thereby reduces the contact angle hysteresis and tilt angle [42.55, 56].

The Influence of Hierarchical Structure with Wax Tubules on the Superhydrophobicity

Morphological Characterization, Wettability, and Adhesion Forces of T. majus Tubules. Figure 42.54a shows SEM micrographs of the nanostructure and hierarchical structure fabricated with two different masses (0.6 and 0.8 μg/mm^2) of *T. majus* [42.57]. SEM micrographs show an increase in the amount of tubules on flat and microstructure surfaces after deposition of higher masses of wax. The tubules of *T. majus* wax grown in an ethanol atmosphere are comparable to the wax morphology found on the leaves of *T. majus*. Surfaces show a homogenous distribution of the wax mass on the specimen surfaces, and tubules provide the desired nanostructure of three-dimensional tubules on flat and microstructure surfaces. The tubule morphology of *T. majus* wax is shown in detail in Fig. 42.54b. The tubular crystals are hollow structures, randomly orientated on the surface and embedded into an amorphous wax layer. They are randomly distributed on the surface, and their shapes and sizes show some variations. As shown in Fig. 42.54b, and based on additional specimens, the tubular diameter varied between 100 and 300 nm, and their length varied between 300 and 1200 nm.

Atomic force microscopy (AFM) was used to characterize the nanostructure fabricated using *T. majus* wax of 0.8 μg/mm^2 after storage at 50 °C with ethanol vapor [42.57]. Statistical parameters of the nanostructure [RMS height, peak-to-valley height, and summit density (η)] were calculated and are presented in Table 42.9 [42.3, 4]. A summit is defined as a point whose height is greater than of its four nearest-neighboring points, above a threshold value of 10% of RMS height to avoid measurement errors. The measurement results were reproducible within ±5%.

To study the effect of structures with various length scales on superhydrophobicity, static contact angle, contact angle hysteresis and tilt angle, and adhesive forces of four structures produced using *T. majus* wax were measured [42.57]. Nanostructures formed on flat

a)
Nanostructure (0.6 μg/mm^2)
10 μm
2 μm
Hierarchical structure (0.6 μg/mm^2)
10 μm
2 μm
Nanostructure (0.8 μg/mm^2)
10 μm
2 μm
Hierarchical structure (0.8 μg/mm^2)
10 μm
2 μm

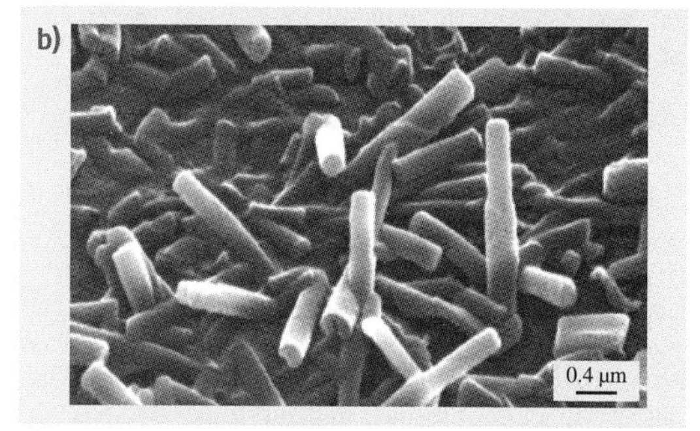

Fig. 42.54a,b SEM micrographs taken at 45° tilt angle (shown using two magnifications) of (**a**) the nanostructure and hierarchical structure fabricated with two different masses (0.6 and 0.8 μg/mm^2) of *T. majus* wax after storage at 50 °C with ethanol vapor. (**b**) Three-dimensional tubule-forming nanostructures on the surface fabricated with 0.8 μg/mm^2 mass of *T. majus* wax after storage at 50 °C with ethanol vapor (after [42.57])

and microstructured surfaces were fabricated using *T. majus* wax of 0.8 μg/mm^2 after storage at 50 °C with ethanol vapor. The data are shown in Fig. 42.55. The static contact angle of a flat surface coated with a film

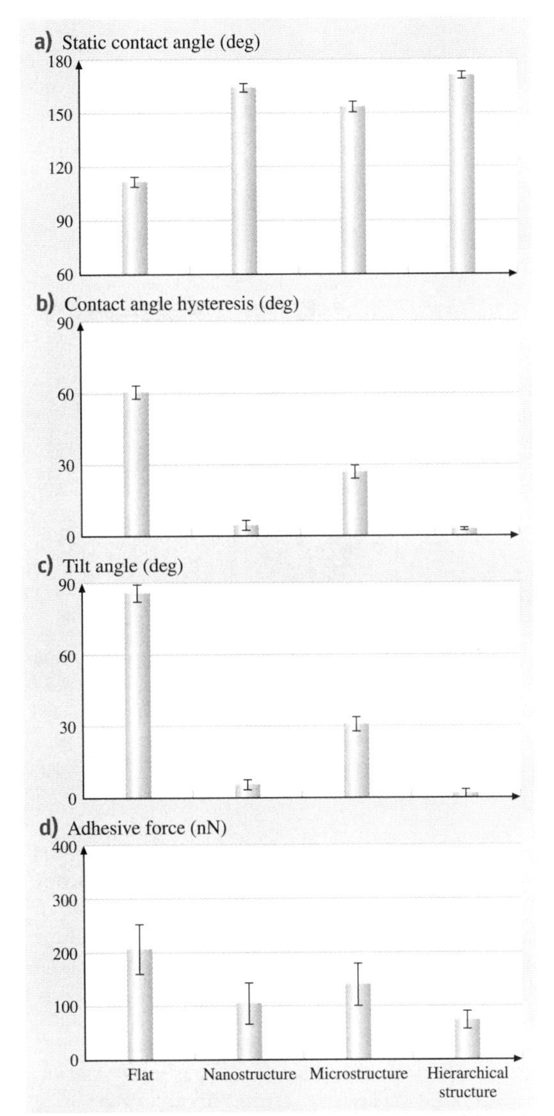

Fig. 42.55a–d Bar chart showing the measured (**a**) static contact angle, (**b**) contact angle hysteresis, and (**c**) tilt angle on various structures fabricated with 0.8 μm/mm^2 mass of *T. majus* wax after storage at 50 °C with ethanol vapor. (**d**) Adhesive forces for various structures, measured using a 15 μm-radius borosilicate tip (after [42.57])

of *T. majus* wax was 112°, and increased to 164° when *T. majus* wax formed a nanostructure of tubules on it. On the flat specimen with a microstructure on it, the static contact angle was 154°, but increased to 171° for the hierarchical surface structure. Contact angle hysteresis and tilt angle for flat, micro-, and nanostructured surfaces show similar trends. Flat surface showed a contact angle hysteresis of 61° and a tilt angle of 86°. The microstructured surface shows a reduction of contact angle hysteresis and tilt angle, but a water droplet still needed a tilt angle of 31° before sliding. As tubules are formed on the flat and microstructured surfaces, the nanostructured and hierarchical structure surfaces have low contact angle hysteresis of 5° and 3°, respectively. These properties are superior to those of plant leaves, including lotus leaves. Adhesive force measured using a 15 μm-radius borosilicate tip in an AFM showed a similar trend to the wetting properties. The adhesion force of the hierarchical surface structure was lower than that of micro- and nanostructured surfaces because the contact between the tip and surface was lower as a result of the contact area being reduced [42.3, 4].

To further verify the effect of hierarchical structure on the propensity for air pocket formation, *Bhushan* et al. [42.57] performed evaporation experiments with a droplet on microstructure and hierarchical structure fabricated with 0.8 μm/mm^2 mass of *T. majus* wax with ethanol vapor at 50 °C. Figure 42.56 shows successive photographs of a droplet evaporating from the two structured surfaces. On the microstructured surface, the light passes below the droplet and air pockets can be seen, so to start with the droplet is in the Cassie–Baxter regime. When the radius of the droplet decreased to 425 μm, the air pockets were not visible anymore, and the droplet was in the Wenzel regime. This transition results from an impalement of the droplet in the patterned surface, characterized by a smaller contact angle. For the hierarchical structure, an air pocket was clearly visible at the bottom area of the droplet throughout, and the droplet was in a hydrophobic state until the droplet evaporated completely. This suggests that a hierarchical structure with nanostructures prevents liquid from filling the gaps between the pillars.

Morphological Characterization, Wettability, and Adhesion Forces of Lotus Tubules. For the development of nanostructures by tubule formation, lotus wax was used [42.30]. Figure 42.57a shows scanning electron microscope (SEM) micrographs of flat surfaces with the tubules nanostructure. The microstructures shown

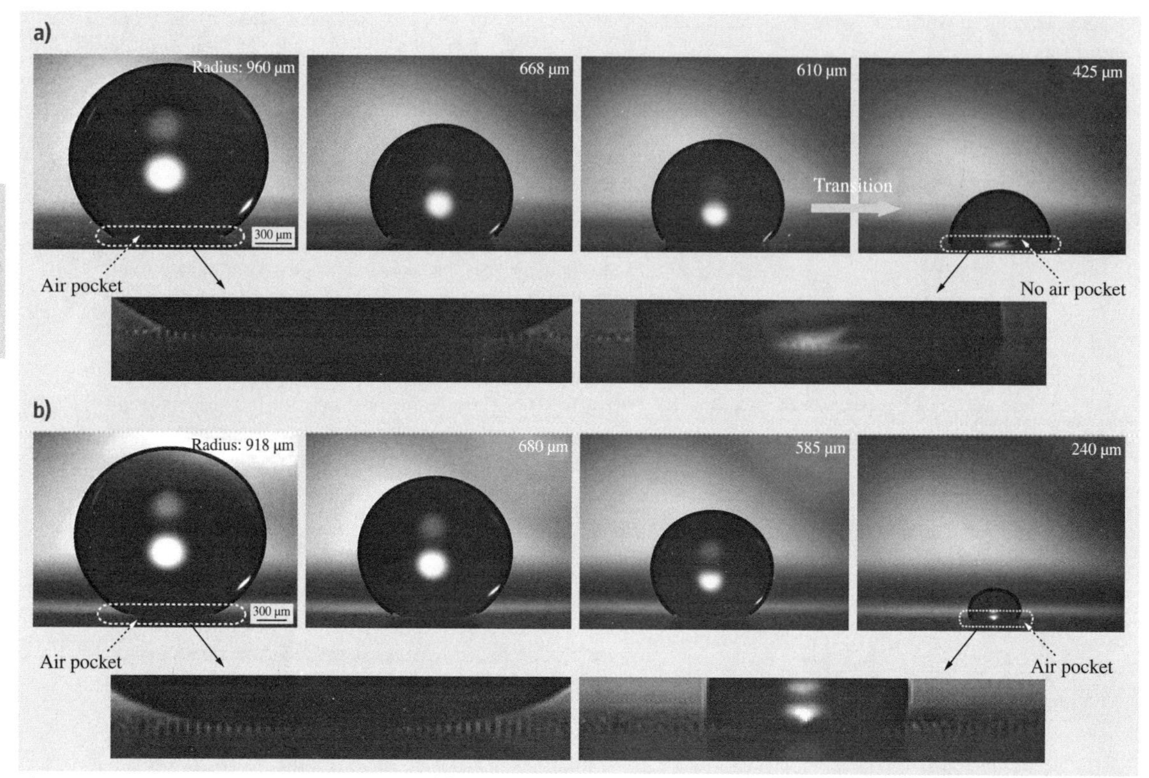

Fig. 42.56a,b Evaporation of a droplet on (**a**) microstructured and (**b**) hierarchical structured surfaces fabricated with $0.8\,\mu m/mm^2$ mass of *T. majus* wax after storage at 50 °C with ethanol vapor. The initial radius of the droplet was $\approx 950\,\mu m$; the time interval between the first two photos was 180 s and between the latter two was 60 s. As the radius of the droplet reached 425 µm (footprint = 836 µm) on the microstructured surface, the transition from the Cassie–Baxter regime to the Wenzel regime occurred, as indicated by the *arrow*. On the hierarchical structured surface, air pockets (visible at the bottom area of the droplet) exist until the droplet evaporated completely (after [42.57])

in Fig. 42.57b are the lotus leaf and the micropatterned Si replica covered with a lotus wax film. Hierarchical structures were fabricated with microstructured lotus leaf replicas and micropatterned Si replicas covered with a nanostructure of lotus wax tubules, as shown in Fig. 42.57c. The SEM micrographs show an overview (left column), a detail in higher magnification (middle column), and a high magnification (right column) of the created flat wax layers and tubules nanostructures. The grown tubules provide the desired nanostructure on flat and microstructured surfaces. The recrystallized lotus wax shows tubular hollow structures, with random orientation on the surfaces. Their shapes and sizes show only a few variations. The tubular diameter varied between 100 and 150 nm, and their length varied between 1500 and 2000 nm. Atomic force microscopy (AFM) was used to characterize the nanostructure of the lotus wax tubules. The statistical parameters of the nanostructure [RMS height, peak-to-valley height, and

Fig. 42.57a–c SEM micrographs taken at 45° tilt angle (shown using three magnifications) of (**a**) nanostructure on flat replica, (**b**) microstructures in lotus replica and micropatterned Si replica, and (**c**) hierarchical structure using lotus and micropatterned Si replicas. Nano- and hierarchical structures were fabricated with $0.8\,\mu g/mm^2$ of lotus wax after storage for 7 days at 50 °C with ethanol vapor (after [42.30]) ▶

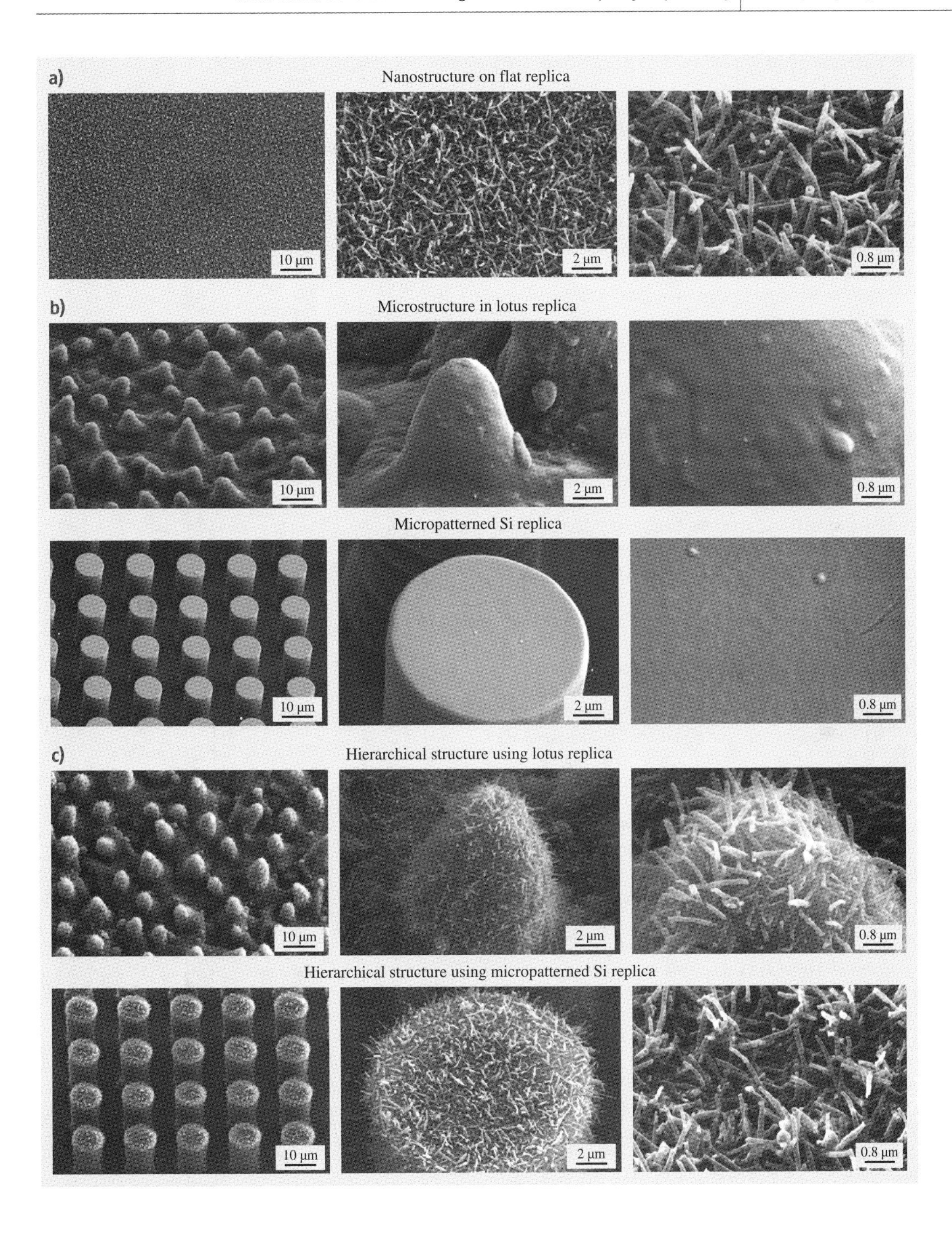
a)
Nanostructure on flat replica
10 μm
2 μm
0.8 μm
b)
Microstructure in lotus replica
10 μm
2 μm
0.8 μm
Micropatterned Si replica
10 μm
2 μm
0.8 μm
c)
Hierarchical structure using lotus replica
10 μm
2 μm
0.8 μm
Hierarchical structure using micropatterned Si replica
10 μm
2 μm
0.8 μm

summit density (η)] were calculated and are presented in Table 42.9.

To study the effect of lotus wax tubule nanostructures on superhydrophobicity, static contact angle, contact angle hysteresis, and tilting angle were measured on flat, microstructured lotus replica, micropatterned Si replica, and hierarchical surfaces. Hierarchical surfaces were made of the lotus leaf replica and micropatterned Si replica with a nanostructure of wax tubules on top. Additionally, fresh lotus leaves were investigated to compare the properties of the fabricated structures with the original biological model.

Figure 42.58 shows that the highest static contact angles of 173°, lowest contact angle hysteresis of 1°, and tilting angle varying between 1° and 2° were found for the hierarchical structured Si replica. The hierarchical structured lotus leaf replica showed a static contact angle of 171°, and the same contact angle hysteresis (2°) and tilt angles of 1–2° as the hierarchical Si replica. Both artificial hierarchical structured surfaces showed similar values to the fresh lotus leaf surface investigated here with a static contact angle of 164°, a contact angle hysteresis of 3°, and a tilting angle of 3°. However, the artificial hierarchical surfaces showed higher static contact angle and lower contact angle hysteresis. Structural differences between the original lotus leaf and the artificial lotus leaf produced here are limited to a difference in wax tubule length, which was 0.5–1 μm longer in the artificial lotus leaf [42.30].

The melting of the wax led to a flat surface with a flat wax film with a much lower static contact angle (119°), a higher contact angle hysteresis (71°), and a high tilting angle of 66°. The data for a flat lotus wax film on a flat replica show that the lotus wax by itself is hydrophobic. However, it has been stated that the wax on the lotus leaf surface, by itself, is weakly hydrophilic [42.188]. We cannot confirm this. The data presented here demonstrate that the native, flat wax of lotus leaves, with a static contact angle of 119°, is hydrophobic but can become superhydrophobic (167°) on increasing the surface roughness after self-assembly into three-dimensional wax tubules. The static contact angle of the lotus wax film is 119°, which is higher than that of the *T. majus* wax film of 112°. However, films made of *n*-hexatriacontane showed static contact angles of only 91°. SEM investigations, made directly after contact angle measurements, revealed no morphological differences between these films. Based on their chemical composition, it should be assumed that the nonpolar *n*-hexatriacontane molecules are more hy-

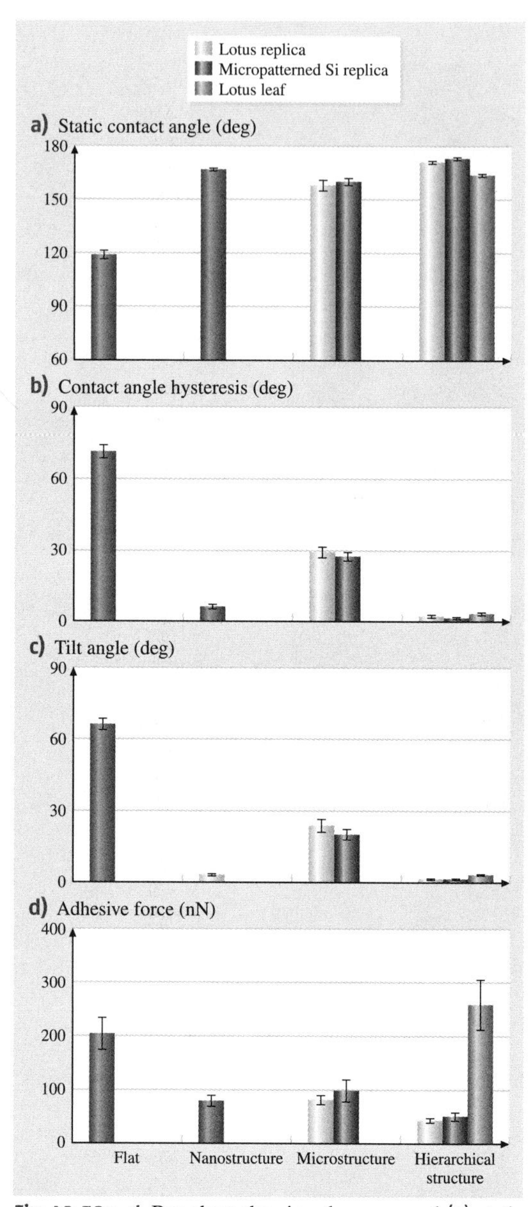

Fig. 42.58a–d Bar chart showing the measured (**a**) static contact angle, (**b**) contact angle hysteresis, and (**c**) tilt angle on various structures fabricated with 0.8 μg/mm^2 of lotus wax after storage for 7 days at 50 °C with ethanol vapor. (**d**) Adhesive forces for various structures, measured using a 15 μm-radius borosilicate tip. The *error bars* represents ±1 standard deviation (after [42.30])

drophobic than the plant waxes, which contain high amounts of oxygen atoms. At this point these differences cannot be explained by structural or chemical differences of the films, but will be the subject of further studies.

Adhesive force measured using a 15 µm-radius borosilicate tip in an AFM also showed a similar trend as the wetting properties for the artificial surfaces (Fig. 42.58) [42.30]. The adhesion force of the hierarchical surface structure was lower than that of micro- and nanostructured and flat surfaces because the contact between the tip and surface was lower as a result of the contact area being reduced. However, for the fresh lotus leaf, there is moisture within the plant material, which causes softening of the leaf, and so when the tip comes into contact with the leaf sample, the sample deforms, and the larger area of contact between the tip and sample causes an increase in the adhesive force [42.3,4].

Self-Cleaning Efficiency of Hierarchical Structured Surfaces

For deposition of contamination on artificial surfaces, various structures were placed in a contamination glass chamber [42.12]. Silicon carbide (SiC) (Guilleaume, Bonn) particles in two different sizes ranges of 1–10 and 10–15 µm were used as contaminants. SiC particles were chosen because of their similarity in shape, size, and hydrophilicity to natural dirt contaminations. The number of particles per area was determined by counting them from a $280 \times 210\,\mu m^2$ image taken by an optical microscope with a camera before and after water cleaning.

For the cleaning test, specimens with the contaminants were subjected to water droplets of ≈ 2 mm diameter, using two microsyringes [42.12]. In order to obtain a relative measure of the self-cleaning ability of hierarchical structures which exhibit the lowest contact angle hysteresis and tilt angle as compared with other structures (flat, nanostructures, and microstructures), the tilt angle chosen for the cleaning tests was slightly above the tilt angle for the hierarchical structures. Thus, experiments were performed at 10° for surfaces covered with n-hexatriacontane and at 3° for surfaces with lotus wax. The water cleaning test was carried out for 2 min (water quantity, 10 ml) with nearly zero kinetic energy of droplets. For watering with nearly zero kinetic energy, the distance between the microsyringes and the surface was set to 0.005 m (nearly zero impact velocity). The chosen impact velocity represents a low value compared with a natural rain shower, in which a water droplet of 2 mm diameter can reach an impact velocity of 6 m/s (measured under controlled conditions) [42.189].

Figure 42.59 shows that none of the investigated surfaces were fully cleaned by water rinsing [42.12]. The data represent the average of five different investigated areas for each experiment. For lotus wax, which forms tubule nanostructures, and n-hexatriacontane, which forms platelet nanostructures, the same tendency of particle removal was found. With the exception of hierarchical structure on all surfaces, larger particles were removed more than small ones. Most particles (70–80%) remained on smooth surfaces, and 50–70% of particles were found on microstructured surfaces. Most particles were removed from the hierarchical structured surfaces, but ≈ 30% of particles remained.

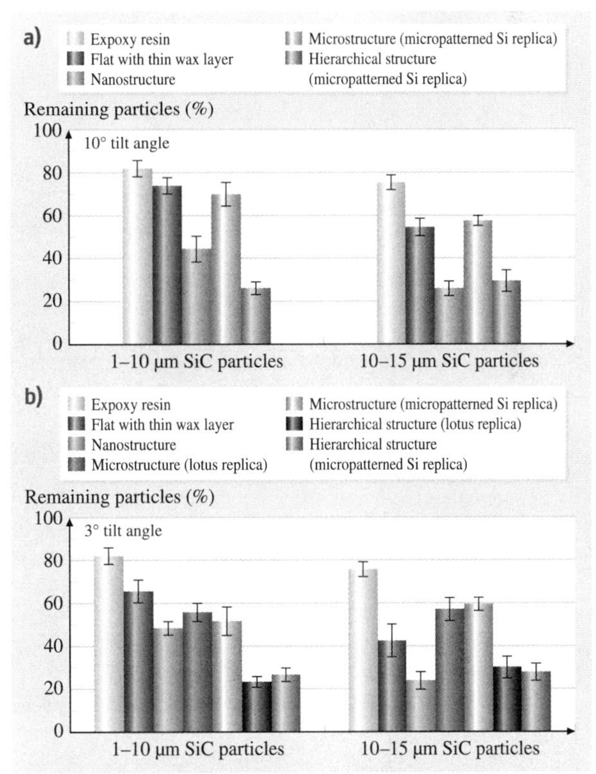

Fig. 42.59a,b Bar charts showing the particles remaining after applying droplets with nearly zero kinetic energy on various structures fabricated using (**a**) n-hexatriacontane and (**b**) lotus wax, using 1–10 and 10–15 µm SiC particles. The experiments on the surfaces with n-hexatriacontane and lotus wax were carried out on stages tilted at 10° and 3°, respectively. The *error bars* represent ±1 standard deviation (after [42.12])

A clear difference in particle removal, independent of particle size, was only found for the flat and nanostructured surfaces, where larger particles were removed with higher efficiency. Observations of droplet behavior during movement on the surfaces showed that droplets were rolling only on the hierarchical structured surfaces. On flat, micro-, and nanostructured surfaces, the first droplets applied did not move, but continuing application of water droplets increased their volume and led to sliding of these large droplets. During this process,

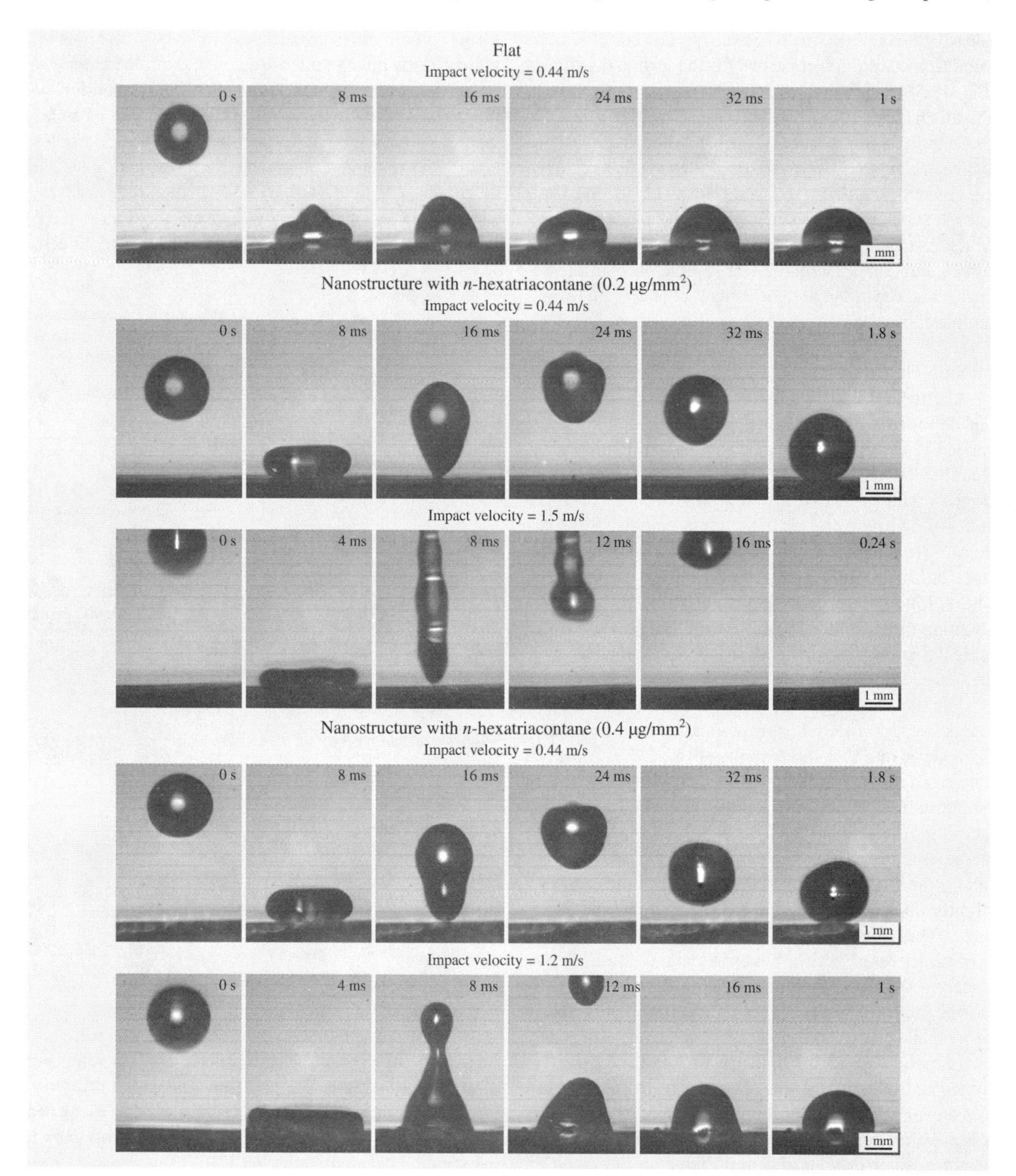

some of the particles were removed from the surfaces. However, the rolling droplets on hierarchical structures did not collect the dirt particles trapped in the cavities of the microstructures. The data clearly show that hierarchical structures have superior cleaning efficiency.

Observation of Transition for a Bouncing Droplet

To observe how the impact velocity influences the transition from the composite solid–air–liquid interface to the homogeneous solid–liquid interface during droplet impact, *Jung* and *Bhushan* [42.62] performed bouncing-droplet experiments on various surfaces with n-hexatriacontane and *T. majus* and lotus waxes. Figure 42.60 shows snapshots of a droplet with 1 mm radius hitting various surfaces fabricated with 0.2 and 0.4 μm/mm^2 of n-hexatriacontane after storage at room temperature. The impact velocity was obtained just prior to the droplet hitting the surface. First, on the flat surface, it was found that the droplet did not bounce

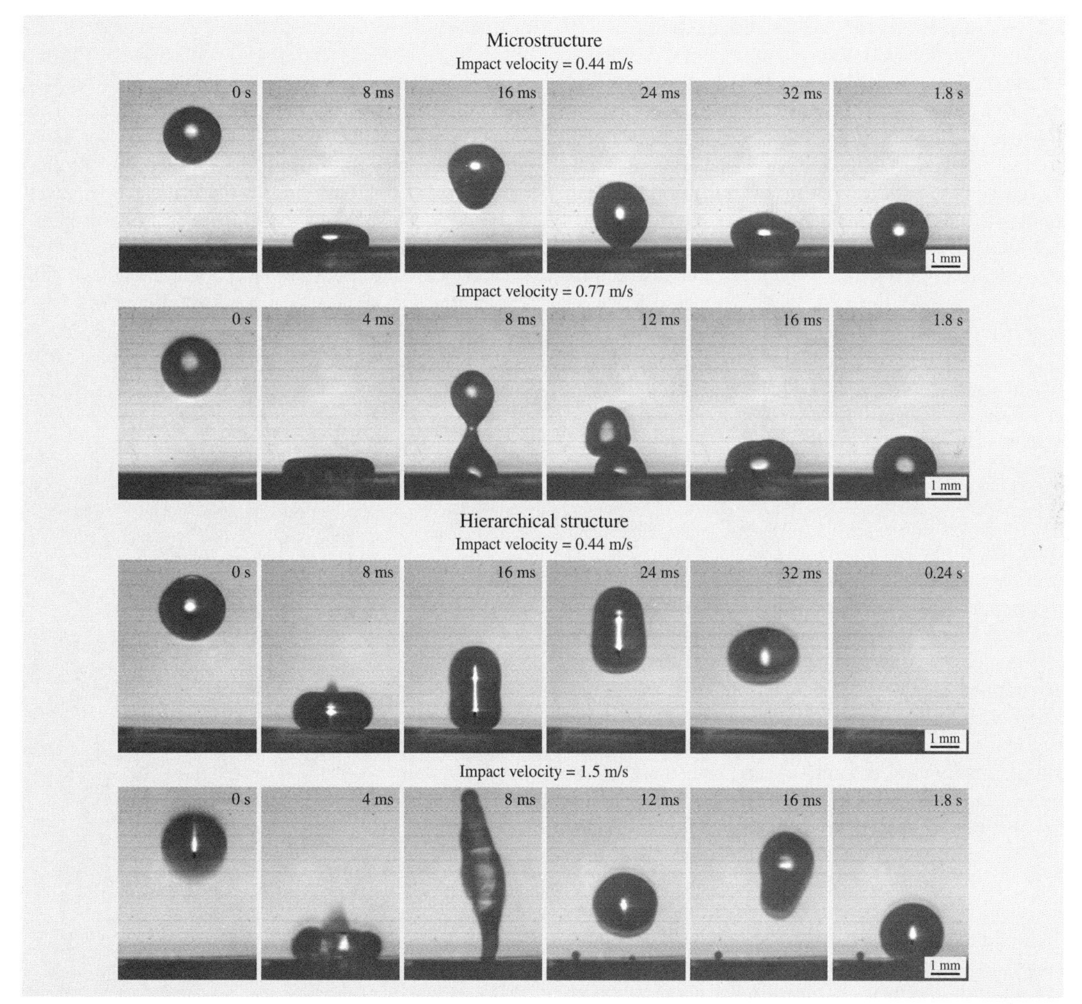

Fig. 42.60 Snapshots of a droplet with 1 mm radius hitting various surfaces fabricated with 0.2 and 0.4 μg/mm^2 of n-hexatriacontane after storage at room temperature. The impact velocity was obtained just prior to the droplet hitting the surface. Pinning of the droplet on the nanostructure with 0.4 μg/mm^2 mass, and on the microstructure occurred at impact velocities of 1.2 and 0.77 m/s, respectively (after [42.62]) ◀ ▲

off, even though the impact velocity applied was up to 1.5 m/s. As shown in the images in the first row for each nano-, micro-, and hierarchical structured surface, a droplet hitting the surface at an impact velocity of 0.44 m/s first deformed and then retracted, and bounced off the surface. Finally, the droplet sat on the surface and had a high contact angle, which suggests the formation of a solid–air–liquid interface. Next, we repeated the impact experiment with an increased impact velocity. As shown in the second row of images, for nanostructure (0.4 $\mu g/mm^2$) and microstructure, bounce off did not occur, and the wetting of the surface (and possibly pinning of the droplet) occurred at impact velocities of 1.2 and 0.77 m/s, respectively, referred to as the critical impact velocity. This is because air pockets do not exist below the droplet as a result of droplet impalement by the structures, characterized by a smaller contact angle. These observations indicate the transition from the composite interface to the homogenous interface. However, as shown in the second row of images, for the nanostructure (0.2 $\mu g/mm^2$) and hierarchical structure, for impact velocity up to 1.5 m/s, bounce off always occurred and wetting of the surface did not occur.

As shown in Fig. 42.61, the critical impact velocity was measured for a droplet with 1 mm radius impacting on various structures on which wetting of the surface (and possibly pinning of the droplet) occurs [42.62]. The arrow indicates that the critical impact velocity can be > 1.5 m/s, or the transition has not occurred. As mentioned earlier, when the mass of n-hexatriacontane increased, the static contact angle of the nanostructure first increased, followed by a decrease at a mass of 0.2 $\mu g/mm^2$. For the critical impact velocity, the same trend was found. It is believed that, if neighboring crystals are separated on a sample with lower crystal density, any trapped air can be squeezed out, whereas if the neighboring crystals are interconnected on a sample with higher density, air remains trapped. At the highest crystal density at a mass of 0.4 $\mu g/mm^2$, there is less open volume compared with at 0.2 $\mu g/mm^2$. For all microstructures with n-hexatriacontane and lotus wax, the critical impact velocities of the droplet are lower than those on the nano- and hierarchical structures due to the larger distance between the pillars, and the solid–air–liquid interface can easily be destabilized due to the dynamic impact on the surface. Based on (42.29), the critical impact velocity of the droplet decreases with the geometric parameter (pitch). The theoretical critical impact velocity for microstructure using (42.29) is 0.5 m/s. This value is lower than the experimental values of critical impact velocity for microstructures by ≈ 30–50% depending on the structured surfaces. In our experiments, when applying impact velocity of up to 1.5 m/s on nano- and hierarchical structures with n-hexatriacontane (0.2 $\mu g/mm^2$ for nanostructure) and lotus wax, wetting of the surface did not occur. The data clearly show that nano- and hierarchical structures are superior to microstructure in terms of maintaining a stable composite solid–air–liquid interface.

To identify whether a homogeneous solid–liquid interface or a composite solid–air–liquid interface exists, the contact angle data in the static condition and after bounce-off were measured on various surfaces, as shown in Fig. 42.62 [42.62]. The contact angles of the left bar for each sample were measured using the droplet with 1 mm radius gently deposited on the surface. The contact angles of the middle and right bars for each sample were measured using the droplet after hitting the surface at 0.44 m/s and at the critical or highest impact velocity. Missing bars mean that the droplet, after hitting the surface, bounced off without coming to sit on the surface. The static contact angle of the droplet after impact at 0.44 m/s is lower than that of the

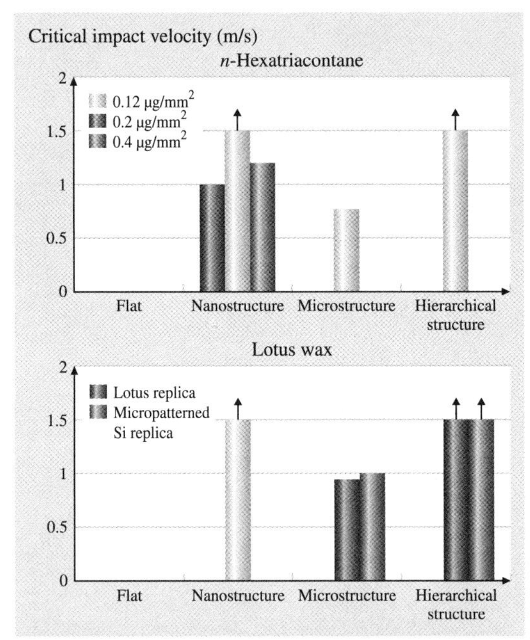

Fig. 42.61 Bar chart showing the measured critical impact velocity of a droplet with 1 mm radius at which transition occurs on various structures. The *arrow* indicates that the critical impact velocity may be more than 1.5 m/s or the transition does not occur (after [42.62])

droplet gently deposited for all of the surfaces with n-hexatriacontane and lotus wax. It can be interpreted that, after impact, the droplet pushes out the entrapped air from the cavities between the pillars under the droplet, resulting in an abrupt increase of the solid–liquid surface area during dynamic impact. As mentioned earlier, after hitting the surface at the critical impact velocity, the contact angles were close to 90° and much lower than that of the droplet gently deposited for nanostructures with n-hexatriacontane (0.12 and 0.4 $\mu g/mm^2$) and microstructures with n-hexatriacontane and lotus wax. Even though the droplet is in the composite interface regime when it is gently deposited on the surface, these observations indicate that the composite solid–air–liquid interface was destroyed due to the dynamic impact on the surface.

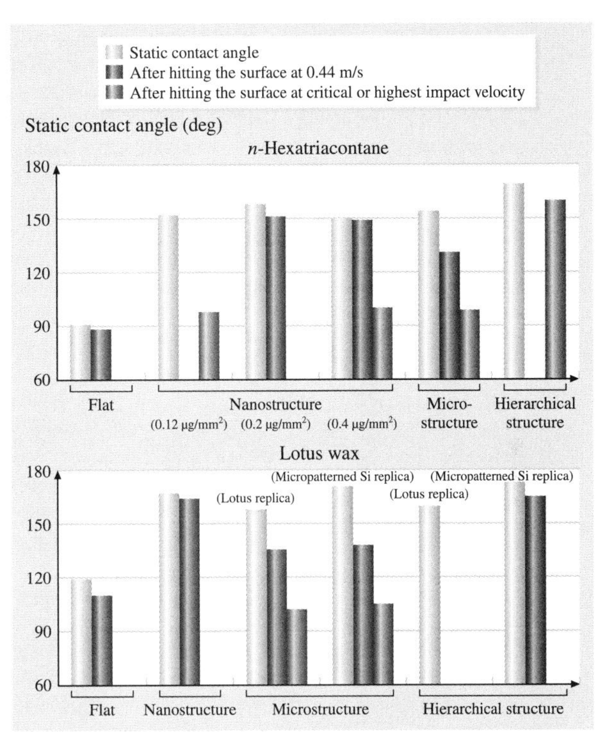

Fig. 42.62 Bar chart showing the measured static contact angle of a droplet on various surfaces. The contact angles of the *left bar* for each sample were measured using a droplet with 1 mm radius gently deposited on the surface. The contact angles of the *middle* and *right bars* for each sample were measured using the droplet after hitting the surface at 0.44 m/s and at the critical or highest impact velocity, respectively (after [42.62])

Observation of Transition for a Vibrating Droplet

Model for the Adhesion and Inertial Forces of a Vibrating Droplet. *Jung* and *Bhushan* [42.62] presented a model for vibration. In this model, they calculate expressions for adhesion force and inertial force as a function of droplet properties and operating conditions. Consider a small droplet of liquid deposited on a surface. The liquid and surface come together under equilibrium at a characteristic angle, called the static contact angle θ, as shown in Fig. 42.63. When a droplet on a surface is vibrated, based on *Lamb* [42.190], a general expression for the resonance frequency f_r of a free oscillating liquid droplet is given as

$$f_r = \sqrt{\frac{n(n-1)(n+2)\gamma}{3\pi\rho V}}\,, \tag{42.37}$$

where V is the volume of the droplet and n represents the number of modes, with 2 being the first mode etc. For a spherical droplet with radius R with a distance between the center of the droplet and the solid of H, the volume of the droplet V is given by

$$\begin{aligned} V &= \tfrac{1}{3}\pi(R+H)^2(2R-H) \\ &= \tfrac{1}{3}\pi R^3(1-\cos\theta)^2(2+\cos\theta)\,. \end{aligned} \tag{42.38}$$

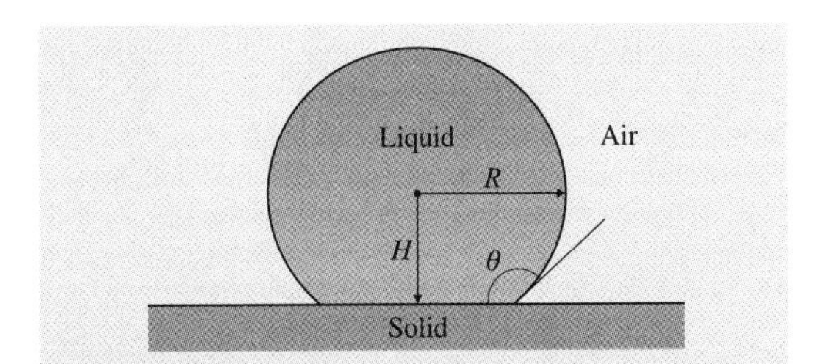

Fig. 42.63 Droplet of liquid in contact with a solid surface – contact angle θ, radius of droplet R, distance between the center of the droplet and the solid H (after [42.62])

Experimental studies were recently carried out by *Noblin* et al. [42.191] and *Celestini* and *Kofman* [42.192]. They showed that the frequency decreases with the volume of the droplet, and the trends are compared with a theoretical model.

When liquid comes into contact with a surface, the energy gained for surfaces coming into contact is greater than the energy required for their separation (or the work of adhesion) by the quantity ΔW, which constitutes the adhesion hysteresis [42.16]. For a surface, the difference between the two values of the interface energy (measured during loading and unloading)

Table 42.11 Summary of the measured resonance frequency at the fixed amplitude value of 0.4 mm, the calculated inertia force of a vibrated droplet, and the adhesive force between the droplet and surface on various surfaces fabricated with n-hexatriacontane and lotus wax. A positive value of ΔF means that the droplet bounced off before the transition occurs. The variation represents ±1 standard deviation (after [42.62])

	Resonance frequency (Hz)		F_A (μN)	F_I (μN)	$\Delta F = F_I - F_A$ (μN)
	First mode ($n=2$)	Second mode ($n=3$)		at transition or bouncing off	
***n*-Hexatriacontane**					
Flat	103	271	260±3.7	–	–
Nanostructure (0.12 μg/mm^2)	63	169	45±2.2	11	−34±2.2 (T)
Nanostructure (0.2 μg/mm^2)	50	147	10±1.6	10	0±1.6 (B)
Nanostructure (0.4 μg/mm^2)	54	149	13±2.0	11	−2±2.0 (B)
Microstructure	86	208	21±1.7	9	−12±1.7 (T)
Hierarchical structure	49	144	0.2±0.06	3.6	3.4±0.06 (B)
Lotus wax					
Flat	92	243	133±3.3	–	–
Nanostructure	47	138	0.8±0.08	3.2	2.4±0.08 (B)
Microstructure (lotus replica)	63	179	14±1.8	11	−3±1.8 (T and B)
Microstructure (micropatterned Si replica)	66	183	10±1.6	10	0±1.6 (B)
Hierarchical structure (lotus replica)	40	139	0.1±0.04	3.6	3.5±0.04 (B)
Hierarchical structure (micropatterned Si replica)	38	137	0.04±0.01	2.7	2.6±0.01 (B)

B – Bouncing off
T – Transition

is given by ΔW. These two values are related to the advancing contact angle θ_{adv} and receding contact angle θ_{rec} of the surface. For example, a model based on Young's equation has been used to calculate the work of adhesion from contact angle hysteresis [42.5, 9, 193, 194]

$$\cos\theta_{adv} - \cos\theta_{rec} = \frac{\Delta W}{\gamma} . \tag{42.39}$$

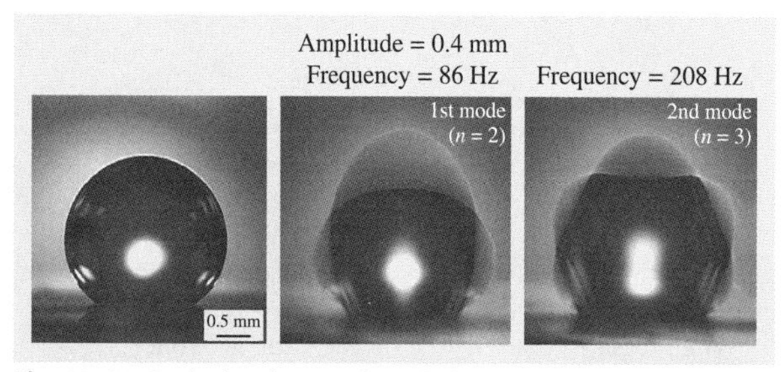

Fig. 42.64 Optical micrographs of droplets on the microstructured surface with n-hexatriacontane before and after vibration at amplitude of 0.4 mm for the first ($n=2$) and second modes ($n=3$) of vibration. The frequencies for the first and second modes were measured to be 86 and 208 Hz, respectively (after [42.62])

From (42.39), the dominant force (adhesion force) responsible for the separation between the droplet and surface is given by [42.195, 196]

$$\begin{aligned} F_A &= L\gamma(\cos\theta_{adv} - \cos\theta_{rec}) \\ &= 2R\sin\theta\gamma(\cos\theta_{adv} - \cos\theta_{rec}) , \end{aligned} \tag{42.40}$$

where L is the length of the triple contact line, referred to as the line of contact of solid, liquid, and air. The radius of a spherical droplet R depends on the contact angle and can be obtained from (42.38).

By applying vertical vibration of a droplet on a surface, the inertial force on the droplet F_I is given by

$$F_I = \rho V A \omega^2 , \tag{42.41}$$

where A and ω are the amplitude and frequency of vibration, respectively. *Bormashenko* et al. [42.111] showed that the transition occurred at a critical value of inertial force acting on the length of a triple line. However, if the inertial force of a droplet vibrated on the surface can overcome the adhesion force between the droplet and surface, i.e., ΔF given by [42.62]

$$\Delta F = F_I - F_A \tag{42.42}$$

is positive, then the droplet can be vertically separated from the surface (bouncing off) before the composite solid–liquid–air interface is destroyed.

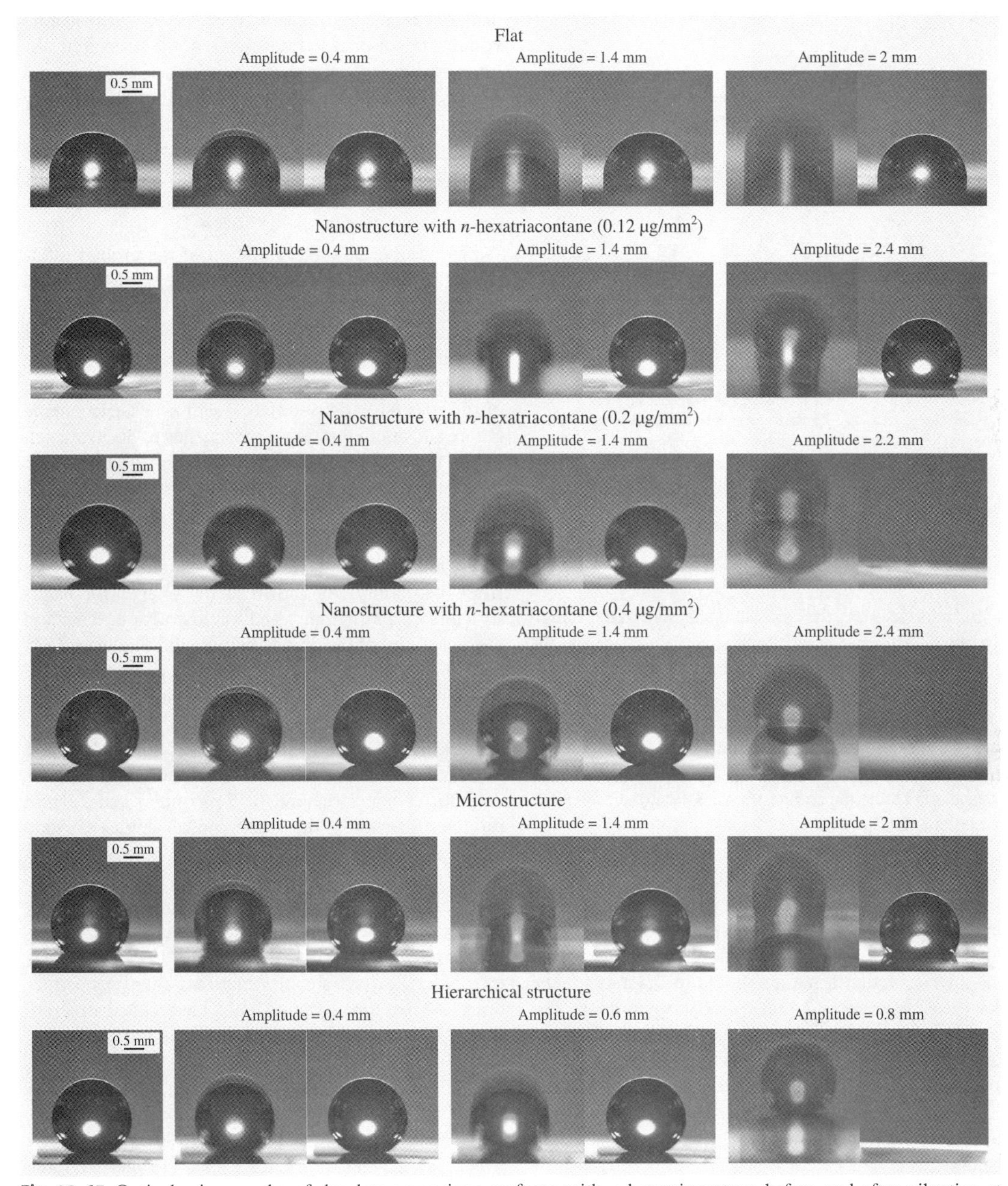

Fig. 42.65 Optical micrographs of droplets on various surfaces with *n*-hexatriacontane before and after vibrating at frequency of 30 Hz. The transition of a droplet on the nanostructure with 0.12 μg/mm^2 mass of *n*-hexatriacontane and on the microstructure occurred at amplitudes of 2.4 and 2.0 mm, respectively (after [42.62])

Vibration Study Results. To estimate the resonance frequency of a droplet, a droplet on various surfaces was vibrated by varying the frequency with a sinusoidal excitation of relatively low amplitude (0.4 mm) [42.62].

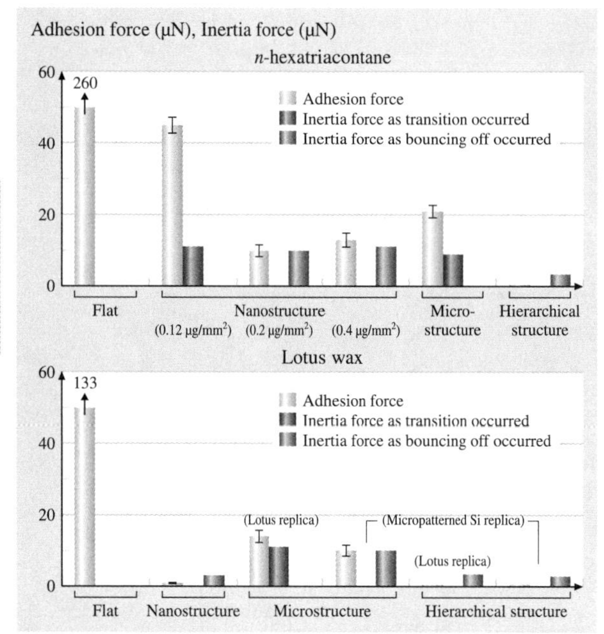

Fig. 42.66 Bar chart showing the calculated adhesive force responsible for the separation between the droplet and the surface and the inertial force of a droplet vibrated on various structures. The inertial forces were obtained as the transition or droplet bounce-off occurred. If the inertial force of the droplet vibrated on the surface can overcome the adhesive force between the droplet and the surface (ΔF is positive), the droplet can be vertically separated from the surface (bounce off) before the composite solid–liquid–air interface is destroyed (after [42.62])

Based on (42.37), the experiments were performed in the first and second modes ($n = 2$ and 3). Figure 42.64 shows optical micrographs of droplets on the microstructured surface with n-hexatriacontane before and after vibration at an amplitude of 0.4 mm for the first ($n = 2$) and second mode ($n = 3$) of vibration. The frequencies for the first and second modes were measured to be 86 and 208 Hz, respectively. The resonance frequencies on various surfaces were measured, and the data are summarized in Table 42.11. For comparison, the theoretical values for the two modes were calculated using (42.37). For calculations, the surface tension of the water–air interface (γ) was taken to be 0.073 N/m, the mass density (ρ) was taken as 1000 kg/m^3 for water, and the volume of the droplet (V) was 5 μl. The theoretical values for the first ($n = 2$) and second modes ($n = 3$) from (42.37) are 110 and 214 Hz, respectively. These values are similar to those of flat surfaces with n-hexatriacontane and lotus wax. However, for the same volume, the resonance frequencies of the structured surfaces are lower than those of flat surfaces. *Celestini* and *Kofman* [42.192] showed that the resonance frequency depends on the contact angle of the structured surfaces, and it decreases with increasing contact angle. The hierarchical structures with highest contact angle have the lowest resonance frequency, consistent with the results of *Celestini* and *Kofman* [42.192].

To observe how the vibration of the droplet influences the transition from the composite solid–air–liquid interface to the homogeneous solid–liquid interface, *Jung* and *Bhushan* [42.62] performed vibrating-droplet experiments on various surfaces with n-hexatriacontane and lotus wax. Figure 42.65 shows optical micrographs of droplets on various surfaces with n-hexatriacontane before and after vibrating at a frequency of 30 Hz, which is less than the resonance frequency for the first mode ($n = 2$). The vibration amplitude was increased until transition, or until the droplet bounced off. First, on the flat surface, it was found that the droplet did not change much after applying vibration at amplitudes ranging from 0 to 3 mm. As shown in the images for nanostructure (0.12 μg/mm^2) and microstructure, the static contact angles of the droplet before vibrating were 152° and 154°, respectively. After vibrating at amplitudes of 0.4 and 1.4 mm, the contact angles still have similar values (151° for nanostructure and 149° for microstructure), which suggests the formation of a solid–air–liquid interface. However, after vibrating at amplitudes of 2.4 mm for nanostructure (0.12 μg/mm^2) and 2.0 mm for microstructures, the static contact angles became 125° and 121°, respectively. This is because air pockets do not exist below the droplet as a result of droplet impalement by the structures, characterized by a smaller contact angle. These observations indicate the transition from the composite interface to the homogenous interface. Observations of vibration on two nanostructures (0.2 and 0.4 μg/mm^2) and a hierarchical structure showed that the transition did not occur, but the droplet started to bounce off the surface above amplitudes of 2.2, 2.4, and 0.8 mm, respectively.

To study the validity of the proposed model (42.42), the adhesive force responsible for the separation between the droplet and surface and the inertial force of a droplet vibrated on various structures were calculated [42.62]. The adhesive force was obtained from (42.40) using the static contact angle and contact angle hysteresis. The inertial forces were obtained from (42.41) using the amplitude and frequency of vibration as the transition or droplet bounce-off occurred.

The data are presented in Fig. 42.66 and summarized in Table 42.11. As shown in Fig. 42.65, it was observed that the transition occurred as a result of droplet impalement by the structures due to the increasing inertial force of the droplet on the surfaces. However, if the inertial force of the droplet vibrated on the surface can overcome the adhesion force between the droplet and the surface (ΔF is positive), the droplet can be vertically separated from the surface (bounce off) before the composite solid–liquid–air interface is destroyed. The experimental results for bouncing off of the droplet appear to exhibit the same trend as the proposed model (42.42). It is shown that hierarchical structures have a positive difference between the inertial force and adhesive force for droplet bounce-off, which is responsible for superior resistance to dynamic effects and the maintenance of a stable composite solid–air–liquid interface.

42.6 Modeling, Fabrication, and Characterization of Oleophobic/Oleophilic Surfaces

Oleophobic surfaces have the potential for self-cleaning and antifouling from biological and organic contaminants in both air and underwater applications. In this section, we discuss a model for predicting the oleophobic/philic nature and experimental measurements of the wetting properties of surfaces.

42.6.1 Modeling of Contact Angle for Various Interfaces

If a water droplet is placed on a solid surface in air, the solid–air and water–air interfaces come together with a static contact angle θ_W. The value of θ_W can be determined from the condition of the total energy of the system being minimized [42.1, 2] and is given by the Young's equation for the contact angle θ_W

$$\cos\theta_W = \frac{\gamma_{SA} - \gamma_{SW}}{\gamma_{WA}} , \quad (42.43)$$

where γ_{SW}, γ_{SA}, and γ_{WA} are the surface tensions of the solid–water, solid–air, and water–air interfaces, respectively. If an oil droplet is placed on a solid surface in air, the Young's equation for the contact angle θ_O can be expressed by

$$\cos\theta_O = \frac{\gamma_{SA} - \gamma_{SO}}{\gamma_{OA}} , \quad (42.44)$$

where γ_{SO}, γ_{SA}, and γ_{OA} are the surface tensions of the solid–oil, solid–air, and oil–air interfaces, respectively. As predicted by (42.44), if γ_{SO} is higher than γ_{SA}, an oleophobic surface can be achieved.

To create an oleophobic surface in water, let us consider the solid–water–oil interface. If an oil droplet is placed on a solid surface in water, the contact angle of an oil droplet in water θ_{OW} is given by the Young equation

$$\cos\theta_{OW} = \frac{\gamma_{SW} - \gamma_{SO}}{\gamma_{OW}} , \quad (42.45)$$

where γ_{SO}, γ_{SW}, and γ_{OW} are the surface tensions of the solid–oil, solid–water, and oil–water interfaces, respectively. Combining (42.43–42.45), the equation for the contact angle θ_{OW} of an oil droplet in water is given as

$$\cos\theta_{OW} = \frac{\gamma_{OA}\cos\theta_O - \gamma_{WA}\cos\theta_W}{\gamma_{OW}} . \quad (42.46)$$

As predicted by (42.46), for a hydrophilic surface ($\gamma_{SA} > \gamma_{SW}$), an oleophobic surface in the solid–water–oil interface can be created if $\gamma_{OA}\cos\theta_O$ is lower than $\gamma_{WA}\cos\theta_W$. Since the surface tension of oil and organic liquids is much lower than that of water, most hydrophilic surfaces can be made oleophobic at a solid–water–oil interface. For a hydrophobic surface ($\gamma_{SA} < \gamma_{SW}$) and an oleophobic surface at a solid–air–oil interface ($\gamma_{SA} < \gamma_{SO}$), an oleophobic surface at a solid–water–oil interface can be created if $\gamma_{OA}\cos\theta_O$ is higher than $\gamma_{WA}\cos\theta_W$, and vice versa. For a hydrophobic and an oleophilic surface at a solid–air–oil interface, an oleophobic surface at the solid–water–oil interface cannot be created. Associated schematics are shown in Fig. 42.67, and a summary of the hydrophilic/phobic nature of various interfaces is shown in Table 42.12. For an oleophobic surface, oil contaminants are washed away when immersed in water. This effect leads to self-cleaning that can be used for marine ship antifouling [42.197].

42.6.2 Experimental Techniques

For the measurement of static contact angle, deionized water was used for water droplets and hexadecane was

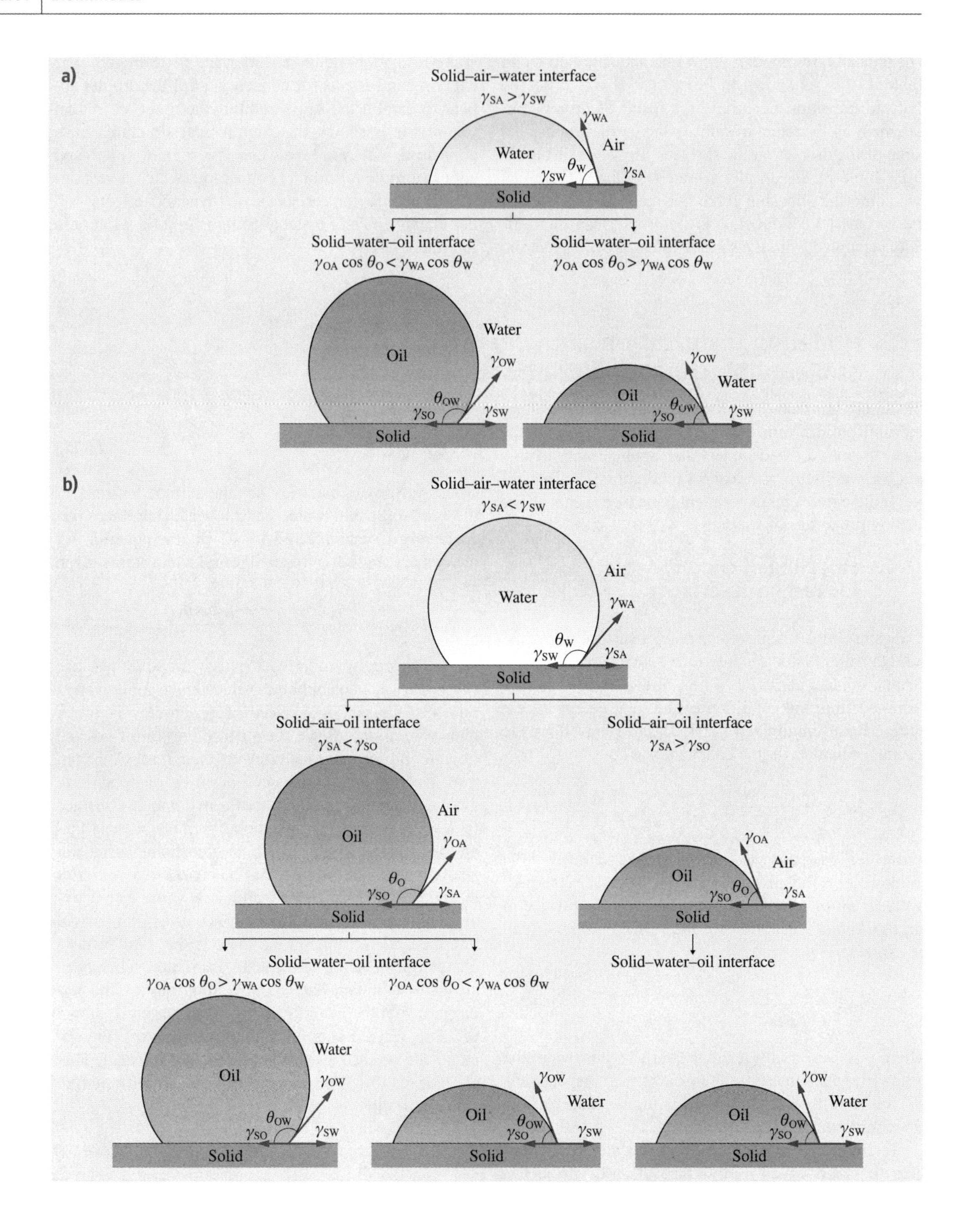
a)
Solid–air–water interface
$\gamma_{SA} > \gamma_{SW}$
γ_{WA}
Water
Air
θ_W
γ_{SW}
γ_{SA}
Solid
Solid–water–oil interface
$\gamma_{OA} \cos\theta_O < \gamma_{WA} \cos\theta_W$
Solid–water–oil interface
$\gamma_{OA} \cos\theta_O > \gamma_{WA} \cos\theta_W$
Water
Oil
γ_{OW}
θ_{OW}
γ_{SO}
γ_{SW}
Solid
b)
Solid–air–water interface
$\gamma_{SA} < \gamma_{SW}$
Air
Water
γ_{WA}
θ_W
γ_{SW}
γ_{SA}
Solid
Solid–air–oil interface
$\gamma_{SA} < \gamma_{SO}$
Solid–air–oil interface
$\gamma_{SA} > \gamma_{SO}$
Air
Oil
γ_{OA}
θ_O
γ_{SO}
γ_{SA}
Solid
Solid–water–oil interface
$\gamma_{OA} \cos\theta_O > \gamma_{WA} \cos\theta_W$
$\gamma_{OA} \cos\theta_O < \gamma_{WA} \cos\theta_W$
Solid–water–oil interface
Water
Oil
γ_{OW}
θ_{OW}
γ_{SO}
γ_{SW}
Solid

Fig. 42.67a,b Schematics of a droplet of liquid exhibiting (**a**) hydrophilic and (**b**) hydrophobic nature at interfaces of three different phases on a surface; θ_W, θ_O, and θ_{OW} are the static contact angles for a water droplet, oil droplet, and oil droplet in water, respectively (after [42.197]) ◀

used for oil droplets [42.197]. The surface tensions of the water–air interface (γ_{WA}), oil–air interface (γ_{OA}), and oil–water interface (γ_{OW}) are 73 mN/m [42.198], 27.5 mN/m [42.198], and 51.4 mN/m [42.199], respectively. The mass densities are 1000 and 773 kg/m^3 for water and hexadecane, respectively. Water and oil droplets of $\approx 5\,\mu l$ in volume (with radius of a spherical droplet ≈ 1 mm) in air environment were gently deposited on the specimen using a microsyringe. The process of wetting behavior of an oil droplet in water was obtained in a solid–water–oil interface system as shown in Fig. 42.68 [42.197]. A specimen was first immersed in the water phase. Then an oil droplet was gently deposited using a microsyringe from the bottom of the system, because the density of the oil used (hexadecane) is lower than that of water. The image of the droplet was obtained by using a digital camcorder (Sony, DCRSR100, Tokyo) with a 10× optical and 120× digital zoom. Images obtained were analyzed for the contact angle using Imagetool software (University of Texas Health Science Center). The measurements were reproducible to within $\pm 2°$.

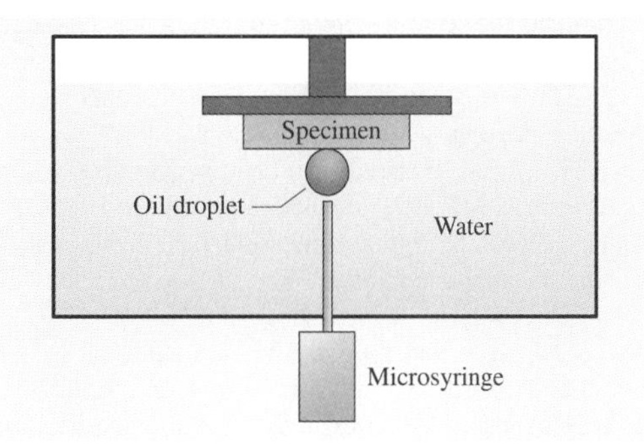

Fig. 42.68 Schematic of a solid–water–oil interface system. The specimen is first immersed in the water phase, and then an oil droplet is gently deposited using a microsyringe, and the static contact angle in the system is measured (after [42.197])

42.6.3 Fabrication and Characterization of Oleophobic Surfaces

A two-step molding process was used to replicate the microstructures with varying pitch values. In this technique, first a negative is generated and then a positive. As a master template for the flat and micropatterned surfaces, a flat Si surface and micropatterned Si surfaces with pillars of 14 μm diameter and 30 μm height with different pitch values (21, 23, 26, 35, 70, 105, 126, 168, and 210 μm), fabricated by photolithography, were used [42.197]. A polyvinylsiloxane dental wax (President Light Body Gel, ISO 4823, polyvinylsiloxane (PLB), Coltene Whaledent, Hamburg) was applied via a dispenser on the surface and immediately pressed down with a glass plate. After complete hardening of the molding mass (3–5 min at room temperature), the silicon master surface and the mold (negative) were separated. After a relaxation time of 30 min for the molding material, the negative replicas were filled with a liquid epoxy resin (Epoxydharz L, no. 236349, Conrad Electronics, Hirschau) with hardener (Harter S, 044-Nr236365, Conrad Electronics, Hirschau). Specimens with microstructures were immediately transferred into a vacuum chamber at 750 mTorr (100 Pa) pressure for 10 s to remove trapped air and to increase resin infiltration through the structures. After hardening at room temperature (24 h at 22 °C), the positive replica was separated from the negative replica. To generate sev-

Table 42.12 Summary of philic/phobic nature at various interfaces (after [42.197])

Solid–air–water interface		Solid–air-oil-interface		Solid–water–oil interface
Hydrophilic ($\gamma_{SA} > \gamma_{SW}$)	→			Oleophobic if $\gamma_{OA}\cos\theta_O < \gamma_{WA}\cos\theta_W$ Oleophilic if $\gamma_{OA}\cos\theta_O > \gamma_{WA}\cos\theta_W$
Hydrophobic ($\gamma_{SA} < \gamma_{SW}$)	→	Oleophobic if $\gamma_{SA} < \gamma_{SO}$	→	Oleophobic if $\gamma_{OA}\cos\theta_O > \gamma_{WA}\cos\theta_W$ Oleophilic if $\gamma_{OA}\cos\theta_O < \gamma_{WA}\cos\theta_W$
		Oleophilic if $\gamma_{SA} > \gamma_{SO}$	→	Oleophilic

eral replicas the second step of replication was repeated 20 times for each surface type.

To study the surfaces with some oleophobicity, a surface coating which has a lower surface tension than that of oil is needed. For this purpose, *Jung* and *Bhushan* [42.197] deposited n-perfluoroeicosane ($C_{20}F_{42}$) (268828, Sigma Aldrich) on the specimen surfaces by thermal evaporation. The surface energy of n-perfluoroeicosane is $6.7\,mJ/m^2$ ($6.7\,mN/m$) [42.200]. The specimens were mounted on a specimen holder with double-sided tape and placed in a vacuum chamber at 30 mTorr (4 kPa pressure), 2 cm above a heating plate loaded with 6000 μg n-perfluoroeicosane [42.58]. The n-perfluoroeicosane was evaporated by heating it to 170 °C. In a vacuum chamber the evaporation from the point source to the substrate occurs in straight line; thus, the amount of sublimated material is equal in a hemispherical region over the point of source [42.184]. In order to estimate the amount of sublimated mass, the surface area of the half sphere was calculated by using the formula $2\pi r^2$, where the radius (r) represents the distance between the specimen to be covered and the heating plate with the substance to be evaporated. The calculated amount of n-perfluoroeicosane deposited on the surfaces was $2.4\,\mu g/mm^2$ (the amount of n-perfluoroeicosane loaded on the heating plate divided by the surface area).

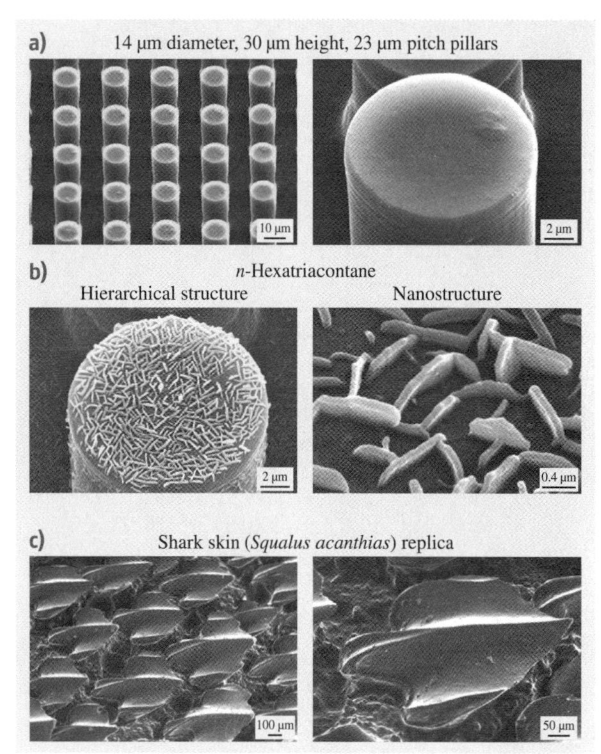

Fig. 42.69a–c SEM micrographs taken at a 45° tilt angle showing two magnifications of (**a**) the micropatterned surface, (**b**) the hierarchical structure and nanostructure with three-dimensional platelets on the surface fabricated with $0.2\,\mu g/mm^2$ mass of n-hexatriacontane, and (**c**) shark-skin (*Squalus acanthias*) replica. The shark-skin replica shows only three ribs on each scale. It is clearly visible that the height of the V-shaped riblets varies between 200–500 μm and their spacing varies between 100–300 μm (after [42.197])

Hierarchical structures were fabricated using a two-step fabrication process, including the production of microstructured surfaces by soft lithography and subsequent development of nanostructures on top by self-assembled n-hexatriacontane at an amount of $0.2\,\mu g/mm^2$ deposited by thermal evaporation, as described above [42.54, 58].

Jung and *Bhushan* [42.197] created a shark-skin replica using a shark, *Squalus acanthias* L. (Squalidae). The shark was conserved in formaldehyde–acetic acid–ethanol (FAA) solution. A shark is an aquatic animal, and its skin is permanently exposed to contamination from marine organisms, e.g., bacteria and algae. Before replicating the conserved shark skin, the area of interest was first cleaned with acetone and then washed with deionized water; this process was repeated twice. The cleaned skin was placed in air for 1 h for drying. For negative replica, a polyvinylsiloxane dental wax was applied via a dispenser on the upper side of the shark skin and immediately pressed down with a glass plate. After complete hardening of the molding mass (at room temperature 3–5 min), the master surface and the mold (negative) were separated. The first negative replica was made only to remove any remaining contaminations from the shark surface by embedding the dirt into the replica material. A second and third replica of the same area were made to obtain negatives without contamination. For positive replica, the process was followed as described above, and epoxy was used.

Figure 42.69a shows SEM micrographs taken at a 45° tilt angle, showing two magnifications of the micropatterned surface. Figure 42.69b shows the hierarchical structures and nanostructures covered with n-hexatriacontane platelets. The nanostructure is formed by three-dimensional platelets of n-hexatriacontane. Platelets are flat crystals, grown perpendicular to the substrate surface. The platelet thickness varied between 50 and 100 nm, and their length varied between 500 and

1000 nm. As shown in Fig. 42.69c, the shark skin replica shows only three ribs on each scale. It is clearly visible that the height of the V-shaped riblets varies in the range 200–500 μm, and their spacing varies in the range 100–300 μm [42.197].

Wetting Behavior on Flat and Micropatterned Surfaces

To observe the wetting behavior of water and oil droplets for philic/phobic nature at three-phase interfaces, *Jung* and *Bhushan* [42.197] performed experiments with droplets on both hydrophilic and hydrophobic, and oleophilic surfaces in air. Figure 42.70 shows optical micrographs of droplets at interfaces of three different phases on flat epoxy resin and micropatterned surfaces. At a solid–air–water interface, the water droplet was hydrophilic for the flat epoxy resin and superhydrophobic for the micropatterned surface with 23 μm pitch. It is known that air pocket formation between the pillars leads to a high static contact angle for the micropatterned surface. However, at a solid–air–oil interface, the oil droplet was oleophilic for both surfaces. In the solid–water–oil interface system, in which the oil droplet sits on water trapped in the pillars, it was observed that the oil droplet in water was oleophobic and had contact angles of 109° and 151° for flat epoxy resin and micropatterned surface with 23 μm pitch, respectively.

To study optimization of oleophobicity at the two solid–air–water and solid–air–oil interfaces, static contact angles for water and oil droplets were measured on the micropatterned surfaces [42.197]. Figure 42.71 shows the measured static contact angle as a function of pitch between the pillars for a water droplet (circles) and an oil droplet (crosses) in air. The data are compared with predicted static contact angle values obtained using the Wenzel and Cassie–Baxter equations (42.24) and (42.25) (solid lines) with a measured value of θ_0 for the micropatterned surfaces. At a solid–air–water interface for a water droplet, the flat epoxy resin showed a static contact angle of 76°. The static contact angle on micropatterned surfaces is higher than that for the flat surfaces. It first increases with increasing pitch values, then starts to drop rapidly to a value slightly higher than that for the flat surfaces. In the first portion, it jumps to a high value of 150° corresponding to a superhydrophobic surface and continues to increase to 160° at a pitch of 26 μm because open air space increases with increasing pitch, responsible for the propensity for air pocket formation. The sudden drop at a pitch value of ≈ 30 μm corresponds to the transition from the Cassie–Baxter to the Wenzel regime. The experimental observations for the transition are comparable to the value predicted from the Wenzel and Cassie–Baxter equations.

At a solid–air–oil interface for an oil droplet, the flat epoxy resin showed a static contact angle of 13°. As shown in Fig. 42.71, the oil droplets on all micropatterned surfaces were oleophilic, and the contact angle was lower than that of the flat surfaces. It increases with increasing pitch values, as predicted by the Wenzel equation. As mentioned earlier, the surface tension of the oil–air interface is very low for hexadecane. Therefore, it is observed from (42.46) that the surface tension of the solid–oil interface (γ_{SO}) is lower than that of the solid–water interface (γ_{SW}), resulting in an oleophilic state for all micropatterned surfaces.

To study optimization of oleophobicity at a solid–water–oil interface, the static contact angles for oil droplets in water were measured on the micropatterned surfaces [42.197]. Figure 42.71 shows the measured static contact angle as a function of pitch between the pillars for an oil droplet in water (triangles). The data are compared with the predicted static contact angle values obtained using the Wenzel and Cassie–Baxter equations (42.28) and (42.25) (solid lines), with a measured value of θ_0 for the micropatterned surfaces. At a solid–water–oil interface, the oil droplet on the flat epoxy resin was oleophobic and had a static contact angle of 109°. The static contact angle of micropatterned surfaces at the solid–water–oil interface showed a similar trend to that at the solid–air–water interface. As the pitch increases to 26 μm, the static contact angle first increases gradually from 146° to 155° because the oil droplet sits on water trapped in the pillars, and open space increases with increasing pitch. Then, the contact angle starts to decrease rapidly due to the transition from the Cassie–Baxter to the Wenzel regime. The experimental observations for the transition are comparable to the values predicted from the Wenzel and Cassie–Baxter equations. The micropatterned surfaces studied here were either hydrophilic or hydrophobic, and both were oleophilic. At the solid–water–oil interface, they were oleophobic. As shown in Fig. 42.67 and Table 42.12, the data are not consistent with the model for hydrophobic surfaces. However, hydrophilic surfaces became oleophobic at the solid–water–oil interface because $\gamma_{OA} \cos\theta_O$ is higher than $\gamma_{WA} \cos\theta_W$.

Wetting Behavior on Flat and Micropatterned Surfaces with $C_{20}F_{42}$

To study the surfaces with some oleophobicity, *n*-perfluoroeicosane ($C_{20}F_{42}$), which has a lower surface

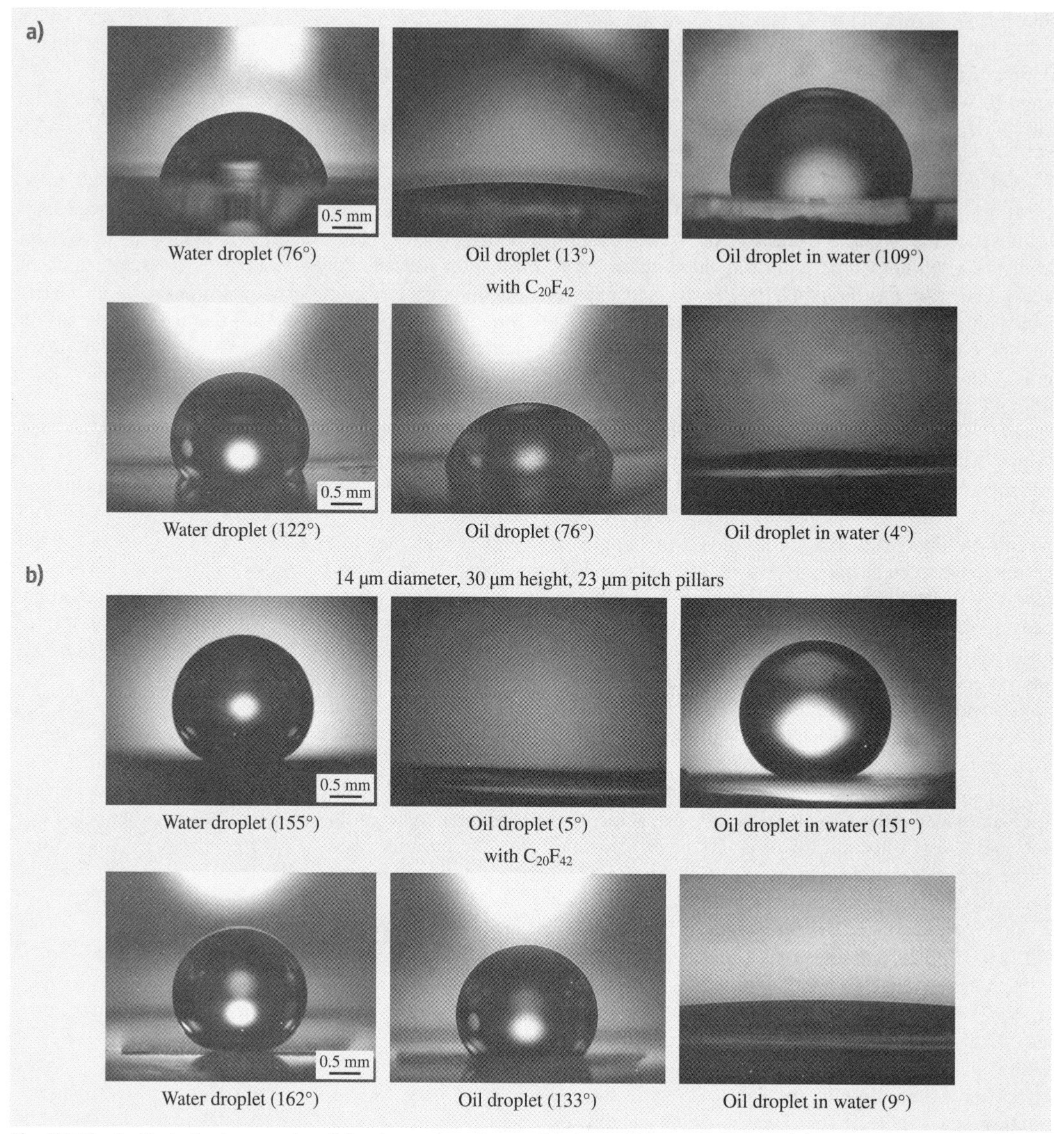

Fig. 42.70a,b Optical micrographs of droplets at interfaces with three different phases on (**a**) flat epoxy resin and (**b**) micropatterned surface without and with $C_{20}F_{42}$. *Left image*: a water droplet is placed on the surface in air. *Middle image*: an oil droplet is placed on the surface in air. *Right image*: an oil droplet is placed on the solid surface in water (after [42.197])

tension than oil, was deposited on the surfaces, and experiments with droplets on hydrophobic and both oleophilic and oleophobic surfaces in air were performed [42.197]. Figure 42.70 shows optical micrographs of droplets in three different phase interfaces on a flat epoxy resin and a micropatterned surface with $C_{20}F_{42}$. At a solid–air–water interface and a solid–air–oil interface, the water droplet and oil droplet showed contact angles of 122° and 76° for the flat epoxy resin with $C_{20}F_{42}$ and contact angles of 162° and 133° for the micropatterned surface with 23 μm pitch with $C_{20}F_{42}$, respectively. However, at a solid–water–oil interface,

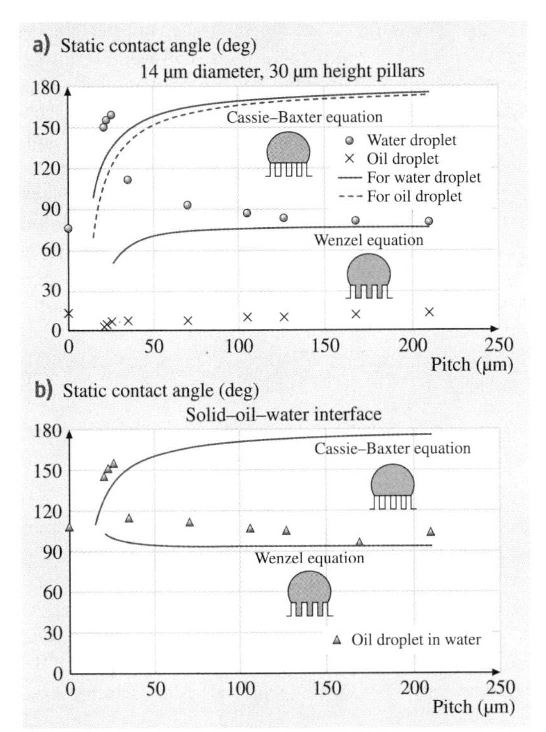

Fig. 42.71a,b Static contact angle as a function of geometric parameters for (**a**) a water droplet (*circles*) and an oil droplet (*crosses*) in air, and (**b**) an oil droplet in water (*triangles*) compared with predicted static contact angle values obtained using the Wenzel and Cassie–Baxter equations (*solid lines*) with a measured value of θ_0 for the micropatterned surfaces (after [42.197])

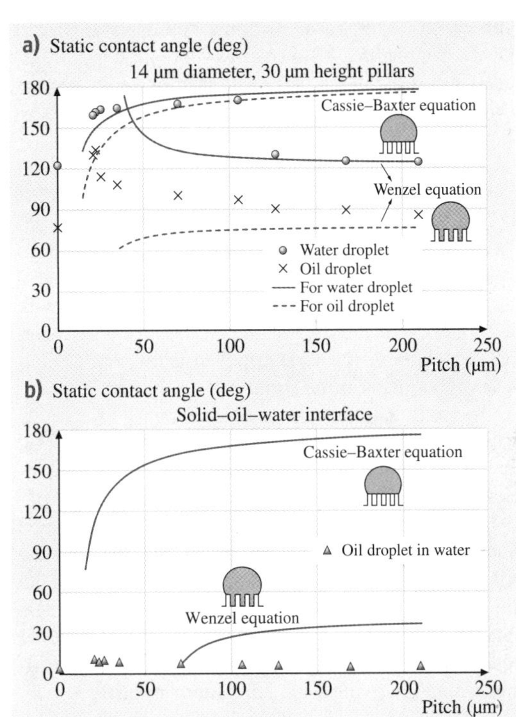

Fig. 42.72a,b Static contact angle as a function of geometric parameters for (**a**) a water droplet (*circles*) and an oil droplet (*crosses*) in air, and (**b**) an oil droplet in water (*triangles*) compared with predicted static contact angle values obtained using the Wenzel and Cassie–Baxter equations (*solid lines*) with a measured value of θ_0 for the micropatterned surfaces with $C_{20}F_{42}$ (after [42.197])

the oil droplet in water was oleophilic and had contact angles of 4° and 9° for both surfaces, respectively. To explain why surfaces that were oleophobic in air became oleophilic in water, the theoretical values for both surfaces were calculated using (42.46). For these calculations, the surface tensions of the water–air interface (γ_{WA}), oil–air interface (γ_{OA}), and oil–water interface (γ_{OW}) were taken to be 73, 27.5, and 51.4 mN/m, and the contact angles for water and oil droplets in air were taken from the measured values. The theoretical values for the flat epoxy resin and the micropatterned surface with 23 μm pitch with $C_{20}F_{42}$ are 28° and 10°, respectively. These values are similar to those found in the experiments. This indicates that the oleophobic surfaces become oleophilic in water.

To study optimization of oleophobicity at two solid–air–water and solid–air–oil interfaces, the static contact angles for water and oil droplets were measured on the micropatterned surfaces with different pitch values and with $C_{20}F_{42}$ [42.197]. Figure 42.72 shows the measured static contact angle as a function of pitch between the pillars for a water droplet (circles) and an oil droplet (crosses) in air. The data are compared with the predicted static contact angle values obtained using the Wenzel and Cassie–Baxter equations (42.24) and (42.25) (solid lines) with a measured value of θ_0 for the micropatterned surfaces with $C_{20}F_{42}$. At a solid–air–water interface for the water droplet, the flat epoxy resin with $C_{20}F_{42}$ showed a static contact angle of 122°. The static contact angle of micropatterned surfaces with $C_{20}F_{42}$ first increases from 158° to 169° with increasing pitch values, then starts to drop rapidly at a pitch value of 110 μm. From comparison of the experimental data with the Wenzel and Cassie–Baxter equations, this corresponds to the transition from the Cassie–Baxter to the Wenzel regime. All surfaces with $C_{20}F_{42}$ had an increased contact angle, and the transition took place at a higher pitch value than for the micropatterned surfaces (Fig. 42.71).

At a solid–air–oil interface for an oil droplet, the flat epoxy resin with $C_{20}F_{42}$ showed a static contact angle of 76°. As shown in Fig. 42.72, the highest contact angle of micropatterned surfaces with $C_{20}F_{42}$ was 133° at a pitch value of 23 μm. Then, it decreases with increasing pitch values, and these values are comparable with the values predicted by the Wenzel equations. The contact angles of all micropatterned surfaces with $C_{20}F_{42}$ were higher than for flat surfaces.

To study optimization of oleophobicity in a solid–water–oil interface, the static contact angles for oil droplets in water were measured on the micropatterned surfaces with different pitch values and with $C_{20}F_{42}$ [42.197]. Figure 42.72 shows the measured static contact angle as a function of pitch between the pillars for an oil droplet in water (triangles). The data are compared with the predicted static contact angle values obtained using the Wenzel and Cassie–Baxter equations (42.24) and (42.25) (solid lines) with a measured value of θ_0 for the micropatterned surfaces with $C_{20}F_{42}$. At a solid–water–oil interface, the flat epoxy resin with $C_{20}F_{42}$ was oleophilic and had a static contact angle of 4°. All micropatterned surfaces with $C_{20}F_{42}$ were oleophilic and had contact angles lower than 10°. The reason why surfaces that are hydrophobic and oleophobic in air became oleophilic in water can be explained from Fig. 42.67 and Table 42.12. The contact angle for a water droplet is higher than that for an oil droplet on all surfaces with $C_{20}F_{42}$, and the surface tension of the water–air interface (γ_{WA}) is higher than that of the oil–air interface (γ_{OA}). Therefore, it is observed that $\gamma_{WA} \cos\theta_W$ is higher than $\gamma_{OA} \cos\theta_O$, and so the surfaces become oleophilic at the solid–water–oil interface.

Wetting Behavior on Nano- and Hierarchical Structures and Shark-Skin Replica

To observe the wetting behavior of water and oil droplets for nano- and hierarchical structures found from lotus plant surfaces, experiments with the droplets on the surfaces were performed at the three-phase interface [42.197]. Figure 42.73 shows optical micrographs of droplets at interfaces with three different phases on a nanostructure and a hierarchical structure fabricated with 0.2 μg/mm^2 mass of n-hexatriacontane.

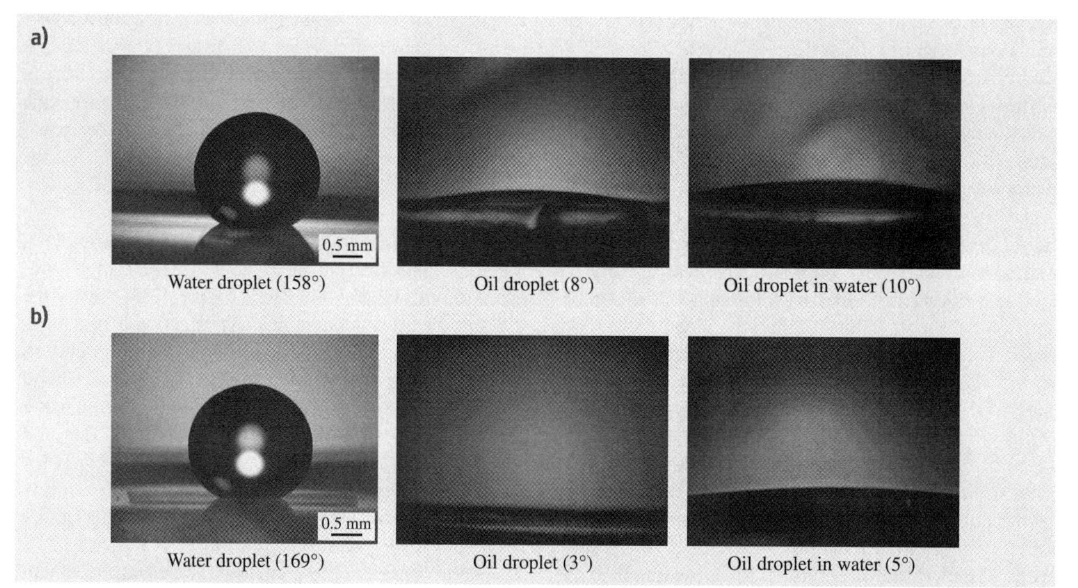

Fig. 42.73a,b Optical micrographs of droplets at interfaces with three different phases on **(a)** nanostructure and **(b)** hierarchical structure fabricated with 0.2 μg/mm^2 mass of n-hexatriacontane. *Left image*: a water droplet is placed on the surface in air. *Middle image*: an oil droplet is placed on the surface in air. *Right image*: an oil droplet is placed on the solid surface in water (after [42.197])

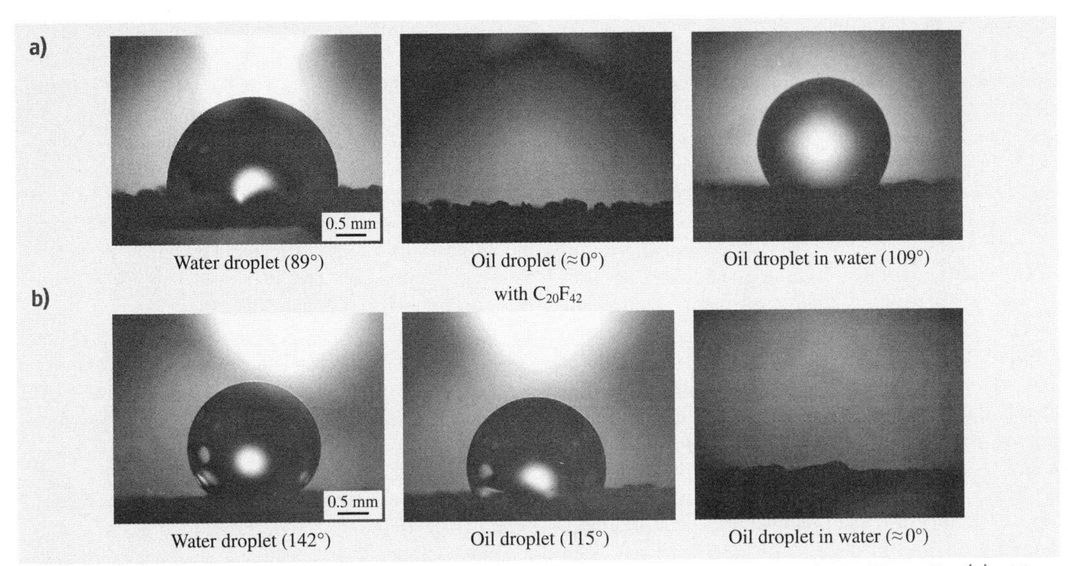

Fig. 42.74a,b Optical micrographs of droplets at interfaces with three different phases on shark-skin replica (**a**) without and (**b**) with $C_{20}F_{42}$. *Left image*: a water droplet is placed on the surface in air. *Middle image*: an oil droplet is placed on the surface in air. *Right image*: an oil droplet is placed on the solid surface in water (after [42.197])

Both nano- and hierarchical structures were superhydrophobic and had a static contact angle of 158° and 169° at the solid–air–water interface, respectively. However, they are oleophilic at the solid–air–oil interface because the surface energy of *n*-hexatriacontane is 31.4 mJ/m^2 (31.4 mN/m) [42.201], which is higher than that of an oil droplet (hexadecane). At the solid–water–oil interface, nano- and hierarchical structures had a static contact angle of 10° and 5°, respectively. As shown in Fig. 42.67 and Table 42.12, both surfaces are oleophilic at the solid–water–oil interface.

To study the surface structure of an aquatic animal, experiments with water and oil droplets on the shark-skin replica were performed at a three-phase interface [42.197]. Figure 42.74 shows optical micrographs of droplets at interfaces with three different phases on a shark-skin replica without and with $C_{20}F_{42}$. First, the shark-skin replica had contact angles of 89° and ≈ 0° for water and oil droplets, respectively. After the surface was coated with $C_{20}F_{42}$, the contact angles of water and oil droplets became 142° and 115°, respectively. At the solid–water–oil interface, the oil droplet in water on the shark-skin replica became oleophobic and had a contact angle of 109°. Based on (42.46), the calculated value was 59° for an oil droplet in water on a shark-skin replica. This difference may result from the open space under the scales of the shark skin replica, responsible for the propensity for trapped water pocket formation, as shown in Fig. 42.69. Shark-skin replica with $C_{20}F_{42}$ was oleophilic and had a contact angle of ≈ 0°. This state is the same as the micropatterned surfaces with $C_{20}F_{42}$, as shown in Fig. 42.67 and Table 42.12.

42.7 Conclusions

In this chapter, the theoretical basis of superhydrophobicity and self-cleaning and characterization of natural and artificial superhydrophobic surfaces are presented. While the theoretical foundations of the wetting of rough surfaces were developed decades ago, emerging applications and the need to design nonstick surfaces have led to the intensive modeling and experimental research in the area. In particular, such problems as the transition between the Wenzel and Cassie–Baxter wetting regimes, contact angle hysteresis, the role ofhierarchical roughness, and the possibility of the creation of reversible hydrophobicity have been very actively investigated in the past decade. We have presented various experimental and theoretical approaches to understand these phenomena. These findings provide new insights into the fundamental mechanisms of wetting and are leading to the creation of successful nonadhesive and self-cleaning surfaces, and some with low adhesion.

Oleophobic surfaces have the potential for self-cleaning and antifouling against biological and organic contaminants in both air and underwater applications. We have presented a model for predicting the contact angle of water and oil droplets. The surface tension of oil and organic liquids is lower than that of water. So, to make the surface oleophobic at a solid–air–oil interface, a material with surface energy lower than that of oil should be used. The wetting behavior of water and oil droplets for hydrophobic/philic and oleophobic/philic surfaces at three-phase interfaces has been investigated. For underwater applications, we have presented the oleophobicity/philicity of oil droplets in water on surfaces with different surface energies for various interfaces and contact angles of water and oil droplets in air.

On the application side, there are several ways to manufacture superhydrophobic surfaces, and new methods continue to emerge. Some methods (such as lithography) allow scientists to create patterned surfaces with clearly defined and controlled geometrical features. These features have a typical size ranging from 1 to 100 μm. Other (and often cheaper) methods lead to self-assembled or random rough surfaces. These methods include, for example, lithography, etching, deposition, and self-assembly. Hierarchical structures are typical for superhydrophobic surfaces in nature. Several technologies are available to produce them. For example, hierarchical surfaces can be produced by replication of a micropattern and self-assembly of hydrophobic alkanes. The fabrication technique is a low-cost two-step process, which provides flexibility in the fabrication of a variety of hierarchical structures.

Proper control of roughness constitutes the main challenge in producing a reliable superhydrophobic surface. If the initial material is hydrophilic, a surface treatment or coating is required to decrease the surface energy. While two factors – roughness and low surface energy – are required for superhydrophobicity, self-cleaning, and low adhesion, the role of roughness clearly dominates. The function of the hierarchical roughness still remains to be further investigated as well as the mechanisms that trigger Cassie–Baxter to Wenzel regime transition.

References

42.1 A.V. Adamson: *Physical Chemistry of Surfaces* (Wiley, New York 1990)

42.2 J.N. Israelachvili: *Intermolecular and Surface Forces*, 2nd edn. (Academic, London 1992)

42.3 B. Bhushan: *Principles and Applications of Tribology* (Wiley, New York 1999)

42.4 B. Bhushan: *Introduction to Tribology* (Wiley, New York 2002)

42.5 M. Nosonovsky, B. Bhushan: *Multiscale Dissipative Mechanisms and Hierarchical Surfaces: Friction, Superhydrophobicity, and Biomimetics* (Springer, Berlin, Heidelberg 2008)

42.6 C.W. Extrand: Model for contact angle and hysteresis on rough and ultraphobic surfaces, Langmuir **18**, 7991–7999 (2002)

42.7 J. Kijlstra, K. Reihs, A. Klami: Roughness and topology of ultra-hydrophobic surfaces, Colloids Surf. A **206**, 521–529 (2002)

42.8 B. Bhushan, Y.C. Jung: Wetting, adhesion and friction of superhydrophobic and hydrophilic leaves and fabricated micro-/nanopatterned surfaces, J. Phys. Condens. Matter **20**, 225010 (2008)

42.9 M. Nosonovsky, B. Bhushan: Multiscale friction mechanisms and hierarchical surfaces in nano- and bio-tribology, Mater. Sci. Eng. R **58**, 162–193 (2007)

42.10 M. Nosonovsky, B. Bhushan: Roughness-induced superhydrophobicity: A way to design nonadhesive surfaces, J. Phys. Condens. Matter **20**, 225009 (2008)

42.11 M. Nosonovsky, B. Bhushan: Biologically-inspired surfaces: Broadening the scope of roughness, Adv. Funct. Mater. **18**, 843–855 (2008)

42.12 B. Bhushan, Y.C. Jung, K. Koch: Self-cleaning efficiency of artificial superhydrophobic surfaces, Langmuir **25**, 3240–3248 (2009)

42.13 M. Nosonovsky, B. Bhushan: Superhydrophobicity for energy conversion and conservation applications, J. Adhes. Sci. Technol. **22**, 2105–2115 (2008)

42.14 M. Nosonovsky, B. Bhushan: Multiscale effects and capillary interactions in functional biomimetic surfaces for energy conversion and green engineering, Philos. Trans. R. Soc. A **367**, 1511–1539 (2009)

42.15 M. Nosonovsky, B. Bhushan: Superhydrophobic surfaces and emerging applications: Nonadhesion, energy, green engineering, Curr. Opin. Colloid Interface Sci. **14**(4), 270–280 (2009)

42.16 B. Bhushan: Adhesion and stiction: Mechanisms, measurement techniques and methods for reduction, J. Vac. Sci. Technol. B **21**, 2262–2296 (2003)

42.17 B. Bhushan: *Nanotribology and Nanomechanics – An Introduction*, 2nd edn. (Springer, Berlin, Heidelberg 2008)

42.18 B. Bhushan, J.N. Israelachvili, U. Landman: Nanotribology: Friction, wear and lubrication at the atomic scale, Nature **374**, 607–616 (1995)

42.19 B. Bhushan: *Tribology and Mechanics of Magnetic Storage Systems*, 2nd edn. (Springer, New York 1996)

42.20 B. Bhushan: *Tribology Issues and Opportunities in MEMS* (Kluwer, Dordrecht 1998)

42.21 B. Bhushan: *Modern Tribology Handbook* (CRC, Boca Raton 2001)

42.22 C. Neinhuis, W. Barthlott: Characterization and distribution of water-repellent, self-cleaning plant surfaces, Ann. Bot. **79**, 667–677 (1997)

42.23 W. Barthlott, C. Neinhuis: Purity of the sacred Lotus, or escape from contamination in biological surfaces, Planta **202**, 1–8 (1997)

42.24 P. Wagner, R. Fürstner, W. Barthlott, C. Neinhuis: Quantitative assessment to the structural basis of water repellency in natural and technical surfaces, J. Exp. Bot. **54**, 1295–1303 (2003)

42.25 Z. Burton, B. Bhushan: Surface characterization and adhesion and friction properties of hydrophobic leaf surfaces, Ultramicroscopy **106**, 709–719 (2006)

42.26 B. Bhushan, Y.C. Jung: Micro- and nanoscale characterization of hydrophobic and hydrophilic leaf surface, Nanotechnology **17**, 2758–2772 (2006)

42.27 B. Bhushan: Biomimetics: Lessons from nature – An overview, Philos. Trans. R. Soc. A **367**, 1445–1486 (2009)

42.28 K. Koch, B. Bhushan, W. Barthlott: Diversity of structure, morphology, and wetting of plant surfaces, Soft Matter **4**, 1943–1963 (2008), (invited)

42.29 K. Koch, B. Bhushan, W. Barthlott: Multifunctional surface structures of plants: An inspiration for biomimetics, Prog. Mater. Sci. **54**, 137–178 (2009), (invited)

42.30 K. Koch, B. Bhushan, Y.C. Jung, W. Barthlott: Fabrication of artificial Lotus leaves and significance of hierarchical structure for superhydrophobicity and low adhesion, Soft Matter **5**, 1386–1393 (2009)

42.31 X.F. Gao, L. Jiang: Biophysics: Water-repellent legs of water striders, Nature **432**, 36 (2004)

42.32 X. Gao, X. Yan, X. Yao, L. Xu, K. Zhang, J. Zhang, B. Yang, L. Jiang: The dry-style antifogging properties of mosquito compound eyes and artificial analogues prepared by soft lithography, Adv. Mater. **19**, 2213–2217 (2007)

42.33 M. Liu, S. Wang, Z. Wei, Y. Song, L. Jiang: Bioinspired design of a superoleophobic and low adhesive water/solid interface, Adv. Mater. **21**, 665–669 (2009)

42.34 D.W. Bechert, M. Bruse, W. Hage: Experiments with three-dimensional riblets as an idealized model of shark skin, Exp. Fluids **28**, 403–412 (2000)

42.35 J. Genzer, K. Efimenko: Recent developments in superhydrophobic surfaces and their relevance to marine fouling: A review, Biofouling **22**, 339–360 (2006)

42.36 R.N. Wenzel: Resistance of solid surfaces to wetting by water, Ind. Eng. Chem. **28**, 988–994 (1936)

42.37 A. Cassie, S. Baxter: Wettability of porous surfaces, Trans. Faraday Soc. **40**, 546–551 (1944)

42.38 R.E. Johnson, R.H. Dettre: Contact angle hysteresis. In: *Contact Angle, Wettability, and Adhesion*, Adv. Chem. Ser., Vol. 43, ed. by F.M. Fowkes (American Chemical Society, Washington 1964) pp. 112–135

42.39 J. Bico, U. Thiele, D. Quéré: Wetting of textured surfaces, Colloids Surf. A **206**, 41–46 (2002)

42.40 A. Marmur: Wetting on hydrophobic rough surfaces: To be heterogeneous or not to be?, Langmuir **19**, 8343–8348 (2003)

42.41 A. Marmur: The Lotus effect: Superhydrophobicity and metastability, Langmuir **20**, 3517–3519 (2004)

42.42 A. Lafuma, D. Quéré: Superhydrophobic states, Nat. Mater. **2**, 457–460 (2003)

42.43 N.A. Patankar: Transition between superhydrophobic states on rough surfaces, Langmuir **20**, 7097–7102 (2004)

42.44 B. He, N.A. Patankar, J. Lee: Multiple equilibrium droplet shapes and design criterion for rough hydrophobic surfaces, Langmuir **19**, 4999–5003 (2003)

42.45 S. Herminghaus: Roughness-induced nonwetting, Europhys. Lett. **52**, 165–170 (2000)

42.46 N.A. Patankar: Mimicking the Lotus effect: Influence of double roughness structures and slender pillars, Langmuir **20**, 8209–8213 (2004)

42.47 M. Sun, C. Luo, L. Xu, H. Ji, Q. Ouyang, D. Yu, Y. Chen: Artificial Lotus leaf by nanocasting, Langmuir **21**, 8978–8981 (2005)

42.48 S. Shibuichi, T. Onda, N. Satoh, K. Tsujii: Super-water-repellent surfaces resulting from fractal structure, J. Phys. Chem. **100**, 19512–19517 (1996)

42.49 M. Nosonovsky, B. Bhushan: Roughness optimization for biomimetic superhydrophobic surfaces, Microsyst. Technol. **11**, 535–549 (2005)

42.50 M. Nosonovsky, B. Bhushan: Stochastic model for metastable wetting of roughness-induced superhydrophobic surfaces, Microsyst. Technol. **12**, 231–237 (2006)

42.51 M. Nosonovsky, B. Bhushan: Wetting of rough three-dimensional superhydrophobic surfaces, Microsyst. Technol. **12**, 273–281 (2006)

42.52 B. Bhushan, Y.C. Jung: Wetting study of patterned surfaces for superhydrophobicity, Ultramicroscopy **107**, 1033–1041 (2007)

42.53 B. Bhushan, M. Nosonovsky, Y.C. Jung: Towards optimization of patterned superhydrophobic surfaces, J. R. Soc. Interface **4**, 643–648 (2007)

42.54 B. Bhushan, K. Koch, Y.C. Jung: Nanostructures for superhydrophobicity and low adhesion, Soft Matter **4**, 1799–1804 (2008)

42.55 B. Bhushan, K. Koch, Y.C. Jung: Biomimetic hierarchical structure for self-cleaning, Appl. Phys. Lett. **93**, 093101 (2008)

42.56 B. Bhushan, K. Koch, Y.C. Jung: Fabrication and characterization of the hierarchical structure for superhydrophobicity, Ultramicroscopy **109**, 1029–1034 (2009)

42.57 B. Bhushan, Y.C. Jung, A. Niemietz, K. Koch: Lotus-like biomimetic hierarchical structures developed by the self-assembly of tubular plant waxes, Langmuir **25**, 1659–1666 (2009)

42.58 B. Bhushan, Y.C. Jung, K. Koch: Micro-, nano- and hierarchical structures for superhydrophobicity, self-cleaning and low adhesion, Philos. Trans. R. Soc. A **367**, 1631–1672 (2009)

42.59 Y.C. Jung, B. Bhushan: Wetting transition of water droplets on superhydrophobic patterned surfaces, Scr. Mater. **57**, 1057–1060 (2007)

42.60 Y.C. Jung, B. Bhushan: Wetting behavior during evaporation and condensation of water microdroplets on superhydrophobic patterned surfaces, J. Microsc. **229**, 127–140 (2008)

42.61 Y.C. Jung, B. Bhushan: Dynamic effects of bouncing water droplets on superhydrophobic surfaces, Langmuir **24**, 6262–6269 (2008)

42.62 Y.C. Jung, B. Bhushan: Dynamic effects induced transition of droplets on biomimetic superhydrophobic surfaces, Langmuir **25**(16), 9208–9218 (2009)

42.63 M. Nosonovsky, B. Bhushan: Hierarchical roughness makes superhydrophobic surfaces stable, Microelectron. Eng. **84**, 382–386 (2007)

42.64 M. Nosonovsky, B. Bhushan: Hierarchical roughness optimization for biomimetic superhydrophobic surfaces, Ultramicroscopy **107**, 969–979 (2007)

42.65 M. Nosonovsky, B. Bhushan: Patterned nonadhesive surfaces: Superhydrophobicity and wetting regime transitions, Langmuir **24**, 1525–1533 (2008)

42.66 M. Nosonovsky, B. Bhushan: Capillary effects and instabilities in nanocontacts, Ultramicroscopy **108**, 1181–1185 (2008)

42.67 M. Nosonovsky, B. Bhushan: Do hierarchical mechanisms of superhydrophobicity lead to self-organized criticality?, Scr. Mater. **59**, 941–944 (2008)

42.68 M. Nosonovsky, B. Bhushan: Energy transitions in superhydrophobicity: Low adhesion, easy flow and bouncing, J. Phys. Condens. Matter **20**, 395005 (2008)

42.69 C.W. Extrand: Criteria for ultralyophobic surfaces, Langmuir **20**, 5013–5018 (2004)

42.70 M. Nosonovsky, B. Bhushan: Biomimetic superhydrophobic surfaces: Multiscale approach, Nano Lett. **7**, 2633–2637 (2007)

42.71 J. Bico, C. Marzolin, D. Quéré: Pearl drops, Europhys. Lett. **47**, 220–226 (1999)

42.72 D. Oner, T.J. McCarthy: Ultrahydrophobic surfaces. Effects of topography length scales on wettability, Langmuir **16**, 7777–7778 (2000)

42.73 Z. Yoshimitsu, A. Nakajima, T. Watanabe, K. Hashimoto: Effects of surface structure on the hydrophobicity and sliding behavior of water droplets, Langmuir **18**, 5818–5822 (2002)

42.74 D. Richard, C. Clanet, D. Quéré: Contact time of a bouncing drop, Nature **417**, 811 (2002)

42.75 D. Bartolo, F. Bouamrirene, E. Verneuil, A. Buguln, P. Silberzan, S. Moulinet: Bouncing or sticky droplets: Impalement transitions on superhydrophobic micropatterned surfaces, Europhys. Lett. **74**, 299–305 (2006)

42.76 M. Reyssat, A. Pepin, F. Marty, Y. Chen, D. Quéré: Bouncing transitions on microtextured materials, Europhys. Lett. **74**, 306–312 (2006)

42.77 F.G. Yost, J.R. Michael, E.T. Eisenmann: Extensive wetting due to roughness, Acta Metall. Mater. **45**, 299–305 (1995)

42.78 S. Semal, T.D. Blake, V. Geskin, M.L. de Ruijter, G. Castelein, J. De Coninck: Influence of surface roughness on wetting dynamics, Langmuir **15**, 8765–8770 (1999)

42.79 H.Y. Erbil, A.L. Demirel, Y. Avci: Transformation of a simple plastic into a superhydrophobic surface, Science **299**, 1377–1380 (2003)

42.80 Z. Burton, B. Bhushan: Hydrophobicity, adhesion, and friction properties of nanopatterned polymers and scale dependence for micro- and nanoelectromechanical systems, Nano Lett. **5**, 1607–1613 (2005)

42.81 Y.C. Jung, B. Bhushan: Contact angle, adhesion, and friction properties of micro- and nanopatterned polymers for superhydrophobicity, Nanotechnology **17**, 4970–4980 (2006)

42.82 C. Bourges-Monnier, M.E.R. Shanahan: Influence of evaporation on contact angle, Langmuir **11**, 2820–2829 (1995)

42.83 S.M. Rowan, M.I. Newton, G. McHale: Evaporation of microdroplets and the wetting of solid surfaces, J. Phys. Chem. **99**, 13268–13271 (1995)

42.84 H.Y. Erbil, G. McHale, M.I. Newton: Drop evaporation on solid surfaces: Constant contact angle mode, Langmuir **18**, 2636–2641 (2002)

42.85 G. McHale, S. Aqil, N.J. Shirtcliffe, M.I. Newton, H.Y. Erbil: Analysis of droplet evaporation on a superhydrophobic surface, Langmuir **21**, 11053–11060 (2005)

42.86 G.D. Danilatos, J.V. Brancik: Observation of liquid transport in the ESEM, Proc. 44th Annu. Meet. EMSA (1986) pp. 678–679
42.87 N.A. Stelmashenko, J.P. Craven, A.M. Donald, E.M. Terentjev, B.L. Thiel: Topographic contrast of partially wetting water droplets in environmental scanning electron microscopy, J. Microsc. **204**, 172–183 (2001)
42.88 P.G. de Gennes, F. Brochard-Wyart, D. Quéré: *Capillarity and Wetting Phenomena* (Springer, Berlin, Heidelberg 2003)
42.89 J.N. Israelachvili, M.L. Gee: Contact angles on chemically heterogeneous surfaces, Langmuir **5**, 288–289 (1989)
42.90 O.N. Tretinnikov: Wettability and microstructure of polymer surfaces: Stereochemical and conformational aspects. In: *Apparent and Microscopic Contact Angles*, ed. by J. Drelich, J.S. Laskowski, K.L. Mittal (VSP, Utrecht 2000) pp. 111–128
42.91 W. Li, A. Amirfazli: A thermodynamic approach for determining the contact angle hysteresis for superhydrophobic surfaces, J. Colloid Interface Sci. **292**, 195–201 (2006)
42.92 B.V. Derjaguin, N.V. Churaev: Structural component of disjoining pressure, J. Colloid Interface Sci. **49**, 249–255 (1974)
42.93 A. Checco, P. Guenoun, J. Daillant: Nonlinear dependence of the contact angle of nanodroplets on contact line curvatures, Phys. Rev. Lett. **91**, 186101 (2003)
42.94 M.A. Anisimov: Divergence of Tolman's length for a droplet near the critical point, Phys. Rev. Lett. **98**, 035702 (2007)
42.95 T. Pompe, A. Fery, S. Herminghaus: Measurement of contact line tension by analysis of the three-phase boundary with nanometer resolution. In: *Apparent and Microscopic Contact Angles*, ed. by J. Drelich, J.S. Laskowski, K.L. Mittal (VSP, Utrecht 2000) pp. 3–12
42.96 L. Boruvka, A.W. Neumann: Generalization of the classical theory of capillarity, J. Chem. Phys. **66**, 5464–5476 (1977)
42.97 M. Nosonovsky: On the range of applicability of the Wenzel and Cassie equations, Langmuir **23**, 9919–9920 (2007)
42.98 L. Gao, T.J. McCarthy: How Wenzel and Cassie were wrong, Langmuir **23**, 3762–3765 (2007)
42.99 C.W. Extrand: Contact angle hysteresis on surfaces with chemically heterogeneous islands, Langmuir **19**, 3793–3796 (2003)
42.100 L. Barbieri, E. Wagner, P. Hoffmann: Water wetting transition parameters of perfluorinated substrates with periodically distributed flat-top microscale obstavles, Langmuir **23**, 1723–1734 (2007)
42.101 F.E. Bartell, J.W. Shepard: Surface roughness as related to hysteresis of contact angles, J. Phys. Chem. **57**, 455–458 (1953)
42.102 P. Gupta, A. Ulman, F. Fanfan, A. Korniakov, K. Loos: Mixed self-assembled monolayer of alkanethiolates on ultrasmooth gold do not exhibit contact angle hysteresis, J. Am. Chem. Soc. **127**, 4–5 (2005)
42.103 N. Eustathopoulos, M.G. Nicholas, B. Drevet: *Wettability at High Temperatures* (Pergamon, Amsterdam 1999)
42.104 M. Nosonovsky: Model for solid-liquid and solid-solid friction for rough surfaces with adhesion hysteresis, J. Chem. Phys. **126**, 224701 (2007)
42.105 Y.T. Cheng, D.E. Rodak, A. Angelopoulos, T. Gacek: Microscopic observations of condensation of water on lotus leaves, Appl. Phys. Lett. **87**, 194112 (2005)
42.106 A. Tuteja, W. Choi, M. Ma, J.M. Mabry, S.A. Mazzella, G.C. Rutledge, G.H. McKinley, R.E. Cohen: Designing superoleophobic surfaces, Science **318**, 1618–1622 (2007)
42.107 H. Kamusewitz, W. Possart, D. Paul: The relation between Young's equilibrium contact angle and the hysteresis on rough paraffin wax surfaces, Colloids Surf. A **156**, 271–279 (1999)
42.108 T.N. Krupenkin, J.A. Taylor, T.M. Schneider, S. Yang: From rolling ball to complete wetting: The dynamic tuning of liquids on nanostructured surfaces, Langmuir **20**, 3824–3827 (2004)
42.109 V. Bahadur, S.V. Garimella: Electrowetting-based control of static droplet states on rough surfaces, Langmuir **23**, 4918–4924 (2007)
42.110 X.J. Feng, L. Feng, M.H. Jin, J. Zhai, L. Jiang, D.B. Zhu: Reversible super-hydrophobicity to super-hydrophilicity transition of aligned ZnO nanorod films, J. Am. Chem. Soc. **126**, 62–63 (2004)
42.111 E. Bormashenko, R. Pogreb, G. Whyman, M. Erlich: Cassie–Wenzel wetting transition in vibrated drops deposited on the rough surfaces: Is dynamic Cassie–Wenzel transition 2-D or 1-D affair?, Langmuir **23**, 6501–6503 (2007)
42.112 D. Quéré: Nonsticking drops, Rep. Prog. Phys. **68**, 2495–2535 (2005)
42.113 M. Nosonovsky: Multiscale roughness and stability of superhydrophobic biomimetic interfaces, Langmuir **23**, 3157–3161 (2007)
42.114 C. Ishino, K. Okumura: Nucleation scenarios for wetting transition on textured surfaces: The effect of contact angle hysteresis, Europhys. Lett. **76**, 464–470 (2006)
42.115 E. Bormashenko, Y. Bormashenko, T. Stein, G. Whyman, R. Pogreb, Z. Barkay: Environmental scanning electron microscope study of the fine structure of the triple line and Cassie–Wenzel wetting transition for sessile drops deposited on rough polymer substrates, Langmuir **23**, 4378–4382 (2007)
42.116 E.A. Baker: Chemistry and morphology of plant epicuticular waxes. In: *The Plant Cuticle*, ed. by D.F. Cutler, K.L. Alvin, C.E. Price (Academic, London 1982) pp. 139–165

42.117 R. Jetter, L. Kunst, A.L. Samuels: Composition of plant cuticular waxes. In: *Biology of the Plant Cuticle*, ed. by M. Riederer, C. Müller (Blackwell, Oxford 2006) pp. 145–181

42.118 K. Koch, A. Dommisse, W. Barthlott: Chemistry and crystal growth of plant wax tubules of Lotus (*Nelumbo nucifera*) and Nasturtium (*Tropaeolum majus*) leaves on technical substrates, Cryst. Growth Des. **6**, 2571–2578 (2006)

42.119 C.Y. Poon, B. Bhushan: Comparison of surface roughness measurements by stylus profiler, AFM and noncontact optical profiler, Wear **190**, 76–88 (1995)

42.120 V.N. Koinkar, B. Bhushan: Effect of scan size and surface roughness on microscale friction measurements, J. Appl. Phys. **81**, 2472–2479 (1997)

42.121 N.S. Tambe, B. Bhushan: Scale dependence of micro-/nanofriction and adhesion of MEMS/NEMS materials, coatings and lubricants, Nanotechnology **15**, 1561–1570 (2004)

42.122 R. Fürstner, W. Barthlott, C. Neinhuis, P. Walzel: Wetting and self-cleaning properties of artificial superhydrophobic surfaces, Langmuir **21**, 956–961 (2005)

42.123 L. Gao, T.J. McCarthy: The Lotus effect explained: Two reasons why two length scales of topography are important, Langmuir **22**, 2966–2967 (2006)

42.124 L. Xu, W. Chen, A. Mulchandani, Y. Yan: Reversible conversion of conducting polymer films from superhydrophobic to superhydrophilic, Angew. Chem. Int. Ed. Engl. **44**, 6009–6012 (2005)

42.125 N.J. Shirtcliffe, G. McHale, M.I. Newton, C.C. Perry, P. Roach: Porous materials show superhydrophobic to superhydrophilic switching, Chem. Commun. **25**, 3135–3137 (2005)

42.126 S. Wang, H. Liu, D. Liu, X. Ma, X. Fang, L. Jiang: Enthalpy driven three state switching of a superhydrophilic/superhydrophobic surfaces, Angew. Chem. Int. Ed. Engl. **46**, 3915–3917 (2007)

42.127 T.N. Krupenkin, J.A. Taylor, E.N. Wang, P. Kolodner, M. Hodes, T.R. Salamon: Reversible wetting-dewetting transitions on delectrically tunable superhydrophobic nanostructured surfaces, Langmuir **23**, 9128–9133 (2007)

42.128 A. Nakajima, A. Fujishima, K. Hashimoto, T. Watanabe: Preparation of transparent superhydrophobic boehmite and silica films by sublimation of aluminum acetylacetonate, Adv. Mater. **11**, 1365–1368 (1999)

42.129 E. Martines, K. Seunarine, H. Morgan, N. Gadegaard, C.D.W. Wilkinson, M.O. Riehle: Superhydrophobicity and superhydrophilicity of regular nanopatterns, Nano Lett. **5**, 2097–2103 (2005)

42.130 B. Cappella, E. Bonaccurso: Solvent-assisted nanolithography on polystyrene surfaces using the atomic force microscope, Nanotechnology **18**, 155307 (2007)

42.131 C. Martín, G. Rius, X. Borrisé, F. Pérez-Murano: Nanolithography on thin layers of PMMA using atomic force microscopy, Nanotechnology **16**, 1016–1022 (2005)

42.132 M. Ma, R.M. Hill: Superhydrophobic surfaces, Curr. Opin. Colloid Interface Sci. **11**, 193–202 (2006)

42.133 H. Jansen, M. de Boer, R. Legtenberg, M. Elwenspoek: The black silicon method: A universal method for determining the parameter setting of a fluorine-based reactive ion etcher in deep silicon trench etching with profile control, J. Micromech. Microeng. **5**, 115–120 (1995)

42.134 S.R. Coulson, I. Woodward, J.P.S. Badyal, S.A. Brewer, C. Willis: Super-repellent composite fluoropolymer surfaces, J. Phys. Chem. B **104**, 8836–8840 (2000)

42.135 J. Shiu, C. Kuo, P. Chen, C. Mou: Fabrication of tunable superhydrophobic surfaces by nanosphere lithography, Chem. Mater. **16**, 561–564 (2004)

42.136 K. Teshima, H. Sugimura, Y. Inoue, O. Takai, A. Takano: Transparent ultra water-repellent poly(ethylene terephthalate) substrates fabricated by oxygen plasma treatment and subsequent hydrophobic coating, Appl. Surf. Sci. **244**, 619–622 (2005)

42.137 M.T. Khorasani, H. Mirzadeh, Z. Kermani: Wettability of porous polydimethylsiloxane surface: Morphology study, Appl. Surf. Sci. **242**, 339–345 (2005)

42.138 B. Qian, Z. Shen: Fabrication of superhydrophobic surfaces by dislocation-selective chemical etching on aluminum, copper, and zinc substrates, Langmuir **21**, 9007–9009 (2005)

42.139 J.L. Zhang, J.A. Li, Y.C. Han: Superhydrophobic PTFE surfaces by extension, Macromol. Rapid Commun. **25**, 1105–1108 (2004)

42.140 H. Yabu, M. Shimomura: Single-step fabrication of transparent superhydrophobic porous polymer films, Chem. Mater. **17**, 5231–5234 (2005)

42.141 M. Ma, R.M. Hill, J.L. Lowery, S.V. Fridrikh, G.C. Rutledge: Electrospun poly(styrene-block-dimethylsiloxane) block copolymer fibers exhibiting superhydrophobicity, Langmuir **21**, 5549–5554 (2005)

42.142 E. Bormashenko, T. Stein, G. Whyman, Y. Bormashenko, E. Pogreb: Wetting properties of the multiscaled nanostructured polymer and metallic superhydrophobic surfaces, Langmuir **22**, 9982–9985 (2006)

42.143 W. Lee, M. Jin, W. Yoo, J. Lee: Nanostructuring of a polymeric substrate with well-defined nanometer-scale topography and tailored surface wettability, Langmuir **20**, 7665–7669 (2004)

42.144 N. Chiou, C. Lu, J. Guan, L.J. Lee, A.J. Epstein: Growth and alignment of polyaniline nanofibres with superhydrophobic, superhydrophilic and other properties, Nat. Nanotechnol. **2**, 354–357 (2007)

42.145 N.J. Shirtcliffe, G. McHale, M.I. Newton, G. Chabrol, C.C. Perry: Dual-scale roughness produces unusually water-repellent surfaces, Adv. Mater. **16**, 1929–1932 (2004)
42.146 M. Hikita, K. Tanaka, T. Nakamura, T. Kajiyama, A. Takahara: Superliquid-repellent surfaces prepared by colloidal silica nanoparticles covered with fluoroalkyl groups, Langmuir **21**, 7299–7302 (2005)
42.147 H.M. Shang, Y. Wang, S.J. Limmer, T.P. Chou, K. Takahashi, G.Z. Cao: Optically transparent superhydrophobic silica-based films, Thin Solid Films **472**, 37–43 (2005)
42.148 Y. Zhao, M. Li, Q. Lu, Z. Shi: Superhydrophobic polyimide films with a hierarchical topography: Combined replica molding and layer-by-layer assembly, Langmuir **24**, 12651–12657 (2008)
42.149 L. Zhai, F.C. Cebeci, R.E. Cohen, M.F. Rubner: Stable superhydrophobic coatings from polyelectrolyte multilayers, Nano Lett. **4**, 1349–1353 (2004)
42.150 R.J. Klein, P.M. Biesheuvel, B.C. Yu, C.D. Meinhart, F.F. Lange: Producing super-hydrophobic surfaces with nano-silica spheres, Z. Metallkd. **94**, 377–380 (2003)
42.151 W. Ming, D. Wu, R. van Benthem, G. de With: Superhydrophobic films from raspberry-like particles, Nano Lett. **5**, 2298–2301 (2005)
42.152 X. Zhang, S. Feng, X. Yu, H. Liu, Y. Fu, Z. Wang, L. Jiang, X. Li: Polyelectrolyte multilayer as matrix for electrochemical deposition of gold clusters: Toward superhydrophobic surface, J. Am. Chem. Soc. **126**, 3064–3065 (2004)
42.153 K.K.S. Lau, J. Bico, K.B.K. Teo, M. Chhowalla, G.A.J. Amaratunga, W.I. Milne, G.H. McKinley, K.K. Gleason: Superhydrophobic carbon nanotube forests, Nano Lett. **3**, 1701–1705 (2003)
42.154 L. Huang, S.P. Lau, H.Y. Yang, E.S.P. Leong, S.F. Yu: Stable superhydrophobic surface via carbon nanotubes coated with a ZnO thin film, J. Phys. Chem. **109**, 7746–7748 (2005)
42.155 M.E. Abdelsalam, P.N. Bartlett, T. Kelf, J. Baumberg: Wetting of regularly structured gold surfaces, Langmuir **21**, 1753–1757 (2005)
42.156 L. Zhu, Y. Xiu, J. Xu, P.A. Tamirisa, D.W. Hess, C. Wong: Superhydrophobicity on two-tier rough surfaces fabricated by controlled growth of aligned carbon nanotube arrays coated with fluorocarbon, Langmuir **21**, 11208–11212 (2005)
42.157 Y. Zhao, T. Tong, L. Delzeit, A. Kashani, M. Meyyappan, A. Majumdar: Interfacial energy and strength of multiwalled-carbon-nanotube-based dry adhesive, J. Vac. Sci. Technol. B **24**, 331–335 (2006)
42.158 N. Zhao, Q.D. Xie, L.H. Weng, S.Q. Wang, X.Y. Zhang, J. Xu: Superhydrophobic surface from vapor-induced phase separation of copolymer micellar solution, Macromolecules **38**, 8996–8999 (2005)
42.159 X. Wu, L. Zheng, D. Wu: Fabrication of superhydrophobic surfaces from microstructured ZnO-based surfaces via a wet-chemical route, Langmuir **21**, 2665–2667 (2005)
42.160 J.T. Han, Y. Jang, D.Y. Lee, J.H. Park, S.H. Song, D.Y. Ban, K. Cho: Fabrication of a bionic superhydrophobic metal surface by sulfur-induced morphological development, J. Mater. Chem. **15**, 3089–3092 (2005)
42.161 E. Hosono, S. Fujihara, I. Honma, H. Zhou: Superhydrophobic perpendicular nanopin film by the bottom-up process, J. Am. Chem. Soc. **127**, 13458–13459 (2005)
42.162 F. Shi, Y. Song, J. Niu, X. Xia, Z. Wang, X. Zhang: Facile method to fabricate a large-scale superhydrophobic surface by galvanic cell reaction, Chem. Mater. **18**, 1365–1368 (2006)
42.163 M.T. Northen, K.L. Turner: A batch fabricated biomimetic dry adhesive, Nanotechnology **16**, 1159–1166 (2005)
42.164 M.A.S. Chong, Y.B. Zheng, H. Gao, L.K. Tan: Combinational template-assisted fabrication of hierarchically ordered nanowire arrays on substrates for device applications, Appl. Phys. Lett. **89**, 233104 (2006)
42.165 Y. Wang, Q. Zhu, H. Zhang: Fabrication and magnetic properties of hierarchical porous hollow nickel microspheres, J. Mater. Chem. **16**, 1212–1214 (2006)
42.166 D. Kim, W. Hwang, H.C. Park, K.H. Lee: Superhydrophobic micro- and nanostructures based on polymer sticking, Key Eng. Mater. **334/335**, 897–900 (2007)
42.167 A. del Campo, C. Greiner: SU-8: A photoresist for high-aspect-ratio and 3-D submicron lithography, J. Micromech. Microeng. **17**, R81–R95 (2007)
42.168 B. Cortese, S.D. Amone, M. Manca, I. Viola, R. Cingolani, G. Gigli: Superhydrophobicity due to the hierarchical scale roughness of PDMS surfaces, Langmuir **24**, 2712–2718 (2008)
42.169 C.Y. Kuan, M.H. Hon, J.M. Chou, I.C. Leu: Wetting characteristics on micro-/nanostructured zinc oxide coatings, J. Electrochem. Soc. **156**, J32–J36 (2009)
42.170 Y.C. Jung, B. Bhushan: Technique to measure contact angle of micro/nanodroplets using atomic force microscopy, J. Vac. Sci. Technol. A **26**, 777–782 (2008)
42.171 M. Brugnara, C. Della Volpe, S. Siboni, D. Zeni: Contact angle analysis on polymethylmethacrylate and commercial wax by using an environmental scanning electron microscope, Scanning **28**, 267–273 (2006)
42.172 J.P. Cleveland, S. Manne, D. Bocek, P.K. Hansma: A nondestructive method for determining the spring constant of cantilevers for scanning force microscopy, Rev. Sci. Instrum. **64**, 403–405 (1993)
42.173 B. Bhushan, G.S. Blackman: Atomic force microscopy of magnetic rigid disks and sliders and

Part F | 42

its applications to tribology, ASME J. Tribol. **113**, 452–457 (1991)
42.174 N. Chen, B. Bhushan: Atomic force microscopy studies of conditioner thickness distribution and binding interactions on the hair surface, J. Microsc. **221**, 203–215 (2006)
42.175 R.A. Lodge, B. Bhushan: Surface characterization of human hair using tapping mode atomic force microscopy and measurement of conditioner thickness distribution, J. Vac. Sci. Technol. A **24**, 1258–1269 (2006)
42.176 S.E. Choi, P.J. Yoo, S.J. Baek, T.W. Kim, H.H. Lee: An ultraviolet-curable mold for sub-100 nm lithography, J. Am. Chem. Soc. **126**, 7744–7745 (2004)
42.177 B. Bhushan, D. Hansford, K.K. Lee: Surface modification of silicon and polydimethylsiloxane surfaces with vapor-phase-deposited ultrathin fluorosilane films for biomedical nanodevices, J. Vac. Sci. Technol. A **24**, 1197–1202 (2006)
42.178 T. Pompe, S. Herminghaus: Three-phase contact line energetics from nanoscale liquid surface topographies, Phys. Rev. Lett. **85**, 1930–1933 (2000)
42.179 D. Quéré: Surface wetting model droplets, Nat. Mater. **3**, 79–80 (2004)
42.180 T. Kasai, B. Bhushan, G. Kulik, L. Barbieri, P. Hoffmann: Micro-/nanotribological study of perfluorosilane SAMs for antistiction and low wear, J. Vac. Sci. Technol. B **23**, 995–1003 (2005)
42.181 X. Zhang, S. Tan, N. Zhao, X. Guo, X. Zhang, Y. Zhang, J. Xu: Evaporation of sessile water droplets on superhydrophobic natural lotus and biomimetic polymer surfaces, ChemPhysChem **7**, 2067–2070 (2006)
42.182 K. Koch, A. Dommisse, W. Barthlott, S. Gorb: The use of plant waxes as templates for micro- and nanopatterning of surfaces, Acta Biomater. **3**, 905–909 (2007)
42.183 K. Koch, A.J. Schulte, A. Fischer, S.N. Gorb, W. Barthlott: A fast and low-cost replication technique for nano- and high-aspect-ratio structures of biological and artificial materials, Bioinspir. Biomim. **3**, 046002 (2008)
42.184 R.F. Bunshah: *Handbook of Deposition Technologies for Films and Coatings: Science, Technology and Applications* (Applied Science, Westwood 1994)
42.185 A. Niemietz, W. Barthlott, K. Wandelt, K. Koch: Thermal evaporation of multi-component waxes and thermally activated formation of nanotubules for superhydrophobic surfaces, Prog. Org. Coat. **66**(3), 221–227 (2009)
42.186 D.L. Dorset, W.A. Pangborn, A.J. Hancock: Epitaxial crystallization of alkane chain lipids for electron diffraction analysis, J. Biochem. Biophys. Methods **8**, 29–40 (1983)
42.187 K. Koch, W. Barthlott, S. Koch, A. Hommes, K. Wandelt, W. Mamdouh, S. De-Feyter, P. Broekmann: Structural analysis of wheat wax (*Triticum aestivum*, c.v. *Naturastar* L.): From the molecular level to three dimensional crystals, Planta **223**, 258–270 (2006)
42.188 A. Tuteja, W. Choi, G.H. McKinley, R.E. Cohen, M.F. Rubner: Design parameters for superhydrophobicity and superoleophobicity, Science **318**, 1618 (2008)
42.189 A.I.J.M. van Dijk, L.A. Bruijnzeel, C.J. Rosewell: Rainfall intensity–kinetic energy relationships: A critical literature appraisal, J. Hydrol. **261**, 1–23 (2002)
42.190 H. Lamb: *Hydrodynamics* (Cambridge Univ. Press, Cambridge 1932)
42.191 X. Noblin, A. Buguin, F. Brochard-Wyart: Vibrated sessile drops: Transition between pinned and mobile contact line oscillations, Eur. Phys. J. E **14**, 395–404 (2004)
42.192 F. Celestini, R. Kofman: Vibration of submillimeter-size supported droplets, Phys. Rev. E **73**, 041602 (2006)
42.193 R.J. Good: A thermodynamic derivation of Wenzel's modification of Young's equation for contact angles. Together with a theory of hysteresis, J. Am. Chem. Soc. **74**, 5041–5042 (1952)
42.194 Y.L. Chen, C.A. Helm, J.N. Israelachvili: Molecular mechanisms associated with adhesion and contact angle hysteresis of monolayer surfaces, J. Phys. Chem. **95**, 10736–10747 (1991)
42.195 J.F. Joanny, P.G. de Gennes: A model for contact angle hysteresis, J. Chem. Phys. **81**, 552–562 (1984)
42.196 S.-W. Lee, P.E. Laibinis: Directed movement of liquids on patterned surfaces using noncovalent molecular adsorption, J. Am. Chem. Soc. **122**, 5395–5396 (2000)
42.197 Y.C. Jung, B. Bhushan: Wetting behavior of water and oil droplets in three-phase interfaces for hydrophobicity/hydrophilicity and oleophobicity/oleophilicity, Langmuir **25**, 14165–14173 (2009)
42.198 D.R. Lide: *CRC Handbook of Chemistry and Physics*, 89th edn. (CRC, Boca Raton 2009)
42.199 K. Tajima, T. Tsutsui, H. Murata: Thermodynamic relation of interfacial tensions in three fluid phases, Bull. Chem. Soc. Jpn. **53**, 1165–1166 (1980)
42.200 T. Nishino, M. Meguro, K. Nakamae, M. Matsushita, Y. Ueda: The lowest surface free energy based on $-CF_3$ alignment, Langmuir **15**, 4321–4323 (1999)
42.201 S. Wu: Surface-tension of solids-equation of state analysis, J. Colloid Interface Sci. **71**, 605–609 (1979)

43. Biological and Biologically Inspired Attachment Systems

Stanislav N. Gorb

Many species of animals and plants are supplied with diverse attachment devices, which morphology depends on the species biology and on particular function, in which the attachment device is involved. Many functional solutions have evolved independently in different lineages of animals and plants. Based on the original and literature data, we have proposed classification of biological attachment systems according to several principles:

1. Fundamental physical mechanism, on which the system operates
2. Biological function of the attachment device
3. Duration of the contact.

In more detail, we discuss here locomotory attachment devices capable of multiple bonding and debonding cycles. Finally, we show a biomimetic potential of studies on biological attachment devices.

43.1 Foreword

We are always impressed by the extraordinary locomotory abilities of living creatures, especially if a similar type of locomotion cannot be performed by humans. One such example is walking on the ceiling, which is at least as impressive as flying. During their evolution, many animal groups, such as insects, spiders, and lizards, have developed this ability (Fig. 43.1). Interestingly, there is no simple explanation for this mechanism. Rather, it is a combination of micro- and nanostructured surfaces, viscoelastic materials, biphasic fluids and their transporting systems, as well as the manner of movement itself. Some of these properties are trivial from the physical point of view; others are highly complex and need further experimental studies and theoretical considerations. In addition to locomotory attachment devices, there is an enormous diversity of attachment systems with nonlocomotory function.

During the last decade, interest in biological attachment devices has been renewed, because of the newest experimental techniques enabling high-speed video recordings, force measurements, and elaborate microscopy methods. Biological attachment structures turned out to possess important characters for evolutionary and ecological studies. Additionally, detailed

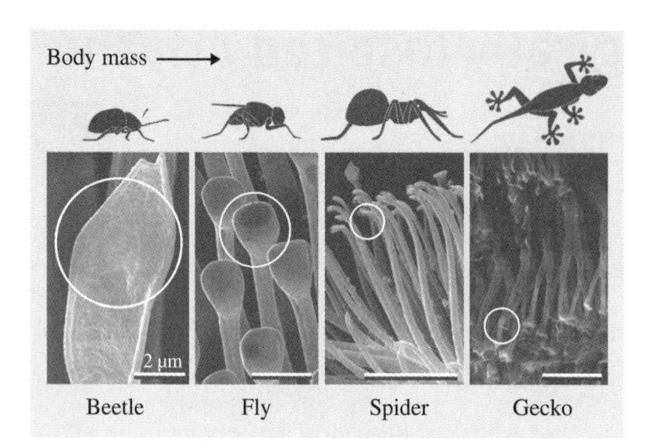

Fig. 43.1 An example of micro- and nanostructures responsible for the attachment ability of different animals. *Circles* indicate terminal contact elements (so-called spatulae) involved in contact formation with the substrate ◀

information about attachment structures and mechanisms has huge potential for biomimetic applications. Because of this growing interest in biological attachment mechanisms, this special chapter is devoted to this freshly emerging topic. It contains information on the micro- and nanostructure of biological attachment devices, and their function, biomechanics, and biomimetic implications.

43.2 Attachment Systems

Attachment devices are functional systems, the purpose of which is either temporary or permanent attachment of an organism to the substrate surface, to another organism, or temporary interconnection of body parts within an organism. Their structure varies enormously and is subject to different functional loads [43.1, 2]. Many species of animals and plants are supplied with diverse attachment devices, whose morphology depends on the species biology and particular biological function of the device. There is no doubt that many functional solutions have evolved independently in different animal lineages [43.3–5]. The evolutionary background and biology of each species influence the specific composition of the attachment systems in each particular organism.

The diversity of biological structures is huge, and the amount of literature published on the morphology of biological attachment devices is also rather large. By comparing various attachment devices in biology [43.1, 2, 6, 7], we have found that biological attachment systems can be subdivided into several groups according to the following principles:

1. The fundamental physical mechanism according to which the system operates
2. The biological function of the attachment device
3. The duration of contact

Eight fundamental attachment mechanisms have been previously recognized (Fig. 43.2):

1. Hooks
2. Lock or snap
3. Clamp
4. Spacer
5. Suction
6. Wet adhesion (capillarity, viscosity, glue/cement)
7. Dry adhesion (van der Waals forces, electrostatic interactions)
8. Friction [43.2, 8]

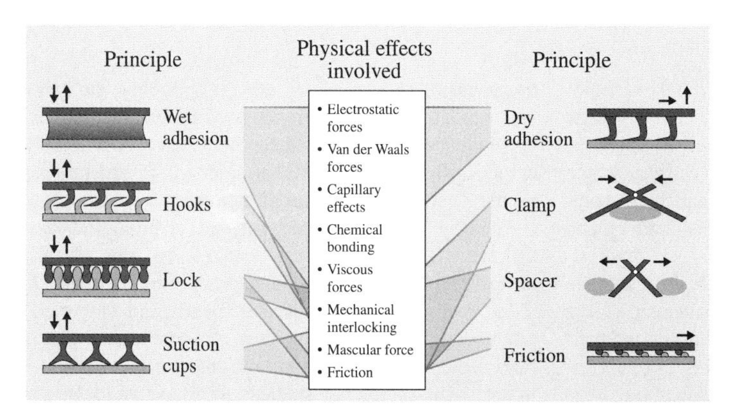

Fig. 43.2 Functional principles of biological attachment systems and physical effects involved

Detailed analysis of these will be given in Sect. 43.5.

According to their time scale of operation, different systems can be subdivided into three main groups: permanent, temporary, and transitory [43.8, 9]. In subsequent sections, we will discuss these classifications and draw conclusions about the general relationships between the attachment mechanism and the functional load of a biological attachment system.

43.3 Biological Functions of Attachment

Attachment is required to fulfil a number of biological functions (Fig. 43.3):

1. Maintenance of position, i. e., attachment to the substrate to stay in place
2. Locomotion requiring strong adhesion and friction with a large number of rapid attachment–detachment cycles
3. Attachment to an animal or plant host for feeding or phoresy (dispersal)
4. Prey capture and a firm hold of the captured prey
5. Temporary attachment between two body parts
6. Maintenance of mechanical contact with the mating partner during copulation
7. Particle manipulation during grooming, sampling, and filtering

Below, we provide a short overview of these functions with some selected examples.

Many sessile animals use various physical principles to attach themselves to a substrate in order to occupy territory. Especially marine organisms, such as mussels (Fig. 43.3d), barnacles, sea anemones, sea urchins, and many others, must withstand rather strong forces generated by waves at the seashore [43.9–12]. Sea anemones and sea urchins may stay in place, attached for quite a long time; however, they are not sessile organisms in the strict sense, because they are capable of movement over the substrate. Also, many climbing plants have developed adhesive structures to secure themselves to the surface of stones and other plants [43.7] (Fig. 43.3c). In general, one of the many functions of plant roots is anchorage in the soil [43.13]. Spiders use secretion of piriform glands to adhere their safety threads (draglines) to the substrate. By so doing, they secure themselves against occasional falls or ensure a short cut to return to their initial location during hunting or exploratory behavior [43.14] (Fig. 43.3b). Many insects and fish attach their eggs to substrates, on which their larvae can potentially feed [43.15–17] (Fig. 43.3a). Long-term attachment to the substrate usually relies on mechanical interlocking or on the use of cements/glues, but a combination of both principles is observed in most systems.

In order to stay attached to the surface, but simultaneously move over it, many elaborate locomotory systems have evolved. They may employ worm-like locomotion or different modes of pedal locomotion, but most of them are equipped with structures or surfaces that must generate strong contact forces (mechanical interlocking, friction, and adhesion) on various

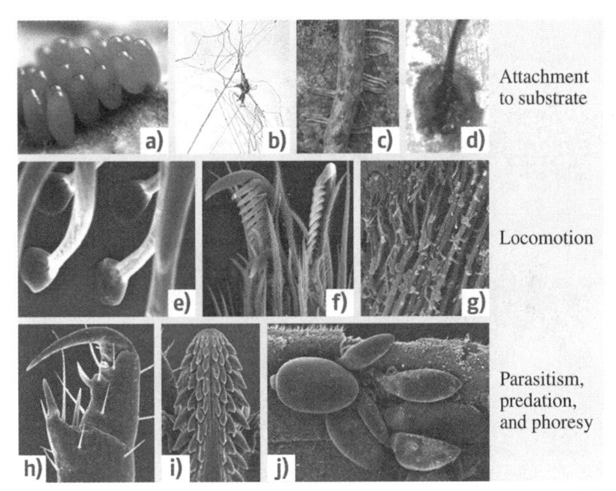

Fig. 43.3a–j Examples of attachment systems according to their biological functions. **(a–d)** Attachment to substrate, maintenance of position: **(a)** eggs of the butterfly *Pieris brassicae* cemented to the cabbage surface; **(b)** dragline of the spider *Lycosa* sp. attached to the glass by sticky threads produced by piriform glands; **(c)** roots of the ivy plant (*Hedera helix*) adhering to tree stem bark; **(d)** plaque of the byssal thread of the mussel *Mytilus edulis*. **(e–g)** Attachment devices used in locomotion: **(e)** tenent hairs of the leg of the fly *Calliphora vicina* adhering to glass; **(f)** claws of the spider *Oxyopes heterophthalmus* adapted for attaching to and walking on threads; **(g)** smooth adhesive pad of the grasshopper *Tettigonia viridissima* (freezing fracture). **(h–j)** Parasitism, predation, and phoresy: **(h)** leg clamp of the lice *Pediculus humanus*; **(i)** ventral aspect of the mite *Ixodes persulcatus* mouthparts used in anchoring in the host tissue; **(j)** mites anchoring in the cuticle of an aquatic insect

substrates. These contact forces are absolutely necessary for the generation of propulsion [43.18]. Insects, spiders, and some lizards (geckos and skinks) bear attachment devices of either hairy (Figs. 43.1 and 43.3e) or smooth type (Fig. 43.3g), which provide proper contact formation with almost any kind of smooth natural substrates and thus generate strong friction and adhesion [43.3, 19–25]. Some of these attachment pads are supplemented with various kinds of fluids (wet adhesion) [43.26–30] whereas others are not (dry adhesion) [43.20, 31–33]. Most of these locomotory systems also bear stiff, pointed hairs and/or claws (Fig. 43.3f), which provide mechanical interlocking to substrates with either strongly corrugated profiles, compliant surfaces or substrates consisting of threads or fibers (spider threads and plant trichomes) [43.34–36].

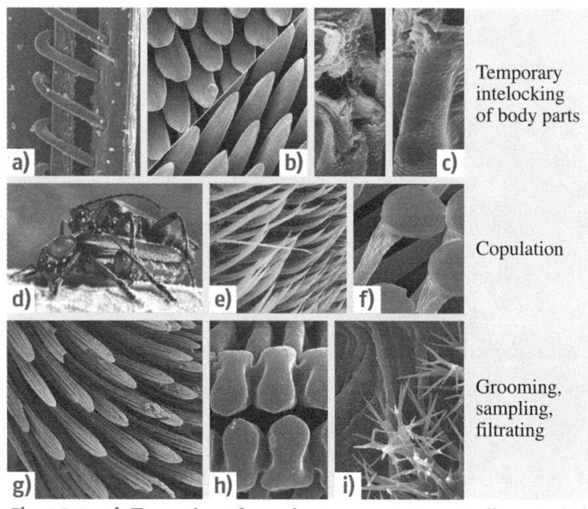

Fig. 43.4a–i Examples of attachment systems according to their biological functions. (**a–c**) Interlocking of body parts: (**a**) temporary locking mechanism between fore- and hindwings in sawfly (*Hymenoptera*); (**b**) two corresponding surfaces covered with microstructures in the head-arresting mechanism in the damselfly *Perissolestes romulus*; (**c**) double tongue-and-groove joint between the right and left covering wings in the beetle *Tenebrio molitor*. (**d–f**) Attachment devices used during copulation; (**d**) copulating beetles *Cantharis fusca*; (**e**) brush-like coverage on the medial surface of abdominal appendages in the male fly *Dolichopus ungulatus* used for attachment to the female; (**f**) specialized tarsal setae of the male beetle *Leptinotarsa decemlineata*. (**g–i**) Grooming, sampling, filtering (particle manipulation); (**g**) cleaning organ on the foreleg of the ant *Formica polyctena*; (**h**) pseudotrachea of the labellum in the fly *Calliphora vicina*; (**i**) filter system of the spiracle in the tenebrionid beetle *Tenebrio molitor*

Although parasitism, predation, and phoresy have slightly different requirements in terms of attachment structures, these systems taken together can still be recognized as providing long-term attachment to another animal (host or prey). Another case of animal–animal attachment is copulation, which, due to its very specialized nature and strong sexual dimorphism, is considered separately here (see later in this section). Most parasites require the ability to form long-term attachments to their hosts. This function is fulfilled by the action of suction cups (some crustaceans and mites), mechanical interlocking (most cases; Fig. 43.3h–j), and by the use of glues or cements. The situation is slightly different in predators, where strong contact forces must be developed very rapidly during prey capture [43.37]. Many predatory animals bear specialized surface structures for mechanical interlocking, some of which rely on microstructures enhancing friction (e.g., aquatic bugs: *Nepa*, *Ranatra*, and *Belostoma*), whereas others use adhesive hairs (spiders [43.38], reduviid bugs [43.39]). In a few cases, suction cups (cephalopod mollusks) or adhesive secretions are involved in capturing prey [43.40–42]. Passive sticky traps are widespread among orb web spiders [43.43–45].

Other interesting cases of attachment structures have been previously described at the interfaces between two different parts within the same organism. Some articulations have to be mobile most of the time, but they have to be firmly interconnected in some behavioral situations. For example, covering wings of beetles and bugs are separated from the body in flight, but locked during walking, in order to provide mechanical integrity of the entire body, to stabilize the origin sites of some leg muscles, and/or to prevent water loss [43.46–50] (Fig. 43.4c). The dragonfly head arrester (Fig. 43.4b) mechanically secures the head at two additional points during pairing or feeding [43.51] and provides high mobility of the head during flight, when the head is involved in the flight control as a mechanosensory organ [43.52]. Some animals attach their forewings to the hindwings in flight, in order to be supported by a larger area in flight (functional diptery) [43.53]. Probably the most interesting aspect of the last example is that such an articulation remains immobile in one direction but provides sliding of the forewing relatively to the hindwing in the other direction (Fig. 43.4a). In birds, keratinous microhooks provide the structural integrity of feathers [43.1, 7]. The principal mechanisms responsible for additional interconnection of body parts are mechanical interlocking and friction [43.54].

In the course of evolution, males of various animal species have developed a number of specializations to attach to the surface of a female during copulation (Fig. 43.4e). In many animals, male and female are attached to one another during copulation with a kind of lock-and-key mechanism, which has not only mechanical, but also sensory function, and is also involved in the recognition of an interspecific mate. In the case of aquatic beetles, the specialized attachment system of males relies on a suction-cup mechanism and is under strong natural selection pressure [43.55–57]. There are some examples where such specialization is so strongly *overdeveloped* (Fig. 43.4f) that males can even be hindered from performing normal walking on smooth surfaces [43.2, 36, 58]. Operation of these systems is based on mechanical interlocking, friction (Fig. 43.4e), suction, and/or capillary adhesion (Fig. 43.4f).

Many attachment devices are specialized for manipulation of particles. These structures are enormously diverse, and include systems used in grooming (cleaning), sampling of food particles, and filtration. Grooming is a very important function for animals, especially when they live in dirty environments. This function is definitely related to adhesion because contamination is usually removed by specialized structures that attract both dust particles and dirt more strongly than other unspecialized surfaces; for example, many ants, wasps, and bees bear flattened leg spines specialized for the cleaning of antennae [43.59–62] (Fig. 43.4g).

Collecting pollen grains and food particles is also a function related to adhesion. Interestingly, such systems in the region of the mouthparts appeared independently in the evolution of such phylogenetically distantly related animal groups as insects, crustaceans, mollusks, and even some vertebrates [43.63]. In bees (*Apoidea*), systems responsible for collecting pollen grains are usually equipped with urticating bristles [43.64,65]. Some beetles bear setae similar to those found on tarsal adhesive pads [43.66]. In aquatic organisms, structures, specialized for sampling, are mostly based on adhesive secretions (tunicates, bivalve mollusks, some polychaetes). Sampling here is closely connected to the function of filtration and burrowing or tube building (some polychaetes and some insect larvae). Filtration systems are usually equipped with long bristles. Such systems are well known from the mouthparts of aquatic invertebrates. The filtering system of insect spiracles is often composed of branched acanthae (Fig. 43.4i). Mouthparts (labellum) in flies (Brachycera, Diptera) bear so-called pseudotrachea [43.67–71], which are able to change the diameter of the filtration sieve, depending on the size of particles in the food (Fig. 43.4h).

43.4 Time Scale of Attachment

Different attachment systems in biology can be subdivided into three main functional groups depending on the time scale over which they operate: permanent, temporary, and transitory attachment [43.7–9, 72] (Fig. 43.5). The first type of these systems is mainly observed in sessile marine organisms, animal eggs, some insects' pupae, some parasitic animals, etc. For this purpose, different kinds of glues/cements or mechanical interlocking are used. Temporary attachment is found in locomotory systems of lizards, tree frogs, some mammals, insects, spiders, cephalopods, echinoderms, and other organisms. There are a number of systems responsible for temporary attachment of a pair of structures within the same organism. Additionally, temporary attachment can be subdivided into two subgroups, each with a different time scale of contact: short term, where the attachment–detachment cycle lasts for at most a few minutes

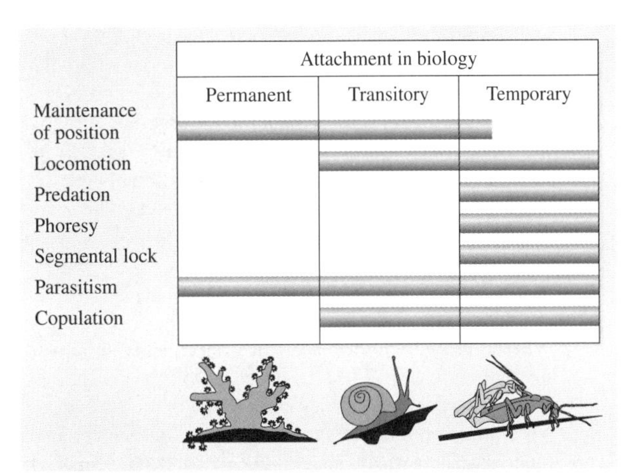

Fig. 43.5 Classification of biological attachment devices according to the time scale of contact

(locomotory devices of geckos, flies, beetles, and spiders), and long term, where the cycle may last from minutes to days (cephalopods, echinoderms, and arresting systems of the head and wings). The short-term systems mainly rely and on van der Waals interactions, capillarity, and viscous forces (Stefan adhesion), whereas the long-term ones usually use a suction-cup mechanism, mechanical interlocking, friction or glues. Transitory attachment is observed in snails, sea anemones, and some flatworms (turbellarians), which are able to move while adhering to the substrate [43.73, 74].

43.5 Principles of Biological Attachment

Attachment devices can join dissimilar materials and, consequently, provide stronger functional flexibility of the entire system [43.10]. They show improved stress distribution in the joint compared with other joining technologies (screw, nail, latch). These features are relevant to the evolution of natural attachment systems and to the development of manmade bonding technologies. As mentioned at the beginning of this chapter, there is a variety of natural attachment systems based on entirely mechanical principles, while others additionally rely on the chemistry of polymers and colloids [43.2, 7, 75] (Fig. 43.2).

Formation of an adhesive bond consists of two phases: contact formation and generation of intrinsic adhesion forces across the joint [43.76]. The action of the adhesive can be supported by mechanical interlocking between irregularities of the surfaces in contact. In the case of using glue, increased surface roughness usually results in increased strength of the bond. Strong adhesion is also possible between two ideally smooth surfaces. If sufficient contact between the substrate and adhesive interface is reached, forces will be set up between atoms and molecules of the two contacting materials, and they will adhere. Van der Waals forces are the most common type of such forces, together with hydrogen bonds. Electrostatic forces may also be involved. Thus, real biological attachment systems usually rely on a number of different functional principles, which are briefly reviewed below.

43.5.1 Interlocking Mechanisms

The hook is a widespread attachment principle across living nature (Fig. 43.3f–i). It has mainly been reported in systems adapted for long-term attachment. Numerous specialized, hook-like structures are found in parasitic animals, adapted for attachment to particular surfaces of their host's body [43.77–79]. A huge variety of hook-like attachment devices has been described for parasitic mites and lice (Fig. 43.3h). The most common example of the hook-like attachment device used for short-term attachment during locomotion is the tarsal claw, which is used to interlock with surface texture (Fig. 43.3f). Prolegs of butterfly caterpillars bear hooks, or crochets, working in concert with the suction devices [43.80–82]. They are not homologous to claws of adults, but serve similar function. Fore- and hindwings of bees, wasps, butterflies, aphids, and other insects can be interlocked with each other with different systems of microhooks [43.81] (Fig. 43.4a). Also plants demonstrate a huge diversity of hook-like microstructures supplementing climbing stems and leaves [43.83]. Barbed seeds and fruits are specialized for dispersal by animals [43.84].

43.5.2 Sucker

A sucker uses the difference between atmospheric pressure and the pressure under the suction cup. Important properties of the cup are its concave shape, flexibility, smooth surface at the edge, and, in some cases,

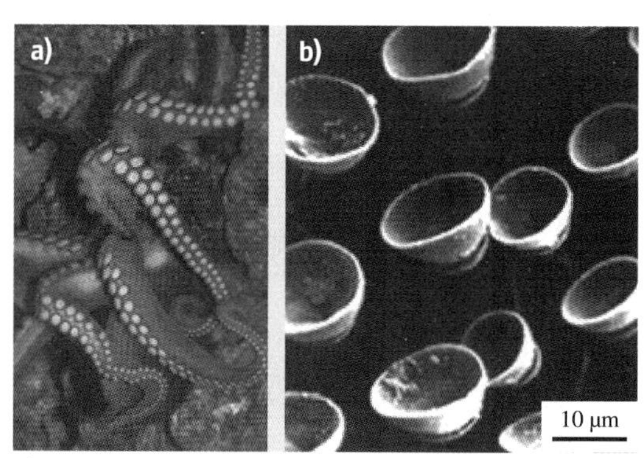

Fig. 43.6a,b Two examples of macroscopical (**a**) and microscopical (**b**) suction cups. (**a**) Arms of *Octopus vulgaris*. (**b**) Microsuckers from tarsal attachment organ of male aquatic beetle *Dytiscus marginatus* are specialized for attachment to the dorsal surface of the female

the presence of muscles or muscular fibers that are responsible for the generation of low pressure. This principle is found widely in soft-bodied animals such as worms and mollusks [43.1] (Fig. 43.6a). However, numerous examples have also been found in insects and crustaceans. Suction cups are widely used by parasites and predators. Some species, such as the aquatic beetle *Dytiscus marginatus*, bear specialized microsuckers for attachment during copulation [43.56] (Fig. 43.6b). True suckers are adapted for attachment to relatively smooth substrata (some stones, plant leaves, and surfaces of other animals). In most functional systems they provide long-term attachment. The sucker requires a muscle to deform the cup and specialized types of resilient tissues. In terrestrial animals, diverse fluids are used to provide better contact of the sucker margin to the substratum. This is essential for long-term maintenance of the lower pressure under the cup.

43.5.3 Snap, Clamp, Spacer

The snap (lock-and-key) attachment principle includes systems with coopted surface profiles: outgrowth and depression. In addition, both surfaces can be covered with tiny surface protuberances or depressions. The snap principle is widely represented in structures related to copulation; for example, male mayflies [43.85] and dragonflies attach to females with extremely specialized lock-and-key devices. The diversity of such devices, even among insects, is huge. These snap-like systems are often supplemented with additional structures, functioning according to hook, clamp, and friction principles. The lock-and-key principle in the design of insect copulation apparatus and its biological significance has been reviewed in detail elsewhere [43.86]. Snap-like mechanisms usually provide long-term fixation and always need two precisely adapted structures. To attach two parts of the snap to each other, or to detach them, muscular force has to be applied.

The clamp principle is usually found in complex mechanical systems, which can attach to or hold onto diverse structures, or substrata, by the use of muscular force. Usually, clamp arms have a particular curvature and are often supplemented by a variety of microstructures. However, they are not necessarily adapted to one particular surface. Devices adapted for prey capture (in predators) or for attachment to the host (in parasites) are often constructed as clamp-like structures. Biting mouth parts of insects can work as a clamp. Many species of crustaceans and insects use their chelae as an organ adapted for capturing, manipulating, and processing prey. Chelae and predatory legs in general, have convergently evolved in different groups of animals [43.87, 88]. Clamp-like mechanisms exist widely in functional systems adapted for copulation. The clamp segment usually flexes during the movement capturing the basal segment [43.89]. The clamp often fits into the groove of the basal segment. The groove margin usually bears rows of spines, or setae, to enhance contact forces. In some crustacean chelae, the muscles belong to the slow type [43.90]. They contract slowly but produce a large force and can remain longer in the contracted condition.

Spacer-like attachment devices rarely occur in biological systems. Such systems are quite similar to the clamp described above, because muscular force or hydraulic mechanisms are involved in both principles. In the case of spacers, however, internal forces are used to transfer the system from the free condition into the working condition, and back. Usually, attachment itself does not require any muscular energy. In many cases, such systems are supplemented with hooked or rough surfaces to interlock with the supporting substratum, or to increase friction with its surface.

43.5.4 Frictional Systems

Probabilistic fasteners are attachment devices composed of two surfaces covered with cuticular micro-outgrowths (Fig. 43.4b). Attachment in such systems

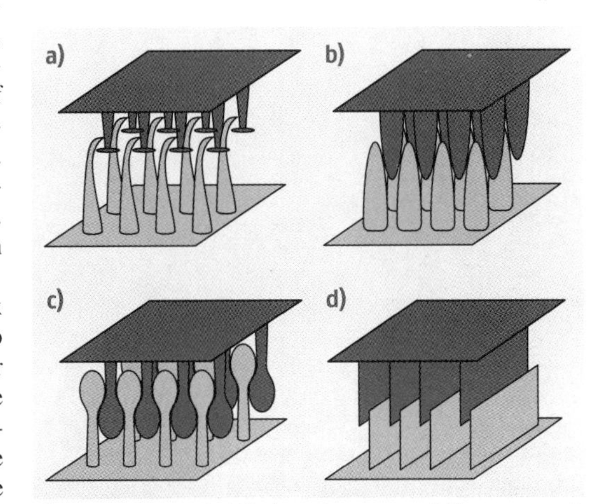

Fig. 43.7a–d Probabilistic fasteners. Microoutgrowth shape in functionally corresponding fields. (**a**) Hook-like and mushroom-like elements. (**b**) Two parabolic elements on both surfaces. (**c**) Clavate elements on both surfaces. (**d**) Plate-like elements on both surfaces

is based on the use of the particular microstructure of two corresponding parts (Fig. 43.7) and the mechanical properties of the materials, the combination of which results in an increase of friction forces in the contact zone. Attachment generated by these systems is fast, precise, and reversible. Such systems have been described in head-arresting systems [43.51], unguitractor plate [43.91], and other intersegmental fixators of antennal and leg joints [43.92], ovipositor valvulae [43.2,93], and wing attachment devices [43.46–50]. The combination of morphological studies of biological systems, experimental data obtained in artificial model systems, and theoretical considerations based on a simple model of the behavior of probabilistic fasteners with parabolic elements has demonstrated that the attachment force in this type of system is strongly dependent on the load force [43.54] (Fig. 43.8). At small loads, the load-to-attachment ratio is rather low, whereas a rapid increase of attachment was detected at higher loads. At very high loads, saturation of the attachment force was revealed (Fig. 43.8c). A simple explanation of the attachment principle is that, with increasing load, single outgrowths of both surfaces slide into gaps in the corresponding part. This results in an increase of lateral load acting on neighboring elements. High lateral forces lead to an increase of friction between single sliding elements.

43.5.5 Adhesion: Glue, Capillary Effects, and Molecular Forces

There is a variety of natural attachment systems based on wet and dry adhesion [capillary interactions, viscous forces (Stefan adhesion), van der Waals forces], whereas others rely on diverse types of glues [43.2, 7, 75]. There are at least three reasons for using adhesives:

1. They join dissimilar materials;
2. They show improved stress distribution within the adhesive bond;
3. They increase design flexibility [43.10].

These reasons are relevant to the evolution of natural attachment systems as well as manmade joining materials. Biological adhesion can be found at all levels of organization of living tissues. Cell contact phenomena have been extensively reviewed in the biomedical and biophysical literature [43.94–96]. The function of attachment appeared very early in evolution; early unicellular organisms had a variety of cellular adaptations for adhesion [43.97, 98]. Many multicellular organisms

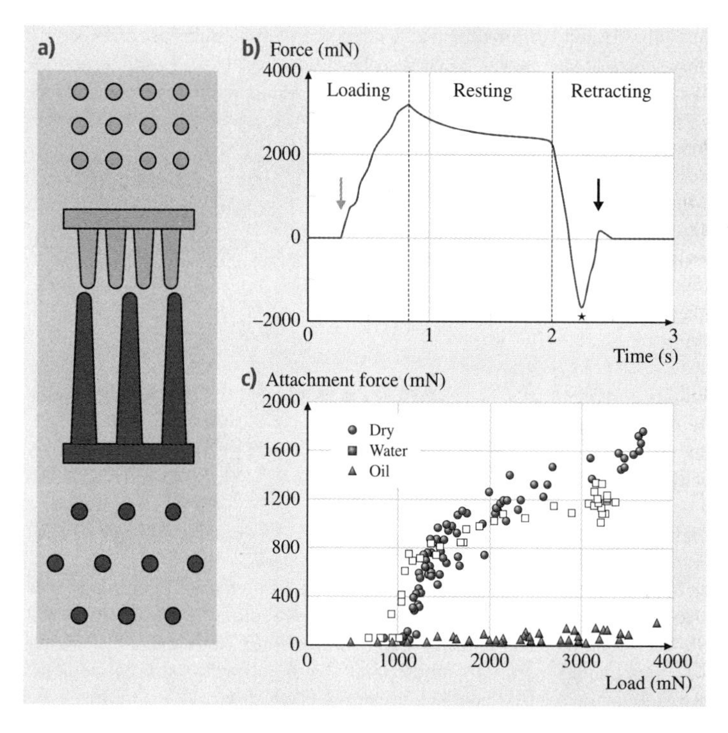

Fig. 43.8a–c Probabilistic fasteners. Results of the force measurements on the artificial fastener system shown in (**a**). (**b**) An example of the force–time curve consisting of three main parts: loading, resting, and retracting. The *gray arrow* indicates contact initiation; the *black arrow* indicates contact breakage; the *star* indicates the beginning of pin sliding at detachment. (**c**) Dependence of the attachment force on the load force for dry, water-covered, and oil-covered samples (after [43.54])

often bear adhesive organs composed of single cells. In many cases, however, cells are specialized into glands, which may be composed of several cell types [43.9, 10].

In the case of glue, adhesive bond formation consists of two phases: contact formation and generation of intrinsic adhesion forces across the joint [43.76, 99]. The action of the adhesive can be supported by mechanical interlocking between irregularities of the contact surfaces. Increased surface roughness usually results in increased strength of the adhesive joint. The last result, however, may be simply explained by increased contact area between contacting surfaces and the glue [43.12]. Strong adhesion is also possible between two ideally smooth surfaces. If sufficient contact area at the interface between the substrate and the adhesive is reached, attractive forces act at the level of atoms and molecules of both contacting materials. van der Waals forces and hydrogen bonds are the most widespread interactions of this type. Electrostatic forces may also be involved: many biological glues are rich in polar and charged residues (amino acids or sugars) and interact with the substrate through electrostatic interactions [43.10, 11]. Electrostatic charges seem to be involved in the adhesion of pollen grains to the surface of insect pollinators [43.65].

Adhesive organs, used for releasable attachment to substrates, as well as those involved in catching prey, demonstrate a huge diversity among living creatures. In their evolution, animals have developed two distinctly different mechanisms to attach themselves to a variety of substrates: with smooth pads or with setose/hairy surfaces [43.2,4–6]. Owing to the flexibility of the material of attachment structures, both mechanisms can maximize the possible contact with the substrate, regardless of its microsculpture. The tips of tenent setae are relatively soft structures [43.100, 101]. In flies, they are usually compressed, widened, and bent at an angle of 60° to the hair shaft [43.102]. When walking on smooth surfaces, these hairs in flies and beetles produce a secretion, which is essential for attachment [43.26–28]. It has been previously hypothesized that capillary adhesion and intermolecular van der Waals forces may contribute to the resulting attachment force [43.103, 104]. The action of intermolecular forces is possible only at very close contact between surfaces [43.20]. The forces increase when the contacting surfaces slide against each other. This may explain why flies sitting on a smooth ceiling always move their legs in a lateral–medial direction [43.105], [43.106]. During these movements, pulvilli slide over the surface, and tape-shaped hair tips spread over the surface and obtain optimal contact.

A contribution of intermolecular interaction to the overall adhesion has been previously shown for geckos [43.20] and beetles [43.104]. In the beetle *Chrysolina polita* (Chrysomelidae), ablation of claws, decrease of air pressure, decrease of relative humidity or electrostatic forces do not influence attachment onto smooth substrata. The resulting attachment force directly depended on the number of single hairs contacting the surface. Recently, the contribution of intermolecular interaction and capillary force has been demonstrated for the fly *Calliphora vicina* in a nanoscale experiment with the use of atomic force microscopy [43.107].

43.6 Locomotory Attachment Pads: Hairy Versus Smooth

Hairy systems consist of fine microstructures (Fig. 43.1), whereas smooth systems are composed of cuticles of unusual inner structure (Fig. 43.9). Due to the flexibility of the material of the attachment pads or fine surface structures, both mechanisms can maximize the possible contact area with a wide range of substrate profiles (Fig. 43.10). These highly specialized structures are not restricted to one particular area of the leg. They may be located on different parts, such as claws, derivatives of the pretarsus, tarsal apex, tarsomeres or tibia. Recent phylogenetic analyses have shown that both types of attachment structures have evolved several times independently in the evolutionary history of insects [43.4, 5, 108, 109]. If we consider other animal groups, such as lizards [43.3, 20, 22, 32, 33] and spiders [43.31, 110], then we can speak about the multiple appearance of the hairy functional solution among animals.

Smooth pads consist of a fibrous material with a specific inner structure. For example, in some orthopterans, tiny filaments are located just under the epicuticle of euplantulae. In grasshoppers *Tettigonia viridissima*, the exocuticle is 45–50 μm thick and consists of the primary filaments, oriented at some angle to the surface [43.111–113] (Fig. 43.9). Such material structure contributes to very specific material properties (soft in compression and strong in tension), which are responsible for the ability of the material to replicate the surface profile. Some functional principles of smooth pads (adaptability, viscoelasticity, pressure sensitivity) are similar to those known from industrial pressure-sensitive adhesives. Hairy attachment pads employ several other features, such as flaw tolerance, and lower sensitivity to contamination or roughness, which make them especially interesting from the biomimetic point of view.

Hairy attachment systems are typical for evolutionarily younger and successful insect groups, such as beetles and flies, and have a huge diversity of forms and ecological niches. This fact may indicate that such a design of adhesive surfaces has an advantage for adhesion enhancement not only in biological systems but also for artificial surfaces with similar geometry. There are several geometrical effects, such as contact splitting, high aspect ratio of single contact structures, and peeling prevention using spatula-like tips of single contact elements, that are responsible for the generation of a strong pull-off force in such attachment devices. These effects found in attachment devices of insects are an important source of information for the further development of biomimetic patterned adhesives. The theoretical background pertaining to these physical effects has been intensively theoretically discussed in several recent publications [43.22, 115–120].

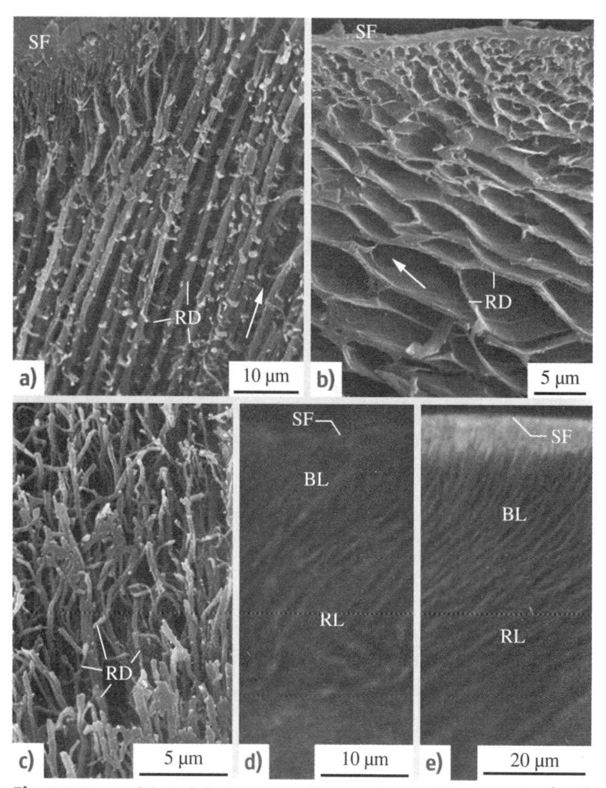

Fig. 43.9a–e Material structure of smooth attachment pads. (**a**,**d**) Grasshopper *Tettigonia viridissima*, euplantula. (**b**) Cicada *Cercopis vulnerata*, arolium. (**c**) Bee *Apis mellifera*, arolium. (**e**) Locust *Locusta migratoria*, euplantula. (**a**–**c**). Scanning electron microscopy (after [43.2]). Freezing-fractured, substituted, dehydrated, and critical-point-dried preparations. (**d**,**e**) Fluorescence microscopy (after [43.114]). BL – layer of branching rods, SF – superficial layer, RD – rods or filaments, RL – layer of primary rods

We have previously shown that the density of hairs strongly increases with increasing body weight [43.7] (Fig. 43.11). This relationship holds because animals cannot increase the area of the attachment devices proportionally to the body weight due to the different scaling rules for mass and surface area [43.121]. Therefore, the increase of the attachment strength in hairy systems is realized by increasing the number of single contact points, i. e., by increasing the hair density. We have explained this general trend by applying the Johnson–Kendall–Roberts (JKR) contact theory [43.122], according to which splitting up the contact into finer subcontacts should be the mechanism for increasing adhesion [43.115]. Since terminal contact elements are mainly spatula-shaped, the adhesion enhancement may be explained by the increase of overall peeling (fracture) line between tape-like spatulae and substrate. However, this trend is presumably different within each single lineage of organisms [43.123, 124]. The fundamental importance of contact splitting for adhesion on smooth and rough substrata has also been explained by a very small

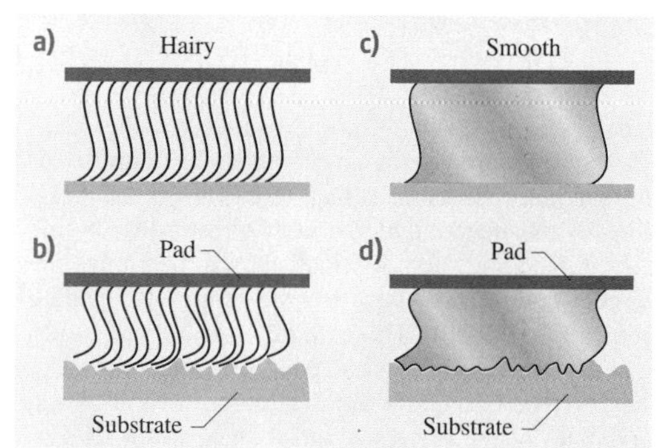

Fig. 43.10a–d Hairy and smooth attachment systems. Diagram of the action of *hairy* (**a**,**b**) and *smooth* (**c**,**d**) pad attachment systems on smooth (**a**,**c**), and structured (**b**,**d**) substrata (after [43.2])

effective elastic modulus of the array of hairs [43.119]. From the scaling analysis, we may suggest that animal lineages relying on dry adhesion (lizards, spiders) usually possess much higher density of terminal contact elements compared with systems using wet adhesive mechanism (insects). Since these effects are based on fundamental physical principles and mostly related to the geometry of the structure, they must hold also for artificial surfaces with similar geometry.

Protuberances on the hairy pads of beetles, earwigs, and flies belong to different types. Setae range in length from a few micrometers to several millimeters. Representatives of the first two lineages have socketed setae on their pads. Fly outgrowths are acanthae: single sclerotized protuberances originating from a single cell [43.125]. Ultrastructural features of adhesive hairs have been previously reported for flies [43.102, 126, 127]: the acanthae are hollow inside, and some of them contain pores under the terminal plate. Such pores presumably deliver an adhesive secretion directly to the contact area. Pore canals at the base of the shaft may additionally transport secretions to the surface. The membranous cuticle of hairy pads is a fibrous composite material with loosely distributed fibers. In coleopterans, the hair bases are embedded in this material, which provides flexibility to the supporting material and helps the pad adapt to a variety of surface profiles [43.2].

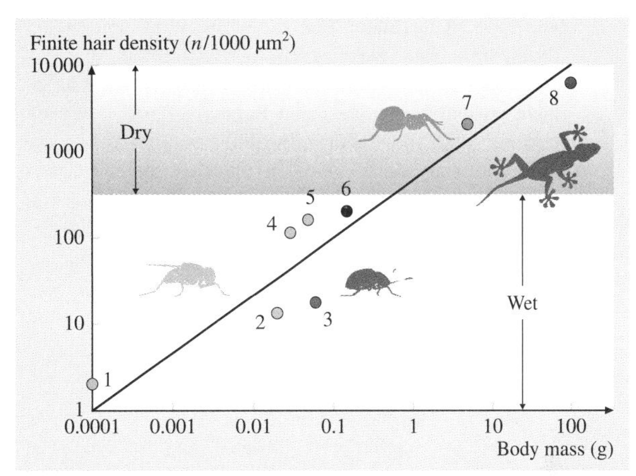

Fig. 43.11 Dependence of the finite hair density of the attachment pads on the body mass in hairy pad systems of representatives from diverse animal groups: 1, 2, 4, 5 – flies; 3 – beetle; 6 – bug; 7 – spider; 8 – gekkonid lizards (after [43.7])

43.7 Dry and Wet Systems

Hairy attachment pads of reduviid bugs [43.128], flies [43.102, 126, 129, 130], and beetles [43.26–28, 36, 37] secrete fluid into the contact area (Fig. 43.12). Such a secretion contains nonvolatile lipid-like substances, but in some species it is a two-phasic emulsion, presumably containing water-soluble and lipid-soluble fractions [43.2, 30]. Hairy attachment systems of the gekkonid lizards and spiders do not produce fluids. In these animals, van der Waals interactions are mainly responsible for the generation of strong attractive forces [43.20]; however, an adsorbed water layer on the surface of solids under ambient conditions can additionally contribute to adhesion in such a *dry* adhesive system [43.31, 33].

In the case of locomotory pads of insect, different basic physical forces contribute to the overall

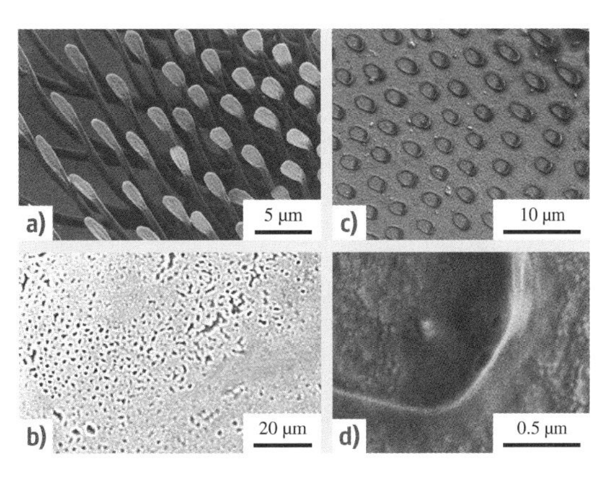

Fig. 43.12a–d Fluid in the wet adhesion system of the fly. **(a)** Tenent setae of the adhesive organ of the fly *Calliphora vicina* (cryo-SEM). **(b)** Footprints viewed in the phase-contrast mode of the light microscope, showing limited ability of the optical microscopy for detailed observation of small volumes of the fluid deposited on the surface. **(c)** Uncoated footprints frozen while the adhesive hairs were in contact with the metalized substrate (cryo-SEM). After freezing, the adhesive organ was removed, allowing observation of fly *footsteps* in the fluid. Fluid menisci between each hair tip and substrate can be seen. **(d)** Meniscus of the print of the single hair tip ▶

adhesion. Attachment was impaired when hairy pads of the bug *Rhodnius prolixus* were treated with organic solvents [43.128]. Experiments with beetles have strongly suggested that cohesive forces, surface tension, and molecular adhesion, mediated by pad secretion, may be involved in the attachment mechanism [43.104]. Recently, multiple local force–volume measurements were carried out on individual terminal plates of the setae of the fly *Calliphora vicina* by application of atomic force microscopy [43.131]. Local adhesion is about two times stronger in the center of the terminal plate than at its border. Adhesion strongly decreases as the volume of the secretion decreases, indicating that a layer of pad secretion, covering the terminal plates, is crucial for the generation of the strong attractive force. These data provide direct evidence that, beside van der Waals and Coulomb forces, attractive capillary forces mediated by pad secretion are a critical factor in the fly's attachment mechanism. One may speculate that the combination of different physical mechanisms is important to generate sufficient adhesion despite variation in the physicochemical properties of the surface (hydrophobic, hydrophilic), surface profile (rough, smooth), and environmental condition (dry, wet).

43.8 Scaling Effects

As we have shown, the mentioned microstructures get smaller and their density higher as the body mass of the animal group increases [43.7, 115] (Fig. 43.11). To test the role of dimensional factors in the generation of friction force by systems of setose attachment pads, six species were studied using light and scanning electron microscopy [43.123]. Flies were selected according to their various body mass and attachment pad dimensions. Variables such as pad area, setal density, the area of a single setal tip, and body mass were individually measured. A centrifugal force tester, equipped with a fiber-optic sensor, was used to measure the friction forces of the pads on a smooth horizontal surface made of plexiglas. It was shown that friction force, which is the resistance force of the insect mass against the sum of the centrifugal and tangential forces, was greater in heavier insects. Although lighter species generated lower forces, the acceleration required to detach an insect was greater in smaller species. The area of attachment pads, setal tip area, and setal density differed significantly in the species studied, and the dependence of these variables on body mass was significant. Frictional properties of the material of the setal tips were not dependent on the dimensions of the fly species. Thus, the properties of the secretion and the mechanical properties of the material of the setal tips are approximately constant among the species studied. It is concluded that differences in friction force must be related mainly to variations in the real contact area generated by different pads on a smooth surface. Although individual variables vary among flies of different dimensions, they usually compensate such that smaller setal tip area is partially compensated for by higher setal density.

It has been previously demonstrated that insects with different pad structure have trouble attaching to microrough plant surfaces [43.132–135]. Hairy pads show a clearly cut minimum of adhesion at certain ranges of substrate roughness. Such a critical range of roughness depends on the relationship between the diameter of single contact elements of the pad and the length scale of the roughness [43.36, 136, 137] (Fig. 43.13). In smooth

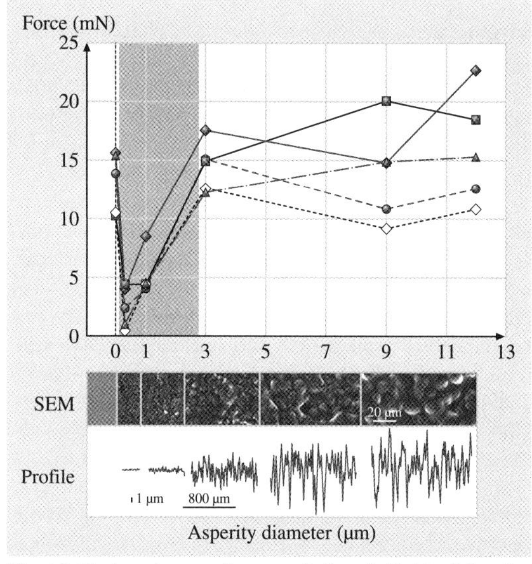

Fig. 43.13 Attachment forces of five individual beetles *Gastrophysa viridula* on substrates with different roughness (asperity diameter)

systems, where contact is not split up into many subcontacts, the adhesion force did not increase with increasing substrate roughness after reaching this minimum [43.138]. This behavior is similar to the adhesive behavior of soft rubber, as previously studied experimentally by Fuller and Tabor [43.139]. In the case of the locust, this effect can be explained by the reduced capability of the interconnected fibers to act independently in response to the local substrate roughness.

43.9 Evolutionary Aspects

In different groups of the same animal lineage, both types of attachment pads can be found. For example, within arachnids, representatives of Solifugae have smooth pads [43.140, 141], whereas most of Araneae are characterized by hairy ones [43.38]. Among reptiles, Gekkonidae and Anolinae have hairy pads [43.3, 19, 20, 32], while some skinks possess smooth ones [43.142]. Numerous insect groups, such as ephemeropterans, orthopterans, plecopterans, and hymenopterans, rely on smooth pads, whereas others (dipterans, coleopterans, megalopterans) have developed hairy structures in their evolution [43.4, 5]. As mentioned above, pads may be located on claws, derivatives of the pretarsus, tarsomeres or tibia. In some cases (sawflies), tarsal spines on the ventral side of the tarsus are transformed into plantae, smooth attachment pads with movable socket-like connection to the tarsus [43.4, 143, 144].

High structural diversity of smooth pads correlates with the difference of their origin. Indeed, recent phylogenetic analysis of hexapods, based on pad characters, processed together with characters of other organ systems, aided in resolving the question of attachment pad evolution in hexapods and showed that smooth pads have evolved several times, independently [43.4, 5, 108, 109]. The number of hexapod groups with smooth pads is much higher compared with those with hairy pads. This fact led us to suggest that smooth systems appeared earlier in insect evolution than did hairy ones. A fibrous composite material such as hexapod cuticle [43.145] was presumably a preadaptation for local development of soft areas of integument due only to a change in fiber architecture and density. Such areas were able to form larger contact areas under the same load conditions, on different substrata.

Smooth and hairy attachment pads are adaptive structures. Their construction and properties may correlate with the preferred substrata used by particular species. That is why we do not see any clear evolutionary pattern in the distribution of lineages with either smooth or hairy types of pads. It has been previously hypothesized that hexapod attachment pads probably evolved to facilitate walking on plant surfaces. On the other side, during their evolution, plants have not only developed structures attracting pollinators, but also a wide variety of structural and chemical attributes of their surfaces related to adhesion prevention in herbivore insects.

43.10 Attachment Devices and Environment

Most of the one million insect species described to date are associated with plants. Such insects should be able to attach successfully to plant surfaces. When dealing with animal attachment to plants, we have to consider a contact problem in which two bodies are involved: animal attachment organs and plant substrates. Both of these parts may have various geometries, and mechanical and chemical properties. Moreover, as shown above, insects produce and deliver a secretory fluid into the contact zone [43.27–30]. The situation is usually even more complex, since many plant substrates bear specialized surface coverage [43.132, 133, 135, 146].

It is well known that insects attach well to macroscopically rough surfaces and use their claws to interlock with surface irregularities [43.147]. To attach to smooth and microrough substrates, insects apply adhesive pads. Since plant surfaces range from rather smooth ones to those covered with trichomes or wax crystals, insect–plant interactions may often rely on the insect's attachment ability to the particular plant surface. In the course of the last 15 years, numerous experiments have been carried out in order to understand insect–plant interactions at the level of the contacting surfaces [43.148–151]. These studies were mainly performed with insects having hairy adhesive pads, the surfaces of which are densely covered with microscopic hairs.

Plant surfaces have a wide range of textures and microsculptures. They may be smooth or structured, i.e., covered with various hairs (trichomes) [43.150] or bear microscopic crystals of epicuticular waxes [43.152–154]. To study the role of different structures of a plant surface in insect attachment, various plant surfaces were screened [43.151].

The attachment ability of the beetle *Chrysolina fastuosa* was tested on 99 surfaces of 83 plant species, belonging to 45 families [43.151]. Insects attached successfully to smooth, dry hairy and felt-like substrates, but they could not attach properly to surfaces covered with wax crystals. To explain the antiadhesive properties of substrates structured with crystalline waxes, four hypotheses were proposed (Fig. 43.14):

- *Roughness hypothesis*: Wax crystals cause microroughness, which considerably decreases the real contact area between the substrate and the setal tips of insect adhesive pads;
- *Contamination hypothesis*: Wax crystals are easily detachable structures that contaminate pads;
- *Wax-dissolution hypothesis*: Insect pad secretion may dissolve wax crystals. This would result in the appearance of a thick layer of fluid, making the substrate slippery;
- *Fluid-absorption hypothesis*: Structured wax coverage may absorb the fluid from the setal surface.

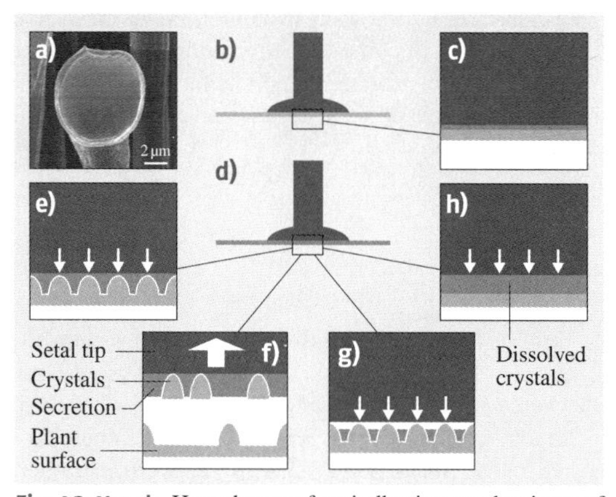

Fig. 43.14a–h Hypotheses of antiadhesive mechanisms of a waxy plant surface. (**a**) Tenent seta of the adhesive pad of the *Chrysolina fastuosa* beetle; (**b,c**) setal contact with a smooth surface; (**d–h**) setal contact with a waxy surface; (**e**) roughness hypothesis; (**f**) contamination hypothesis; (**g**) fluid-absorption hypothesis; (**h**) wax-dissolution hypothesis (after [43.151])

The influence of surface roughness on insect attachment has been tested experimentally in several studies [43.2, 36, 136, 138]. Centrifugal experiments on the attachment ability of insects on artificial substrates with varying surface roughness showed that the fly *Musca domestica* and the beetle *Gastrophysa viridula* generated much higher forces on either smooth or rough surfaces with an asperity size exceeding 3.0 µm (Fig. 43.13). This effect may be explained by spatula-like terminal elements of setae that are able to generate sufficient contact with large surface irregularities. The worst attachment was observed on substrates with roughness of 0.3–1.0 µm, corresponding to the size of plant wax crystals. Because of the small size of wax crystals, the area of real contact between these substrates and the tips of insect spatulae is very small. Since the attachment force depends on the area of real contact, insects are not able to attach successfully to surfaces with such microroughness.

Since many wax crystals are easily erodible structures, they may contaminate adhesive pads, and this may reduce proper functioning of the pads. The contamination hypothesis was tested experimentally for several insects and a series of plant species. The data on the contamination of beetle tarsi by wax crystals of 12 waxy plant surfaces [43.146] showed that the plants differ essentially in their contaminating effects on insect pads. Differences were found in both the nature of the contamination, i.e., the structure of the contaminating material and the presence of recognizable crystals, and the degree of contamination (the portion of setal tip surface covered with contaminating particles, and the portion of setae covered with the wax). These differences were hypothesized to be caused by various micromorphologies of waxes in the plant species studied. Analyzing the relationship between the contaminating ability and geometrical parameters of wax crystals, we found that the contamination is related to both the largest dimension and the largest aspect ratio of the crystals.

Pad contamination was also observed in a study on the waxy surface of the slippery zone in trapping organs (pitchers) of the carnivorous plant *Nepenthes alata* [43.149]. The wax coverage consists of two clearly distinguished superimposed layers, differing in their structure, chemical composition, and mechanical properties. Laboratory experiments with tethered beetles showed that both wax layers essentially reduce the attachment force of insects, however, in different ways. Crystals of the upper wax layer contaminate the insect adhesive pads, whereas the lower wax layer leads to the reduction of the real contact area of the insect feet with the pitcher surface.

43.11 Design Principles

Comparison of a wide variety of animal groups revealed that the size of single contacting points gets smaller and their density increases as the body mass of the animal increases (Figs. 43.1 and 43.11). Additionally, a number of other design principles can be extracted from studies on biological systems (Fig. 43.15). The effective elastic modulus of the fiber arrays is very small, which is of fundamental importance for proper contact formation and adhesion on smooth and rough substrates [43.119, 155]. It is predicted that an additional advantage of patterned surfaces is reliability of contact on various surface profiles [43.24] and increased tolerance to defects at individual contacts. In a real situation, failure of some microcontacts, because of dust particles or mechanical damage of single hairs, would minimally influence adhesion [43.156]. In the case of a solitary contact, even slight damage of the contact due to the presence of contamination or surface irregularities will immediately lead to contact breakage, similar to crack propagation in bulk material [43.157]. These theoretical considerations were proven experimentally: It has been shown that structured surfaces have increased tenacity (adhesion force per unit contact area) [43.156, 158].

Why does a structured surface stick better? The effectiveness of the attachment system may be evaluated by using energy. Contact energy is the difference between adhesion energy and elastic energy, stored in a contact. The structured sample has lower surface rigidity and thus higher flexibility. This increases the ability of the surface to adapt to substrate irregularities, in contrast to a flat sample. A flat sample is able to establish contact only at the tips of substrate irregularities and therefore generates rather low real contact area. Thus, one may formulate an ultimate requirement for an effective attachment system: *The system must reach maximal contact energy at the minimal elastic energy spent for formation of the contact area.*

There are two ways to increase the adhesion energy of a contact. The first is to increase the specific adhesion energy (SAE) for constant contact area. The second is to increase the real contact area for constant SAE. The first approach has some limitations. SAE

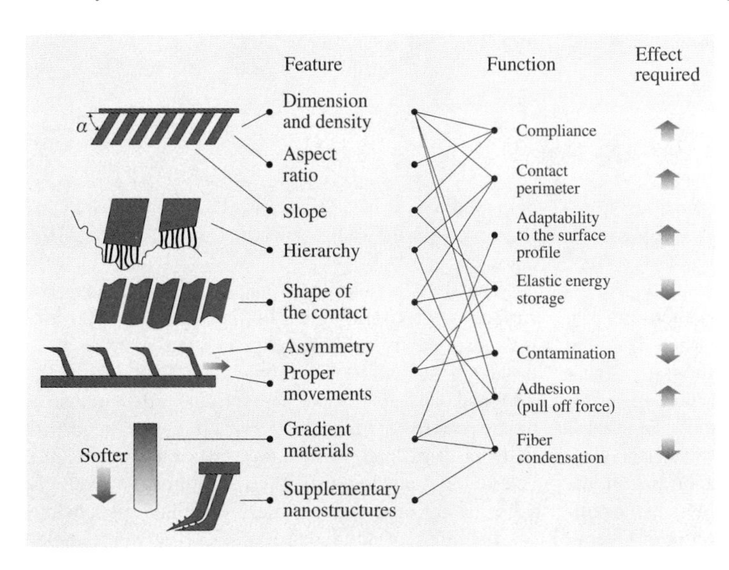

Fig. 43.15 Some functional principles (features) by which reversible biological adhesive systems operate and their relationships with specific functions. The resulting effect required to produce strong adhesion is shown on the *right-hand side* (*up arrow* indicates increase of the effect by a specific design feature). Simultaneous implementation of all these features in one artificial biomimetic system is desirable but hardly possible. However, one principle or a combination of several ones can be implemented, depending on the requirements for a particular biologically inspired material or system

ranges from 50–70 mJ/m^2 (glass, silicon, and GaAs) to 300–400 mJ/m^2 (metals). Additionally, a high SAE results in rather fast passivity of the surface, even under normal conditions. Thus, under natural conditions in the presence of humidity, various gases, and dust, a surface with high SAE will be contaminated very quickly and its SAE will drop rapidly, effecting a decrease of surface stickiness.

Adhesion energy of contact may also be enhanced in a second way, namely by an increase of the real contact area between two surfaces. The relationship between real contact area and geometrical contact area, estimated by measuring the resistance at the interface in metals, is about 1/10 000 [43.159], and it depends mainly on the surface geometry. Surface patterning into structures with high aspect ratio results in higher deformability of the surface, and therefore higher ability to achieve large areas of real contact [43.160], especially with uneven substrates [43.161]. A structured surface has a lower portion of elastic energy stored in the material. Such a decrease of the stored elastic energy of deformation results in an increase of the adhesion energy [43.158]. This approach does not have the disadvantages mentioned above for the first approach. The area of real contact may vary widely, depending on the load applied, and is less dependent on contamination than a surface with high SAE. The presence of several hierarchical levels of structures in biological attachment systems may presumably increase the above effect [43.24, 161].

During pull-off, the fibers may elongate many micrometers before the force in the fiber is large enough to break the bond to the substrate. Since the spring constant associated with a long fiber is very small, the displacement may be very large, leading to very large effective surface energy. Therefore, it was previously suggested that extension of *long bonds* may lead to an increase in the energy dissipation at the interface and thus to an increase in the reliability of the adhesive contact [43.119, 162].

An additional functional principle is related to the spatula-shaped tips of the setae (Figs. 43.1 and 43.12a), which are responsible for proper contact formation with the substrate due to the low bending stiffness of the plates with or without a minimum of normal load [43.120]. While sliding over the surface, thin plate-like spatulae may easily make contact with the surface by adapting to the surface profile and replicating surface irregularities of a certain length scale [43.101].

Some other functional principles according to which reversible biological adhesive systems operate are given in Fig. 43.15. These principles relate to the dimension and density of surface structures, and their aspect ratio and slope [43.163]. Hierarchical design of surface features may enhance adaptability to real surfaces, which normally have fractal roughness. Shape of the contact may aid in tuning pull-off forces at the level of single contact elements. By changing shape one may adjust the adhesive properties of the material to a particular application. Asymmetrical shape of single contact elements in combination with the proper movements may provide a route to switchable adhesives. Additional structured coatings and the use of gradient materials together with properly selected aspect ratio, density, and elastic modulus may prevent condensation of structures [43.155].

43.12 Biomimetics: Where We Are Now

There are several artificial surfaces previously described in the literature that were claimed to improve pull-off forces in contact with a flat surface (Fig. 43.16). These materials have been produced using various micro- and nanofabrication methods ranging from laser technology (Fig. 43.16e,f), carbon nanotube packaging (Fig. 43.16g,h) to various lithography techniques [43.156, 158, 164–168]. Some of these materials do not demonstrate an improvement of adhesion measured in a flat-on-flat scheme. Other materials were strongly limited in the patterned area or/and in the number of adhesion cycles, despite increases in the pull-off forces achieved by the surface patterning. Overall patterned area is usually restricted to a few square centimeters, and the increased adhesion occurs only for a few cycles.

The first large-scale bioinspired fibrillar adhesive surface with overall area in the range of 500 cm^2 was inspired by microstructures found in male beetles from the family Chrysomelidae [43.2, 104, 169]. The pad surface responsible for this effect consists of a pattern of hairs (fibers, pillars) with broad flattened tips and a narrowed flexible region just below the flattened tip. These features, as well as a hexagonal distribution pattern of pillars responsible for the high packing density [43.170], were imple-

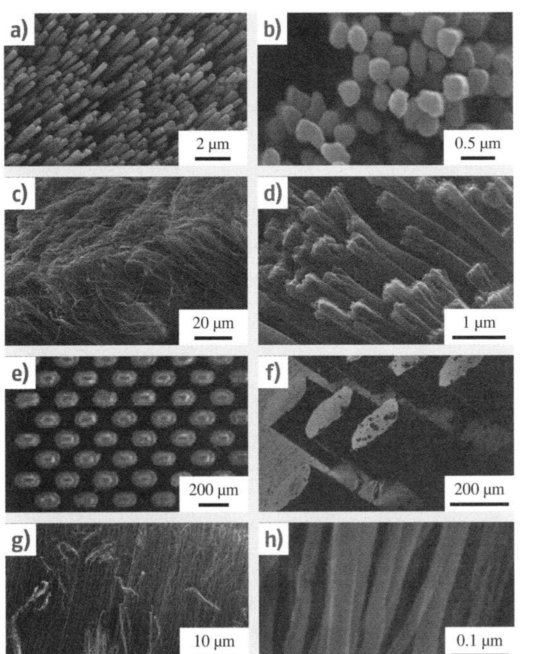

Fig. 43.16a–h Bioinspired fibrillar adhesives of different dimensions. (**a**,**b**) Metallic nanowhiskers; (**c**,**d**) hierarchically branching fiber arrays made of epoxy resin; (**e**,**f**) silicone molds of laser-patterned metal surface; (**g**,**h**) carbon nanotubes ((**a–d**,**g**,**h**) original micrographs) ((**e**,**f**) after [43.158])

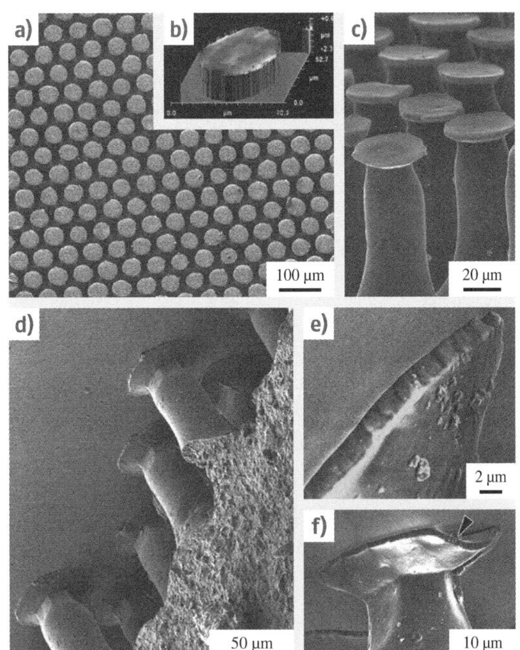

Fig. 43.17a–f Patterned insect-inspired polyvinylsiloxane (PVS) surface. (**a**) Single structures are distributed on the surface according to a hexagonal pattern in order to achieve the highest packaging degree of single pillars (above aspect, SEM image). (**b**) White-light interferometer image of single pillar head demonstrates the almost flat shape of the contacting surface. (**c**) Side aspect of the pillar array. (**d–f**) Behavior of structured PVS surfaces in contact with the glass surface (SEM images). *Black arrowhead* shows a dust particle in contact (after [43.156])

mented in the design of a patterned polymer tape (Fig. 43.17).

The adhesive properties of this tape were characterized using a variety of measurement techniques and compared with those of flat tape made from the same polymer. Compared with the flat tape, the patterned one demonstrated considerably higher adhesion in a peeling test [43.171] and higher pull-off force per unit apparent contact area in measurements according to the flat-to-flat scheme [43.156, 172]. Excellent performance of the patterned polymer tape with similar pillar shape has also been demonstrated elsewhere [43.173]. The structured tape is less sensitive to contamination by dust particles than the flat tape or regular Scotch tape [43.174]. After being contaminated, the structured tape can recover its initial adhesive properties completely, on washing with an aqueous soap solution.

Multiple mechanisms are involved in adhesion enhancement in surfaces patterned in a specific (bioinspired) way. The majority of effects responsible for stronger adhesion of patterned surfaces are related to the enhancement of the contact formation. Lower surface rigidity of structured samples provides greater adaptability to the substrate profile. The presence of several hierarchical levels of structure in biological systems may presumably increase this effect, especially on natural surfaces with naturally rough profiles, where different roughness wavelengths are superimposed. The adaptability of the fibrillar surface can be increased by making single fibers taller, thinner, and softer, but all three of these approaches lead to fiber condensation (lateral collapse, conglutination), which results in adhesion decrease [43.118, 155]. Hierarchical design is the solution that optimizes contact formation with the minimal degree of fiber condensation. Whereas in insects there is only one hierarchical level of outgrowths on the attachment organ, a second level of outgrowths is present in spiders [43.2, 38]. Even more levels have been reported for geckos [43.19, 20, 23, 175]. A flat sample is able to build real contact only at the tips of substrate irregularities and therefore generates only rather low total contact area.

Another functional advantage of patterned adhesion systems is prevention of crack propagation. In a multicontact system, cracks will be stopped at the level of a single contact element, and the new crack has to be formed at each subcontact during detachment. The formation of new cracks requires more energy than crack propagation in a flat-to-flat contact. Therefore, contact breakage of a patterned tape requires higher energy dissipation [43.139]. This behavior is analogous to the fracture mechanics of solids: cracks propagate more easily in uniform materials than in composites [43.176]. Crack propagation behavior on patterned adhesive films leads to significant enhancement of fracture energy [43.117]. In addition to this effect, cracks can be stopped at the level of each single outgrowth in the material, as described above, because of the lip-like rim at the tip of each pillar.

An additional functional principle is related to the spatula- or mushroom-shaped tips of the setae, which are responsible for proper contact formation with the substrate due to the low bending stiffness of the plates, with or without a minimum of a normal load [43.120]. While sliding over the surface, thin plate-like spatulae of insects, spiders or geckos may easily make contact with the surface by adapting to the surface profile and replicating surface irregularities of a certain length scale. Additionally, such thin plate-like structures provide greater adaptability to an uneven surface profile. In the mushroom-shaped biomimetic microstructure, the thin lip-like rim of the pillar tip is able to replicate surface irregularities of a certain length scale and, in combination with a narrowed flexible area below the tip, may adapt to the local slope of the substrate [43.136]. Also the combination of the rim and narrowing below the rim increases contact tolerance against external disturbances. This material demonstrates not only excellent adhesion in flat-to-flat adhesion tests but also rather high tolerance against crack propagation in peeling tests.

Patterned material with mushroom-shaped microstructures has several structural and functional hierarchical levels that together account for its enhanced adhesive properties. The fist level is the thin backing responsible for tape adaptability to surface unevenness. The thickness, and therefore bending stiffness, of the backing has to be as low as possible [43.164]. The second level is represented by pillars, which provide adaptability to surface features at the level of dozens of micrometers. The third level is the flexible narrowing aiding in adjustment of the terminal plate to the local substrate slope. Finally, the thin rim or lip is capable of replication of roughness and dust particles at the level of single micrometers. A certain degree of redundancy of a number of hierarchical levels makes the tape more robust on the nonsmooth profile of the real surface.

The first experimental evidence on contamination reduction in biological hairy adhesive systems has been provided recently for the gecko system [43.177]. This effect was explained by better adhesion of contaminating particles to the substrate than to gecko setae, because of the larger contact between particle and substrate than between particle and gecko spatulae. Similar effects are probably applicable to the patterned polymer surfaces. Another effect reducing contamination of adhesive patterned surface is sinking of dust particles into the gaps between pillars [43.156]. However, another important property of the patterned tape is stronger adhesion even at a relatively strong degree of contamination. This effect relies on the greater adaptability of the flexible rim of each pillar, even if contact formation of single pillars is hampered due to small dust particles. Adaptability of the lip-like margins of terminal elements provides an additional tolerance to contamination by small dust particles. Pressure-sensitive adhesives fail much faster after a number of adhesive cycles, compared with microstructured silicone tape [43.174]. Pressure-sensitive adhesives fail after washing, whereas silicone tape recovers completely.

Patterned polymers have great potential for underwater adhesion. The adhesive strength of patterned tape under water can be additionally enhanced with the aid of soft sticky polymeric coatings [43.178]. Another approach to generate strong adhesion under water is to use a mushroom-like shape at the tip of each pillar [43.179] (Fig. 43.18).

The industry of adhesives is presently following three main goals [43.180]:

1. Increase in the reliability of glued contact
2. Mimicking of natural, environment-friendly glues
3. Development of mechanisms for application of a minute amount of glue to the surface

An additional challenge is the use of substances or mechanisms that allow multiple attachments and detachments, and enable attachment to the broadest variety of surfaces. Many biological attachment devices correspond to some of these requirements. Hairy and smooth leg attachment pads are promising candidates for biomimetics of robot soles adapted for locomotion. Similar principles can be applied to the design of micro-gripper mechanisms with the ability to adapt to a variety of surface profiles.

Robots that could climb smooth and complex inclined terrains like insects and lizards would have many applications such as exploration, inspection or cleaning [43.181, 182]. Walking machines usually use suckers to hold onto vertical surfaces and under a surface. A primary disadvantage of this attachment principle is the large energy consumption required for vacuum maintenance. Novel, biologically inspired materials may enable future robots to walk on smooth surfaces regardless of the direction of gravity. Mini-Whegs [43.183], a small robot (120 g) that uses four-wheel legs for locomotion, was recently converted to a wall-walking robot with compliant, adhesive feet [43.171, 174]. The robot is capable of ascending vertical smooth glass surfaces using micropatterned adhesive. The foot material maintains its properties for nearly twice as many walking cycles before becoming contaminated. Similar efforts to apply knowledge from biological adhesive systems to robotics have been undertaken by Stanford's robotic group. In the Stickybot robot, which is able to walk on the wall using structured adhesive polymers, a gecko-like peeling principle of foot detachment is excellently applied [43.184].

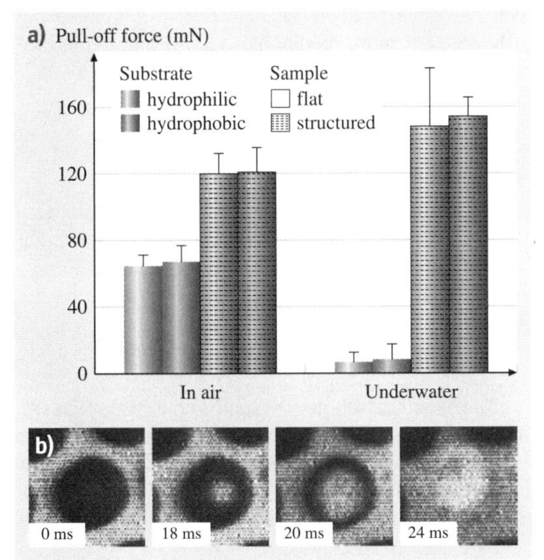

Fig. 43.18a,b Adhesive performance of mushroom-shaped fibrillar microstructure. (**a**) Mean pull-off force of flat and structured samples measured on hydrophilic and hydrophobic substrates in air and deionized water. The *error scales* represent standard deviation from mean values. (**b**) Behavior of a single terminal contact plate detaching from a substrate in air. *Dark areas*, resulting from destructive interference of reflected white light in the glass–silicone interface, visualize the real contact zones (after [43.179])

Observations of insects and geckos have inspired the kinematics of the legs in the Mini-Whegs wall climbing robot. Flies make initial contact with the entire broad, flexible attachment organ (pulvillus) [43.185]. A slight shear component is present in the movement, which provides a preload to the surface of the attachment device. A similar shearing motion has been previously described as part of the attachment mechanism of a single gecko seta [43.20]. Minimal force expenditure during detachment by peeling is also important [43.22, 186, 187]. Disconnecting the entire attachment organ at once requires a strong adhesive force to be overcome, which is energetically disadvantageous for an animal.

There are a lot of activities worldwide aimed at developing surfaces with enhanced or controllable adhesion and friction. Such biomimetic products as Velcro fasteners or 3M Dual-Lock or Scotchmate are well known. Recently, the company Continental has developed a winter tire with honeycomb profiles similar to those on the attachment pads of the grasshopper [43.113, 188] and tree frogs [43.189]. The company promises enhanced wear performance on dry road, less aquaplaning and better braking on wet roads, substantially improved lateral guidance, and better grip and more traction on ice. However, these structures even more effective at the micrometer level, in order to control friction on wet surfaces and prevent stick-and-slip behavior of dry surfaces [43.190].

For materials scientists, results obtained on biological objects emphasize the necessity to couple the inherent material properties of the adhering material with the geometry of the contact [43.155, 191]. The efficiency of natural systems cannot, of course, be copied directly, but some of the concepts can be translated to the materials world to design surfaces with particular properties and functions observed in biological systems. We believe that the huge diversity of biological attachment mechanisms will continuously inspire material scientists and engineers to develop new materials and systems (Fig. 43.19). This is why broad

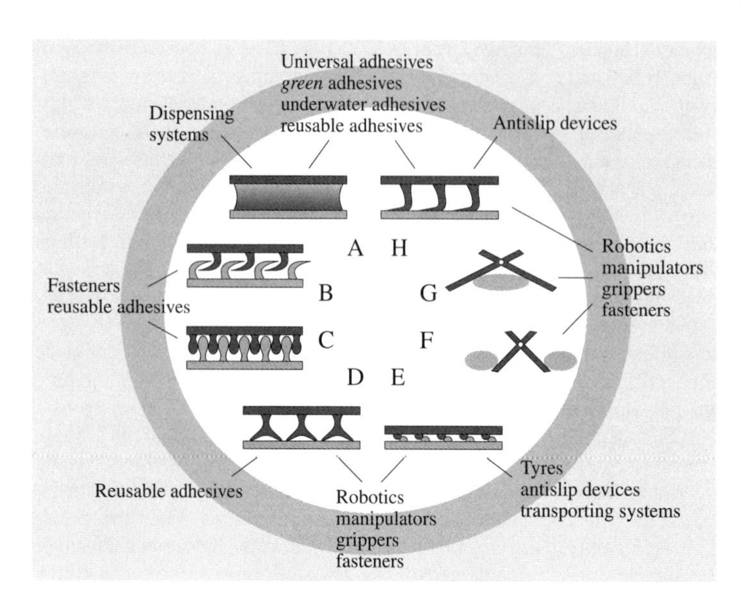

Fig. 43.19 Areas of existing and potential biomimetic applications based on studies of biological attachment devices. A – wet adhesion; B – hooks; C – lock; D – suction cups; E – friction; F – spacer; G – clamp; H – dry adhesion

functional comparative studies on biological surfaces have to be intensified, in order to extract the essential structural, chemical, and mechanical principles behind their functions. Using living nature as an endless source of inspiration might be another reason for saving biological diversity.

43.13 Conclusions

Real biological attachment systems usually rely on a combination of different functional principles based on elaborate micro- and nanostructures.

Attachment systems in biology can be subdivided according to their biological function and the basic physical forces responsible for enhancement of contact forces.

There are three main functional groups of biological attachment devices depending on the time scale over which they operate: permanent, temporary, and transitory attachment.

Reversible attachment devices used for locomotion may be hairy or smooth. Hairy systems consist of fine microstructures, whereas smooth systems are composed of materials of specific inner structure. Due to the flexibility of the material of the attachment pads or fine surface structures, both mechanisms can maximize the possible contact area with a wide range of substrate profiles.

Recent phylogenetic analyses have shown that similar functional solutions for attachment have evolved several times independently in the evolutionary history of animals.

In hairy systems, the density of single contact elements increases strongly with increasing body weight of the animal group. From scaling analysis, we have suggested that animal lineages relying on dry adhesion (lizards, spiders) usually possess much higher density of terminal contact elements compared with systems using wet adhesive mechanism (insects).

The lizard attachment system is based on van der Waals forces and wetting phenomena caused by water absorbed on the surface. In insects, beside van der Waals and Coulomb forces, attractive capillary and viscous forces mediated by pad secretion are a critical factor in their attachment mechanisms.

The ultimate requirement for effective attachment is that the system must achieve maximal contact energy for minimal elastic energy spent for the formation of the contact area.

The effective elastic modulus of fiber arrays is very small, which is of fundamental importance for proper contact formation and adhesion on smooth and rough substrates.

In a multicontact system, cracks will be stopped at the level of a single contact element, and new cracks have to form at each subcontact during detachment. The formation of new cracks affects energy dissipation, which prevents contact breakage.

Hierarchical design of surface features in attachment pads may enhance adaptability to real surfaces, which normally have fractal roughness.

Asymmetrical shape of single contact elements in combination with proper movements may provide a route to switchable adhesives.

Hairy pads show a clear-cut minimum of adhesion at certain ranges of substrate roughness. Such a critical range of roughness depends on the relationship between the diameter of single contact elements of the pad and the length scale of the roughness.

During their evolution, plants have not only developed structures to attract pollinators but also a wide variety of structural and chemical attributes of their surfaces related to preventing the adhesion of herbivores.

To explain the antiadhesive properties of plants structured with crystalline waxes four hypotheses were proposed, based on roughness, contamination, wax dissolution, and fluid absorption.

The self-cleaning effect of fiber arrays is explained by the better adhesion of contaminating particles to the substrate than to the array, because of the larger contact between particle and substrate than between particle and fiber array. Another effect reducing contamination of adhesive patterned surface is sinking dust particles into the gaps between fibers.

Since many functional effects found in biological systems are based on fundamental physical principles and are mostly related to the geometry of the structure, they must also hold for artificial surfaces with similar geometry.

The adhesive strength of patterned tape under water can be additionally enhanced with the aid of soft sticky polymeric coatings. Another way to generate strong adhesion under water is to use a mushroom-like shape at the tip of each pillar.

References

43.1 W. Nachtigall: *Biological Mechanisms of Attachment* (Springer, Berlin, Heidelberg 1974)

43.2 S.N. Gorb: *Attachment Devices of Insect Cuticle* (Kluwer Academic, Dordrecht 2001)

43.3 D.J. Irschick, C.C. Austin, K. Petren, R.N. Fisher, J.B. Losos, O. Ellers: A comparative analysis of clinging ability among pad-bearing lizards, Biol. J. Linn. Soc. **59**, 21–35 (1996)

43.4 R.G. Beutel, S.N. Gorb: Ultrastructure of attachment specializations of hexapods (*Arthropoda*): evolutionary patterns inferred from a revised ordinal phylogeny, J. Zool. Syst. Evol. Res. **39**, 177–207 (2001)

43.5 S.N. Gorb, R.G. Beutel: Evolution of locomotory attachment pads of hexapods, Naturwissenschaften **88**, 530–534 (2001)

43.6 S.N. Gorb: Uncovering insect stickiness: Structure and properties of hairy attachment devices, Am. Entomol. **51**, 31–35 (2005)

43.7 M. Scherge, S.N. Gorb: *Biological Micro- and Nanotribology* (Springer, Berlin, Heidelberg 2001)

43.8 S.N. Gorb: Biological attachment devices: Exploring nature's diversity for biomimetics, Philos. Trans. R. Soc. A **366**, 1557–1574 (2008)

43.9 P. Flammang: Adhesion in echinoderms, Echinoderm Stud. **5**, 1–60 (1996)

43.10 J.H. Waite: Adhesion in byssally attached bivalves, Biol. Rev. **58**, 209–231 (1983)

43.11 J.H. Waite: Mussel byssal adhesion. In: *Marine Biodeterioration. Advanced Techniques Applicable to the Indian Ocean*, ed. by M.F. Thompson, R. Sarojini, R. Nagabhushanam (Balkema, Rotterdam 1988) pp. 35–43

43.12 R. Santos, S.N. Gorb, V. Jamar, P. Flammang: Adhesion of echinoderm tube feet to rough surfaces, J. Exp. Biol. **208**, 2555–2567 (2005)

43.13 A.M. Goodman, M.J. Crook, A.R. Ennos: Anchorage mechanics of the tap root system of winter-sown oilseed rape (*Brassica napus* L.), Ann. Bot. **87**, 397–404 (2001)

43.14 S.N. Gorb, M.A. Landolfa, F.G. Barth: Dragline associated behaviour of the orb web spider *Nephila clavipes* (*Araneoidea*, *Tetragnathidae*), J. Zool. Lond. **244**, 323–330 (1998)

43.15 H. Ohta: Electron microscopic study on adhesive material of Pacific herring (*Clupea pallasi*) eggs, Jpn. J. Ichthyol. **30**, 404–411 (1984)

43.16 R.A. Patzner, R. Glechner: Attaching structures in eggs of native fishes, Limnologica **26**, 179–182 (1996)

43.17 A.R. Panizzi: Possible egg positioning and gluing behavior by ovipositing southern green stink bug *Nezara viridula* (L.) (*Hetereroptera*: *Pentatomidae*), Neotrop. Entomol. **35**, 149–151 (2006)

43.18 V. Radhakrishnan: Locomotion: dealing with friction, Curr. Sci. **74**, 826–840 (1998)

43.19 U. Hiller: Untersuchungen zum Feinbau und zur Funktion der Haftborsten von Reptilien, Z. Morphol. Tiere **62**, 307–362 (1968), in German

43.20 K. Autumn, Y.A. Liang, S.T. Hsieh, W. Zesch, W.P. Chan, T.W. Kenny, R. Fearing, R.J. Full: Adhe-

sive force of a single gecko foot-hair, Nature **405**, 681–685 (2000)
43.21 S.N. Gorb, R. Beutel, E.V. Gorb, Y. Jiao, V. Kastner, S. Niederegger, V.L. Popov, M. Scherge, U. Schwarz, W. Vötsch: Structural design and biomechanics of friction-based releasable attachment devices in insects, Integr. Comp. Biol. **42**, 1127–1139 (2002)
43.22 H. Gao, X. Wang, H. Yao, S.N. Gorb, E. Arzt: Mechanics of hierarchical adhesion structures of geckos, Mech. Mater. **37**, 275–285 (2005)
43.23 G. Huber, S.N. Gorb, R. Spolenak, E. Arzt: Resolving the nanoscale adhesion of individual gecko spatulae by atomic force microscopy, Biol. Lett. **1**, 2–4 (2005)
43.24 B. Bhushan, A.G. Peressadko, T.-W. Kim: Adhesion analysis of two-level hierarchical morphology in natural attachment systems for 'smart adhesion', J. Adhes. Sci. Technol. **20**, 1475–1491 (2006)
43.25 Y. Tian, N. Pesika, H. Zeng, K. Rosenberg, B. Zhao, P. McGuiggan, K. Autumn, J. Israelachvili: Adhesion and friction in gecko toe attachment and detachment, Proc. Natl. Acad. Sci. USA **103**, 19320–19325 (2006)
43.26 S. Ishii: Adhesion of a leaf feeding ladybird *Epilachna vigintioctomaculata* (*Coleoptera*: *Coccinellidae*) on a vertically smooth surface, Appl. Entomol. Zool. **22**, 222–228 (1987)
43.27 A. Kosaki, R. Yamaoka: Chemical composition of footprints and cuticula lipids of three species of lady beetles, Jpn. J. Appl. Entomol. Zool. **40**, 47–53 (1996)
43.28 A.B. Attygalle, D.J. Aneshansley, J. Meinwald, T. Eisner: Defence by foot adhesion in a chrysomelid beetle (*Hemisphaerota cyanea*): characterization of the adhesive oil, Zoology **103**, 1–6 (2000)
43.29 W. Federle, M. Riehle, A.S.G. Curtis, R.J. Full: An integrative study of insect adhesion: mechanics and wet adhesion of pretarsal pads in ants, Integr. Comp. Biol. **42**, 1100–1106 (2002)
43.30 W. Vötsch, G. Nicholson, R. Müller, Y.-D. Stierhof, S.N. Gorb, U. Schwarz: Chemical composition of the attachment pad secretion of the locust *Locusta migratoria*, Insect Biochem. Mol. Biol. **32**, 1605–1613 (2002)
43.31 H. Homann: Haften Spinnen an einer Wasserhaut?, Naturwissenschaften **44**, 318–319 (1957), in German
43.32 K. Autumn, A.M. Peattie: Mechanisms of adhesion in geckos, Integr. Comp. Biol. **42**, 1081–1090 (2002)
43.33 G. Huber, H. Mantz, R. Spolenak, K. Mecke, K. Jacobs, S.N. Gorb, E. Arzt: Evidence for capillarity contributions to gecko adhesion from single spatula nanomechanical measurements, Proc. Natl. Acad. Sci. USA **102**, 16293–16296 (2005)
43.34 S.N. Gorb, F.G. Barth: Locomotor behaviour during prey-capture of a fishing spider *Dolomedes plantarius* (*Araneae*: *Araneidae*): Galloping and stopping, J. Arachnol. **22**, 89–93 (1994)
43.35 Z. Dai, S.N. Gorb, U. Schwarz: Roughness-dependent friction force of the tarsal claw system in the beetle *Pachnoda marginata* (*Coleoptera*, *Scarabaeidae*), J. Exp. Biol. **205**, 2479–2488 (2002)
43.36 D. Voigt, J.M. Schuppert, S. Dattinger, S.N. Gorb: Sexual dimorphism in the attachment ability of the Colorado potato beetle *Leptinotarsa decemlineata* (*Coleoptera*: *Chrysomelidae*) to rough substrates, J. Insect Physiol. **54**, 765–776 (2008)
43.37 O. Betz, G. Kölsch: The role of adhesion in prey capture and predator defence in arthropods, Arthropod Struct. Dev. **33**, 3–30 (2004)
43.38 R. Foelix: *The Biology of Spiders* (Harvard Univ. Press, Cambridge 1982)
43.39 C. Weirauch: Pretarsal structures in *Reduviidae* (*Heteroptera*, *Insecta*), Acta Zool. **86**, 91–110 (2005)
43.40 U. Hintzpeter, T. Bauer: The antennal setal trap of the ground beetle *Loricera pilicornis*: A spezialisation for feeding on *Collembola*, J. Zool. Lond. Ser. A **208**, 615–630 (1986)
43.41 T. Bauer, M. Kredler: Adhesive mouthparts in a ground beetle larva (*Coleoptera*, *Carabidae*, *Loricera pilicornis* F.) and their function during predation, Zool. Anz. **221**, 145–156 (1988)
43.42 O. Betz: Function and evolution of the adhesion-capture apparatus of *Stenus* species (*Coleoptera*, *Staphylinidae*), Zoomorphology **116**, 15–34 (1996)
43.43 B.D. Opell: Changes in spinning anatomy and thread stickiness asociated with the origin of orb-weaving spiders, Biol. J. Linn. Soc. **68**, 593–612 (1999)
43.44 A.C. Hawthorn, B.D. Opell: Evolution of adhesive mechanisms in cribellar spider prey capture thread: Evidence for van der Waals and hygroscopic forces, Biol. J. Linn. Soc. **77**, 1–8 (2002)
43.45 A.C. Hawthorn, B.D. Opell: van der Waals and hygroscopic forces of adhesion generated by spider capture threads, J. Exp. Biol. **206**, 3905–3911 (2003)
43.46 P.M. Hammond: Wing-folding mechanism of beetles, with special reference to investigations of adephagan phylogeny (*Coleoptera*). In: *Carabid Beetles: Their Evolution, Natural History, and Classification*, ed. by T. Ervin, G.E. Ball, D.R. Whitehead (Junk, The Hague 1989) pp. 113–180
43.47 G.A. Samuelson: Binding sites: elytron-to-body meshing structures of possible significance in the higher classification of *Chrysomeloidea*. In: *Chrysomelidae Biology, the Classification, Phylogeny and Genetics*, ed. by P.H.A. Jolivet, M.L. Cox (SPB Academic, Amsterdam 1996) pp. 267–290
43.48 S.N. Gorb: Frictional surfaces of the elytra to body arresting mechanism in tenebrionid beetles (*Coleoptera*: *Tenebrionidae*): Design of co-opted fields of microtrichia and cuticle ultrastructure, Int. J. Insect Morphol. Embryol. **27**, 205–225 (1998)
43.49 S.N. Gorb: Ultrastructure of the thoracic dorso-medial field (TDM) in the elytra-to-body arresting

mechanism in tenebrionid beetles (*Coleoptera*: *Tenebrionidae*), J. Morphol. **240**, 101–113 (1999)

43.50 P.J. Perez Goodwyn, S.N. Gorb: Attachment forces of the hemelytra-locking mechanisms in aquatic bugs (*Heteroptera*: *Belostomatidae*), J. Insect Physiol. **49**, 753–764 (2003)

43.51 S.N. Gorb: Evolution of the dragonfly head-arresting system, Proc. R. Soc. B **266**, 525–535 (1999)

43.52 H. Mittelstaedt: Physiologie des Gleichgewichtssinnes bei fliegenden Libellen, Z. Vergl. Physiol. **32**, 442–463 (1950), in German

43.53 P.J. Perez Goodwyn, S.N. Gorb: Frictional properties of contacting surfaces in the hemelytra-hindwing locking mechanism in the bug *Coreus marginatus* (*Heteroptera*, *Coreidae*), J. Comp. Physiol. A **190**, 575–580 (2004)

43.54 S.N. Gorb, V.L. Popov: Probabilistic fasteners with parabolic elements: biological system, artificial model and theoretical considerations, Philos. Trans. R. Soc. A **360**, 211–225 (2002)

43.55 O. Törne: Die Saugnäpfe der männlichen *Dytiscidae*, Zool. Jahrb. Anat. **29**, 415–448 (1910), in German

43.56 R.B. Aiken, A. Khan: The adhesive strength of the palettes of males of a boreal water beetle, *Dytiscus alaskanus* J. Balfour Browne (*Coleoptera*: *Dytiscidae*), Can. J. Zool. **70**, 1321–1324 (1992)

43.57 K.B. Miller: The phylogeny of diving beetles (*Coleoptera*: *Dytiscidae*) and the evolution of sexual conflict, Biol. J. Linn. Soc. **79**, 359–388 (2003)

43.58 Y. Pelletier, Z. Smilowitz: Specialized tarsal hairs on adult male Colorado potato beetles *Leptinotarsa decemlineata* (Say), hamper its locomotion on smooth surfaces, Can. Entomol. **119**, 1139–1142 (1987)

43.59 K. Schönitzer: Comparative morphology of the antenna cleaner in bees (*Apoidea*), Z. Zool. Syst. Evolutionsforsch. **24**, 35–51 (1986)

43.60 K. Schönitzer, G. Lawitzky: A phylogenetic study of the antenna cleaner in *Formicidae*, *Mutillidae*, and *Tiphiidae* (*Insecta*, *Hymenoptera*), Zoomorphology **107**, 273–285 (1987)

43.61 K. Schönitzer, M. Penner: The function of the antenna cleaner of the honey-bee (*Apis mellifica*), Apidologie **15**, 23–32 (1984)

43.62 A. Francouer, R. Loiselle: Evolution du strigile chez les formicides (Hymenopteres), Nat. Can. **115**, 333–335 (1988), in French

43.63 W. Arens: Comparative functional morphology of the mouthparts of stream animals feeding on epilithic algae, Arch. Hydrobiol. Suppl. **83**, 253–354 (1989)

43.64 J.J. Pasteels, J.M. Pasteels: Les soies cuticulaires des *Megachilidae* (*Apoidea*) vues au microscope electronique a balayage, Acad. R. Belgique, Memoires de la Classe des Sciences, Collection 4 PO S. 2 Ser. (Palais des Academies, Brussels 1972) pp. 1–24, in French

43.65 M. Hesse: Auf welche Weise transportieren Insekten den Blütenstaub?, Linz. Biol. Beitr. **13**, 50 (1981), in German

43.66 H.W. Krenn, J.D. Plant, N.U. Szucsich: Mouthparts of flower-visiting insects, Arthropod Struct. Dev. **34**, 1–40 (2005)

43.67 G.S. Gracham-Smith: Further observations on the anatomy and function of the proboscis of the blow-fly, *Calliphora erythrocephala* L, Parasitology **22**, 47–115 (1930)

43.68 V.G. Dethier: *The Physiology of Insect Senses* (Wiley, New York 1963)

43.69 V.F. Zaytsev: Microstructure of the labella of the fly proboscis. II. Pseudotracheal framework; structure and evolution, Entomol. Rev. **63**, 35–39 (1984)

43.70 R.J. Elzinga, A.B. Broce: Labellar modifications of *Muscomorpha* flies (*Diptera*), Ann. Entomol. Soc. Am. **79**, 150–209 (1986)

43.71 C.A. Driscoll, M.A. Condon: Labellar modifications of *Blepharoneura* (*Diptera*: *Tephritidae*): neotropical fruit flies that damage and feed on plant surfaces, Ann. Entomol. Soc. Am. **87**, 448–453 (1994)

43.72 S. Tyler: The role of function in determination of homology and convergence – Examples from invertebrate adhesive organs, Fortschr. Zool. **36**, 331–347 (1988)

43.73 M.W. Denny: A quantitative model for the adhesive locomotion of the terrestrial slug, *Ariolimax columbianus*, J. Exp. Biol. **91**, 195–217 (1981)

43.74 M.W. Denny, J.M. Gosline: The physical properties of the pedal mucus of the terrestrial slug *Ariolimax columbianus*, J. Exp. Biol. **88**, 375–393 (1980)

43.75 G. Habenicht: *Kleben: Grundlagen, Technologien, Anwendung* (Springer, Berlin, Heidelberg 2002), in German

43.76 M.J. Naldrett: Adhesives. In: *Biomechanics – Materials: A Practical Approach*, ed. by J.F.V. Vincent (IRL, Oxford 1992) pp. 219–240

43.77 Z. Kabata: Copepoda parasitic on Australian fishes. VII. *Shiinoa occlusa* gen. et sp. nov., J. Nat. Hist. **2**, 497–504 (1968)

43.78 Z. Kabata, B. Cousens: The structure of the attachment organ of *Lernaeopodidae* (*Crustacea*: *Copepoda*), J. Fish. Res. Board Can. **29**, 1015–1023 (1972)

43.79 P. Hunter, R.M.T. Rosario: Associations of *Mesostigmata* with other arthropods, Annu. Rev. Entomol. **33**, 393–417 (1988)

43.80 R. Barbier: Morphogenese et evolution de la cuticle et des crochets des fauses-pattes, au cours du developpement larvaire de la *Galleria mellonella* L. (*Lepidoptera*, *Pyralidae*), Bull. Soc. Zool. Fr. **110**, 205–221 (1985)

43.81 E.S. Nielsen, I.F.B. Common: *Lepidoptera*. In: *The Insects of Australia*, ed. by CSIRO (Cornell Univ. Press, New York 1991) pp. 817–916
43.82 I. Hasenfuss: The adhesive devices in larvae of *Lepidoptera* (*Insecta*, *Pterygota*), Zoomorphology **119**, 143–162 (1999)
43.83 F. Gallenmüller, T. Steinbrecher, D. Voigt, C. Weisskopf, T. Speck, S.N. Gorb: Ecobiomechanics of *Galium aparine* L., Fifth Plant Biomech. Conf., Stockholm, Vol. 1 (STFI-Packforsk AB, Stockholm 2006) pp. 37–42
43.84 E.V. Gorb, S.N. Gorb: Contact separation force of the fruit burrs in four plant species adapted to dispersal by mechanical interlocking, Plant Physiol. Biochem. **40**, 373–381 (2002)
43.85 W.P. McCafferty, D.W. Bloodgood: The female and male coupling apparatus in *Tortopus* mayflies, Aquat. Insects **11**, 141–146 (1989)
43.86 A.M. Shapiro, A.H. Porter: The lock-and-key hypothesis: Evolutionary and biosystematic interpretation of insect genitalia, Annu. Rev. Entomol. **34**, 231–245 (1989)
43.87 S.N. Gorb: Design of the predatory legs of water bugs (*Hemiptera*: *Nepidae*, *Naucoridae*, *Notonectidae*, *Gerridae*), J. Morphol. **223**, 289–302 (1995)
43.88 L.I. Frantsevich: The coxal articulation of the striking leg: A comparative study, J. Morphol. **236**, 127–138 (1998)
43.89 R.G. Loxton, I. Nicholls: The functional morphology of the praying mantis forelimb (*Dictyoptera*: *Mantodea*), Zool. J. Linn. Soc. **66**, 185–203 (1979)
43.90 W.J. Costello, C.K. Govind: Contractile proteins of fast and slow fibers during differentiation of lobster claw muscle, Dev. Biol. **104**, 434–440 (1984)
43.91 S.N. Gorb: Design of insect unguitractor apparatus, J. Morphol. **230**, 219–230 (1996)
43.92 S.N. Gorb: The jumping mechanism of cicada *Cercopis vulnerata* (*Auchenorrhyncha*, *Cercopidae*): skeleton-muscle organisation, frictional surfaces, and inverse-kinematic model of leg movements, Arthropod Struct. Dev. **33**, 201–220 (2004)
43.93 A.D. Austin, T.O. Browning: A mechanism for movements of eggs along insect ovipositors, Int. J. Insect Morphol. Embryol. **10**, 93–108 (1981)
43.94 L. Weiss: A biophysical consideration of cell contact phenomena. In: *Adhesion in Biological Systems*, ed. by R.S. Manly (Academic, New York 1970) pp. 1–14
43.95 M.S. Steinberg: Adhesion in development: an historical overview, Dev. Biol. **180**, 377–388 (1996)
43.96 C.J. Strange: Biological ties that bind, Bioscience **47**, 5–8 (1997)
43.97 W.A. Corpe: Attachment of marine bacteria to solid surfaces. In: *Adhesion in Biological Systems*, ed. by R.S. Manly (Academic, New York 1970) pp. 73–87
43.98 J.A. Callow, M.E. Callow: The *Ulva* spore adhesive system. In: *Biological Adhesives*, ed. by A.M. Smith, J.A. Callow (Springer, Berlin, Heidelberg 2006) pp. 63–78
43.99 M.J. Naldrett: The importance of sulphur cross-links and hydrophobic interactions in the polymerization of barnacle cement, J. Mar. Biol. Assoc. UK **73**, 689–702 (1993)
43.100 S. Niederegger, S.N. Gorb, Y. Jiao: Contact behaviour of tenent setae in attachment pads of the blowfly *Calliphora vicina* (*Diptera*, *Calliphoridae*), J. Comp. Physiol. A **187**, 961–970 (2002)
43.101 T. Eimüller, P. Guttmann, S.N. Gorb: Terminal contact elements of insect attachment devices studied by transmission x-ray microscopy, J. Exp. Biol. **211**, 1958–1963 (2008)
43.102 E. Bauchhenss, M. Renner: Pulvillus of *Calliphora erythrocephala* Meig (*Diptera*; *Calliphoridae*), Int. J. Insect Morphol. Embryol. **6**, 225–227 (1977)
43.103 A.F.G. Dixon, P.C. Croghan, R.P. Gowing: The mechanism by which aphids adhere to smooth surfaces, J. Exp. Biol. **152**, 243–253 (1990)
43.104 N.E. Stork: Experimental analysis of adhesion of *Chrysolina polita* (*Chrysomelidae*, *Coleoptera*) on a variety of surfaces, J. Exp. Biol. **88**, 91–107 (1980)
43.105 V.B. Wigglesworth: How does a fly cling to the under surface of a glass sheet?, J. Exp. Biol. **129**, 363–367 (1987)
43.106 S. Niederegger, S.N. Gorb: Tarsal movements in flies during leg attachment and detachment on a smooth substrate, J. Insect Physiol. **49**, 611–620 (2003)
43.107 M.G. Langer, J.P. Ruppersberg, S.N. Gorb: Adhesion forces measured at the level of a terminal plate of the fly's seta, Proc. R. Soc. Lond. Ser. B **271**, 2209–2215 (2004)
43.108 R.G. Beutel, S.N. Gorb: A revised interpretation of attachment structures in *Hexapoda* with special emphasis on *Mantophasmatodea*, Arthropod Syst. Phyl. **64**, 3–25 (2006)
43.109 R.G. Beutel, S.N. Gorb: Evolutionary scenarios for unusual attachment devices of *Phasmatodea* and *Mantophasmatodea* (*Insecta*), Syst. Entomol. **33**, 501–510 (2008)
43.110 S. Niederegger, S.N. Gorb: Friction and adhesion in the tarsal and metatarsal scopulae of spiders, J. Comp. Physiol. A **192**, 1223–1232 (2006)
43.111 U.D. Kendall: The anatomy of the tarsi of *Schistocerca gregaria* Forskål, Z. Zellforsch. **109**, 112–137 (1970)
43.112 B. Henning: Morphologie und Histologie der Tarsen von *Tettigonia viridissima* L. (*Orthoptera*, *Ensifera*), Z. Morphol. Tiere **79**, 323–342 (1974), in German
43.113 S.N. Gorb, Y. Jiao, M. Scherge: Ultrastructural architecture and mechanical properties of attachment pads in *Tettigonia viridissima* (*Orthoptera*, *Tettigoniidae*), J. Comp. Physiol. A **186**, 821–831 (2000)
43.114 P.J. Perez Goodwyn, A. Peressadko, H. Schwarz, V. Kastner, S. Gorb: Material structure, stiffness, and adhesion: Why attachment pads of the

grasshopper (*Tettigonia viridissima*) adhere more strongly than those of the locust (*Locusta migratoria*) (Insecta: Orthoptera), J. Comp. Physiol. A **192**, 1233–1243 (2006)
43.115 E. Arzt, S.N. Gorb, R. Spolenak: From micro to nano contacts in biological attachment devices, Proc. Natl. Acad. Sci. USA **100**, 10603–10606 (2003)
43.116 C.-Y. Hui, N.J. Glassmaker, T. Tang, A. Jagota: Design of biomimetic fibrillar interfaces. 2. Mechanics of enhanced adhesion, J. R. Soc. Lond. Interface **1**, 35–48 (2004)
43.117 J.Y. Chung, M.K. Chaudhury: Roles of discontinuities in bio-inspired adhesive pads, J. R. Soc. Lond. Interface **2**, 55–61 (2005)
43.118 A. Jagota, S.J. Bennison: Mechanics of adhesion through a fibrillar microstructure, Integr. Comp. Biol. **42**, 1140–1145 (2002)
43.119 B.N.J. Persson: On the mechanism of adhesion in biological systems, J. Chem. Phys. **118**, 7614–7621 (2003)
43.120 B.N.J. Persson, S.N. Gorb: The effect of surface roughness on the adhesion of elastic plates with application to biological systems, J. Chem. Phys. **119**, 11437–11444 (2003)
43.121 S.N. Gorb, E.V. Gorb: Ontogenesis of the attachment ability in the bug *Coreus marginatus* (*Heteroptera, Insecta*), J. Exp. Biol. **207**, 2917–2924 (2004)
43.122 K.L. Johnson, K. Kendall, A.D. Roberts: Surface energy and the contact of elastic solids, Proc. R. Soc. Lond. Ser. A **324**, 301–313 (1971)
43.123 S.N. Gorb, E.V. Gorb, V. Kastner: Scale effects on the attachment pads and friction forces in syrphid flies, J. Exp. Biol. **204**, 1421–1431 (2001)
43.124 A.M. Peattie, R.J. Full: Phylogenetic analysis of the scaling of wet and dry biological fibrillar adhesives, Proc. Natl. Acad. Sci. USA **104**, 18595–18600 (2007)
43.125 A.G. Richards, P.A. Richards: The cuticular protuberances of insects, Int. J. Insect Morphol. Embryol. **8**, 143–157 (1979)
43.126 E. Bauchhenss: Die Pulvillen von *Calliphora erythrocephala* Meig. (*Diptera, Brachycera*) als Adhäsionsorgane, Zoomorphologie **93**, 99–123 (1979), in German
43.127 S.N. Gorb: The design of the fly adhesive pad: distal tenent setae are adapted to the delivery of an adhesive secretion, Proc. R. Soc. Lond. Ser. B **265**, 747–752 (1998)
43.128 J.S. Edwards, M. Tarkanian: The adhesive pads of *Heteroptera*: a re-examination, Proc. R. Entomol. Soc. Lond. Ser. A **45**, 1–5 (1970)
43.129 G. Walker, A.B. Yule, J. Ratcliffe: The adhesive organ of the blowfly, *Calliphora vomitoria*: A functional approach (*Diptera: Calliphoridae*), J. Zool. Lond. **205**, 297–307 (1985)
43.130 S.N. Gorb: Fly microdroplets viewed big: A cryo-SEM approach, Microsc. Today **14**, 38–39 (2006)
43.131 M.G. Langer, J.P. Ruppersberg, S.N. Gorb: Adhesion forces measured at the level of a terminal plate of the fly's seta, Proc. R. Soc. Lond. Ser. B **271**, 2209–2215 (2004)
43.132 N.E. Stork: Role of wax blooms in preventing attachment to brassicas by the mustard beetle, *Phaedon cochleariae*, Entomol. Exp. Appl. **28**, 100–107 (1980)
43.133 S.D. Eigenbrode, K.E. Espelie: Effects of plant epicuticular lipids on insect herbivores, Annu. Rev. Entomol. **40**, 171–194 (1995)
43.134 S.D. Eigenbrode: The effects of plant epicuticular waxy blooms on attachment and effectiveness of predatory insects, Arthropod Struct. Dev. **33**, 91–102 (2004)
43.135 W. Federle, K. Rohrseitz, B. Hölldobler: Attachment forces of ants measured with a centrifuge: Better "wax-runners" have a poorer attachment to a smooth surface, J. Exp. Biol. **203**, 505–512 (2000)
43.136 A. Peressadko, S.N. Gorb: Surface profile and friction force generated by insects, Fortschritt-Ber. VDI **249**, 257–263 (2004)
43.137 G. Huber, S.N. Gorb, N. Hosoda, R. Spolenak, E. Arzt: Influence of surface roughness on gecko adhesion, Acta Biomater. **3**, 607–610 (2007)
43.138 S.N. Gorb: Smooth attachment devices in insects. In: *Advances in Insect Physiology*, Vol. 34, ed. by J. Casas, S.J. Simpson (Elsevier/Academic, London 2008) pp. 81–116
43.139 K. Kendall: *Molecular Adhesion and Its Applications* (Kluwer Academic, New York 2001)
43.140 A. Kästner: Ordnung der *Arachnida*: *Solifugae*. In: *Chelicerata*, ed. by T. Krumbach (de Gruyter, Berlin 1941) pp. 204–215, in German
43.141 P.E. Cushing, J.O. Brookhart, H.-J. Kleebe, G. Zito, P. Payne: The suctorial organ of the *Solifugae* (*Arachnida, Solifugae*), Arthropod Struct. Dev. **34**, 397–406 (2005)
43.142 E.E. Williams, J.A. Peterson: Convergent and alternative designs in the digital adhesive pads of scincid lizards, Science **215**, 1509–1511 (1982)
43.143 W. Schedl: *Hymenoptera*: Unterordnung *Symphyta* (Pflanzenwespen). In: *Handbuch der Zoologie*, ed. by M. Fischer (de Gruyter, Berlin 1991) pp. 1–117, in German
43.144 S. Schulmeister: Morphology and evolution of the tarsal plantulae in *Hymenoptera* (*Insecta*), focussing on the basal lineages, Zool. Scr. **32**, 153–172 (2003)
43.145 A.C. Neville: *Biology of Fibrous Composites* (Cambridge Univ. Press, Cambridge 1993)
43.146 E.V. Gorb, S.N. Gorb: Do plant waxes make insect attachment structures dirty? Experimental evidence for the contamination hypothesis. In: *Ecology and Biomechanics – A Mechanical Approach to the Ecology of Animals and Plants*, ed. by A. Herrel, T. Speck, N.P. Rowe (CRC, Boca Raton 2006) pp. 147–162

43.147 Z. Dai, S.N. Gorb, U. Schwarz: Roughness-dependent friction force of the tarsal claw system in the beetle *Pachnoda marginata* (*Coleoptera, Scarabaeidae*), J. Exp. Biol. **205**, 2479–2488 (2002)
43.148 L. Gaume, P. Perret, E.V. Gorb, S.N. Gorb, J.-J. Labat, N. Rowe: How do plant waxes cause flies to slide? Experimental tests of wax-based trapping mechanisms in three pitfall carnivorous plants, Arthropod Struct. Dev. **33**, 103–111 (2004)
43.149 E.V. Gorb, K. Haas, A. Henrich, S. Enders, N. Barbakadze, S.N. Gorb: Composite structure of the crystalline epicuticular wax layer of the slippery zone in the pitchers of the carnivorous plant *Nepenthes alata* and its effect on insect attachment, J. Exp. Biol. **208**, 4651–4662 (2005)
43.150 D. Voigt, E.V. Gorb, S.N. Gorb: Plant surface–bug interactions: *Dicyphus errans* stalking along trichomes, Arthropod-Plant Interact. **1**, 221–243 (2007)
43.151 E.V. Gorb, S.N. Gorb.: Attachment ability of the beetle *Chrysolina fastuosa* on various plant surfaces, Entomol. Exp. Appl. **105**, 13–28 (2002)
43.152 W. Barthlott, C. Neinhuis, D. Cutler, F. Ditsch, I. Meusel, I. Theisen, H. Wilhelmi: Classification and terminology of plant epicuticular waxes, Bot. J. Linn. Soc. **126**, 237–260 (1998)
43.153 W. Barthlott, I. Theisen, T. Borsch, C. Neinhuis: Epicuticular waxes and vascular plant systematics: integrating micromorphological and chemical data. In: *Deep Morphology: Towards a Renessance of Morphology in Plant Systematics*, ed. by T. Stuessy, F. Hörandl, E. Mayer, V. Rugell (Gantner, Liechtenstein 2003) pp. 189–206
43.154 Z. Burton, B. Bhushan: Surface characterization and adhesion and friction properties of hydrophobic leaf surfaces, Ultramicroscopy **106**, 709–719 (2006)
43.155 R. Spolenak, S.N. Gorb, E. Arzt: Adhesion design maps for bio-inspired attachment systems, Acta Biomater. **1**, 5–13 (2005)
43.156 S.N. Gorb, M. Varenberg, A. Peressadko, J. Tuma: Biomimetic mushroom-shaped fibrillar adhesive microstructure, J. R. Soc. Lond. Interface **4**, 271–275 (2007)
43.157 S.N. Gorb, M. Varenberg: Mushroom-shaped geometry of contact elements in biological adhesive systems, J. Adhes. Sci. Technol. **21**, 1175–1183 (2007)
43.158 A. Peressadko, S.N. Gorb: When less is more: Experimental evidence for tenacity enhancement by division of contact area, J. Adhes. **80**, 247–261 (2004)
43.159 F.P. Bowden: Adhäsion und Reibung, Endeavour **16**, 5–18 (1957), in German
43.160 N.J. Glassmaker, A. Jagota, C.-Y. Hui, J. Kim: Design of biomimetic fibrillar interfaces. 1. Making contact, J. R. Soc. Lond. Interface **1**, 23–33 (2004)
43.161 T.W. Kim, B. Bhushan: Adhesion analysis of multilevel hierarchical attachment system contacting with a rough surface, J. Adhes. Sci. Technol. **21**, 1–20 (2007)
43.162 B.L. Smith, T.E. Schäffer, M. Viani, J.B. Thompson, N. Frederick, J. Kindt, A. Belcher, G.D. Stucky, D.E. Morse, P.K. Hansma: Molecular mechanistic origin of the toughness of natural adhesives, fibers and composites, Nature **399**, 761–763 (1999)
43.163 C. Creton, S.N. Gorb: Sticky feet: from animals to materials, MRS Bulletin **32**, 466–468 (2007)
43.164 A.K. Geim, S.V. Dubonos, I.V. Grigorieva, K.S. Novoselov, A.A. Zhukov, S.Y. Shapoval: Microfabricated adhesive mimicking gecko foot-hair, Nat. Mater. **2**, 461–463 (2003)
43.165 M.T. Northen, K.L. Turner: A batch of fabricated dry adhesive, Nanotechnology **16**, 1159–1166 (2005)
43.166 B. Yurdumakan, N.R. Raravikar, P.M. Ajayan, A. Dhinojwala: Synthetic gecko foot-hairs from multiwalled carbon nanotubes, Chem. Commun., 3799–3801 (2005)
43.167 C.S. Majidi, R.E. Groff, R.S. Fearing: Attachment of fiber array adhesive through side contact, J. Appl. Phys. **98**, 103521 (2005)
43.168 C. Greiner, A. del Campo, E. Arzt: Adhesion of bioinspired micropatterned surfaces: Effects of pillar radius, aspect ratio, and preload, Langmuir **23**, 3495–3502 (2007)
43.169 Y. Pelletier, Z. Smilowitz: Specialized tarsal hairs on adult male Colorado potato beetles, *Leptinotarsa decemlineata* (Say), hamper its locomotion on smooth surfaces, Can. Entomol. **119**, 1139–1142 (1987)
43.170 P. Ball: *The Self Made Tapestry: Pattern Formation in Nature* (Oxford Univ. Press, Oxford 2001)
43.171 K.A. Daltorio, S.N. Gorb, A. Peressadko, A.D. Horchler, R.E. Ritzmann, R.D. Quinn: A robot that climbs walls using micro-structured polymer feet, Int. Conf. Climb. Walk. Robots (CLAWAR), London (2005) pp. 131–138
43.172 M. Varenberg, A. Peressadko, S.N. Gorb, E. Arzt: Effect of real contact geometry on adhesion, Appl. Phys. Lett. **89**, 121905 (2006)
43.173 S. Kim, M. Sitti: Biologically inspired polymer microfibers with spatulate tips as repeatable fibrillar adhesives, Appl. Phys. Lett. **89**, 26911 (2006)
43.174 S.N. Gorb, M. Sinha, A. Peressadko, K.A. Daltorio, R.D. Quinn: Insects did it first: a micropatterned adhesive tape for robotic applications, Bioinsp. Biomim. **2**, 117–125 (2007)
43.175 N.W. Rizzo, K.H. Gardner, D.J. Walls, N.M. Keiper-Hrynko, T.S. Ganzke, D.L. Hallahan: Characterization of the structure and composition of gecko adhesive setae, J. R. Soc. Lond. Interface **3**, 441–451 (2006)
43.176 H. Gao, B. Ji, I.L. Jäger, E. Arzt, P. Fratzl: Materials become insensitive to flaws at nanoscale: Lessons from nature, Proc. Natl. Acad. Sci. USA **100**, 5597–5600 (2003)

43.177 W. Hansen, K. Autumn: Evidence for self-cleaning in gecko setae, Proc. Natl. Acad. Sci. USA **102**, 385–389 (2005)

43.178 H. Lee, B.P. Lee, P.B. Messersmith: A reversible wet/dry adhesive inspired by mussels and geckos, Nature **448**, 338–341 (2007)

43.179 M. Varenberg, S.N. Gorb: A beetle-inspired solution for underwater adhesion, J. R. Soc. Interface **5**(20), 383–385 (2008)

43.180 O.-D. Hennemann: Kleben von Kunststoffen: Anwendung, Ausbildung, Trend, Kunststoffe **90**, 184–188 (2000), in German

43.181 C. Menon, M. Murphy, M. Sitti: Gecko inspired surface climbing robots, IEEE ROBIO'04, Shenyang (2004)

43.182 K. Sangbae, A.T. Asbeck, M.R. Cutkosky, W.R. Provancher: Spinybot II: Climbing hard walls with compliant microspines, Int. Conf. Adv. Robot. (ICAR '05), Seattle (2005)

43.183 J.M. Morrey, B.G.A. Lambrecht, A.D. Horchler, R.E. Ritzmann, R.D. Quinn: Highly mobile and robust small quadruped robots, Int. Conf. Intell. Robots Syst. (IROS '03), Las Vegas (2003) pp. 82–87

43.184 S. Kim, M. Spenko, S. Trujillo, B. Heyneman, V. Mattoli, M.R. Cutkosky: Whole body adhesion: hierarchical, directional and distributed control of adhesive forces for a climbing robot, IEEE Int. Conf. Robot. Autom., Rome (2007) pp. 1268–1273

43.185 S. Niederegger, S.N. Gorb: Tarsal movements in flies during leg attachment and detachment on a smooth substrate, J. Insect Physiol. **49**, 611–620 (2003)

43.186 A.P. Russell, V. Bels: Digital hyperextension in *Anolis sagrei*, Herpetologica **57**, 58–65 (2001)

43.187 A.P. Russell: Integrative functional morphology of the gekkotan adhesive system (*Reptilia*: *Gekkota*), Integr. Comp. Biol. **42**, 1154–1163 (2002)

43.188 Y. Jiao, S.N. Gorb, M. Scherge: Adhesion measured on the attachment pads of *Tettigonia viridissima* (*Orthoptera*, *Insecta*), J. Exp. Biol. **203**, 1887–1895 (2000)

43.189 G. Hanna, W.-J.P. Barnes: Adhesion and detachment of the toe pads of tree frogs, J. Exp. Biol. **155**, 103–125 (1990)

43.190 M. Varenberg, S.N. Gorb: Hexagonal surface micropattern for dry and wet friction, Adv. Mater. **20**, 1–4 (2008)

43.191 R. Spolenak, S.N. Gorb, H. Gao, E. Arzt: Effects of contact shape on the scaling of biological attachments, Proc. R. Soc. Lond. Ser. A **461**, 305–319 (2005)

44. Gecko Feet: Natural Hairy Attachment Systems for Smart Adhesion

Bharat Bhushan

The leg attachment pads of several creatures, including many insects, spiders, and lizards, are capable of attaching to a variety of surfaces and are used for locomotion. Geckoes, in particular, have the largest mass and have developed the most complex hairy attachment structures capable of smart adhesion – the ability to cling to different smooth and rough surfaces and detach at will. These animals make use of about three million microscale hairs (setae) (about 14 000 mm^{-2}) that branch off into hundreds of nanoscale spatulae (about three billion spatula on two feet). This so-called division of contacts provides high dry adhesion. This multiple-level hierarchically structured surface construction provides the gecko with the compliance and adaptability to create a large real area of contact with a variety of surfaces. Modeling of the gecko attachment system as a hierarchical spring model has provided insight into the adhesion enhancement generated by this system. van der Waals forces are the primary mechanism utilized to adhere to surfaces, and capillary forces are a secondary effect that can further increase the adhesion force. Preload applied to the setae increases adhesive force. Although a gecko is capable of producing of the order of 20 N of adhesive force, it retains the ability to remove its feet from an attachment surface at will. The adhesive strength of gecko setae is dependent on orientation; maximum adhesion occurs at 30°. During walking, a gecko is able to peel its foot from surfaces by changing the angle at which its setae contact the surface. Manmade fibrillar structures capable of replicating gecko adhesion have the potential for use in dry superadhesive tapes and treads for wall-climbing robots for various applications. These structures can be created using micro/nanofabrication techniques or self-assembly.

44.1 Overview

The leg attachment pads of several animals including many insects, spiders, and lizards, are capable of attaching to and detaching from a variety of surfaces and are used for locomotion, even on vertical walls or across the ceiling [44.1, 2]. Biological evolution over a long period of time has led to the optimization of their leg attachment systems. This dynamic attachment ability is referred to as reversible or smart adhesion [44.3]. Many insects (e.g., beetles and flies) and spiders have been the subject of investigation. However, the attachment pads of geckoes have been the most widely studied due to the fact that they have the highest body mass and exhibit the most versatile and effective adhesion known in nature. As a result, the vast majority of this chapter will be concerned with gecko feet.

Although there are over 1000 species of geckoes [44.4, 5] that have attachment pads of varying morphology [44.6], the Tokay gecko (*Gekko gecko*) which is native to Southeast Asia, has been the main focus of scientific research [44.7–9]. The Tokay gecko is the second largest gecko species, attaining lengths of ≈ 0.3–0.4 and 0.2–0.3 m for males and females, respectively. They have a distinctive blue or gray body with orange or red spots and can weigh up to 300 g [44.10]. These have been the most widely investigated species of gecko due to the availability and size of these creatures.

Almost 2500 years ago, the ability of the gecko to "run up and down a tree in any way, even with the head downwards" was observed by *Aristotle* [44.11, Book IX, Part 9]. Even though the adhesive ability of geckoes has been known since the time of Aristotle, little was understood about this phenomenon until the late 19th century, when the microscopic hairs covering the toes of the gecko were first noted. The development of electron microscopy in the 1950s enabled scientists to view the complex hierarchical morphology that covers the skin on the gecko's toes. Over the past century and a half, scientific studies have been conducted to determine the factors that allow the gecko to adhere to and detach from surfaces at will, including surface structure [44.6, 8, 12–17], the mechanisms of adhesion [44.6, 7, 18–28], and adhesion strength [44.7, 8, 17, 25, 28, 29]. Modeling the gecko attachment system as a system of springs [44.3, 30–33] has provided valuable insight into adhesion enhancement; van der Waals forces are widely accepted in the literature as the dominant adhesion mechanism utilized by hierarchical attachment systems. Capillary forces created by humidity naturally present in the air can further increase the adhesion force generated by the spatulae [44.33]. Both experimental and theoretical work support these adhesion mechanisms.

There is great interest among the scientific community to further study the characteristics of gecko feet in the hope that this information can be applied to the production of micro/nanosurfaces capable of recreating the adhesion forces generated by these lizards [44.2]. Common manmade adhesives such as tape or glue involve the use of wet adhesives that permanently attach two surfaces. However, replication of the characteristics of gecko feet would enable the development of a superadhesive tape capable of clean dry adhesion. These structures can bind components in microelectronics without the high heat associated with various soldering processes. These structures will never dry out in a vacuum – a common problem in aerospace applications. They have the potential for use in everyday objects such as adhesive tapes, fasteners, and toys, and in high technology such as microelectronic and space applications. Replication of the dynamic climbing and peeling ability of geckoes could find use in the treads of wall-climbing robots.

44.2 Hairy Attachment Systems

There are two kinds of attachment pads: relatively smooth and hairy. Relatively smooth pads, so-called arolia and euplantulae, are soft and deformable and are found in tree frogs, cockroaches, grasshoppers, and bugs. The hairy types consist of long deformable setae and are found in many insects (e.g., beetles, flies), spiders, and lizards. The microstructures utilized by beetles, flies, spiders, and geckoes have similar structures, as can be seen in Fig. 44.1a. As the size (mass) of the creature increases, the radius of the terminal attach-

Fig. 44.1 (a) Terminal elements of the hairy attachment pads of a beetle, fly, spider, and gecko shown at two different scales (after [44.17]) and (b) the dependence of terminal element density on body mass (after [44.34]). Data from *Arzt* et al. [44.17] and *Kesel* et al. [44.35] ▸

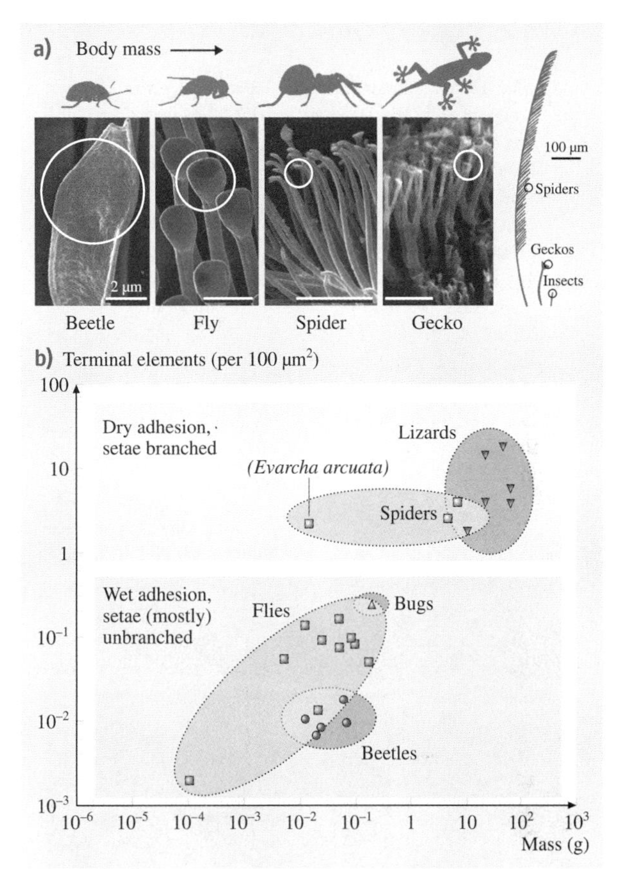

ment elements decreases. This allows a greater number of setae to be packed into an area, hence increasing the linear dimension of contact and the adhesion strength. *Arzt* et al. [44.17] determined that the density of the terminal attachment elements ρ_A /m^2 strongly increases with increasing body mass m in g. In fact, a master curve can be fitted for all the different species (Fig. 44.1b)

$$\log \rho_A = 13.8 + 0.669 \log m \,. \tag{44.1}$$

The correlation coefficient of the master curve is 0.919. Beetles and flies have the largest attachment pads and the lowest density of terminal attachment elements. Spiders have highly refined attachment elements that cover the leg of the spider. Geckoes have both the highest body mass and greatest density of terminal elements (spatulae). Spiders and geckoes can generate high dry adhesion, whereas beetles and flies increase adhesion by secreting liquid stored generally within a spongy layer of cuticle and delivered at the contacting surface through a system of porous channels [44.1, 17, 35]. It should be noted that, in the smooth attachment system discussed earlier, this secretion is essential for attachment.

It should be noted that *Peattie* and *Full* [44.36] have revisited the scaling of terminal attachment elements with body mass using a phylogenetic approach. It their work, a larger set of species (81) over a wider range of body mass and setal morphology were considered. They found that fiber morphology is better predicted by evolutionary history and adhesion mechanism (dry or wet) than by body mass.

Figure 44.2 shows scanning electron micrographs of the end of the legs of two flies – fruit fly (*Drosophila melanogaster*) and syrphid fly. The fruit fly uses setae with flattened tips (spatulae) on two hairy rods for attachment to smooth surfaces and two front claws for attachment to rough surfaces. The front claws are also used for locomotion. The syrphid fly uses setae on the legs for attachment. In both cases, fluid is secreted at the contacting surface to increase adhesion.

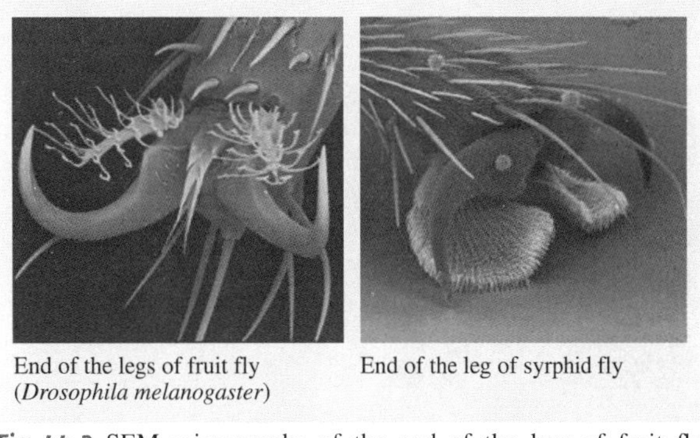

Fig. 44.2 SEM micrographs of the end of the legs of fruit fly (*Drosophila melanogaster*) and syrphid fly (after [44.1])

44.3 Tokay Gecko

44.3.1 Construction of the Tokay Gecko

The explanation for the adhesion properties of gecko feet can be found in the surface morphology of the skin on the toes of the gecko. The skin is comprised of a complex hierarchical structure of lamellae, setae, branches, and spatulae [44.6]. Figure 44.3 shows various scanning electron microscopy (SEM) micrographs of a gecko foot, showing the hierarchical structure down to the nanoscale. Figure 44.4 shows a schematic of the structure and Table 44.1 summarizes the surface characteristics. The gecko attachment system consists of an intricate hierarchy of structures beginning with lamellae, soft ridges 1–2 mm in length [44.6] that are located on the attachment pads (toes) that compress easily so that contact can be made with rough bumpy surfaces. Tiny curved hairs known as setae extend from the lamellae at a density of $\approx 14\,000$ per mm^2 [44.15]. These setae are typically 30–130 μm in length, 5–10 μm in diameter [44.6, 7, 12, 14], and are composed primarily of β-keratin [44.13, 38] with some α-keratin component [44.39]. At the end of each seta, 100–1000 spatulae [44.6, 7], typically 2–5 μm in length and with a diameter of 0.1–0.2 μm [44.6], branch out and form the points of contact with the surface. The tips of the spatulae are ≈ 0.2–0.3 μm in width [44.6], 0.5 μm in length, and 0.01 μm in thickness [44.40] and garner their name from their resemblance to a spatula.

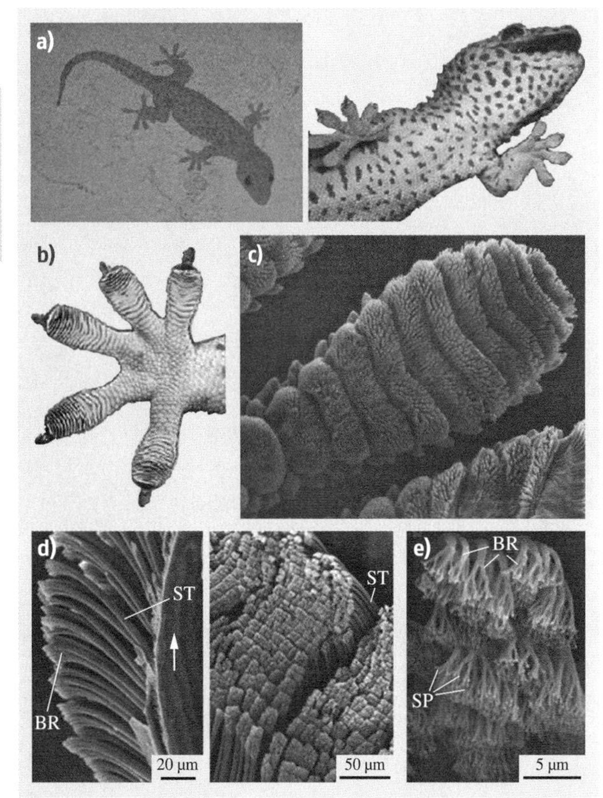

Fig. 44.3 (a) Tokay gecko (after [44.25]). The hierarchical structures of a gecko foot: (b) a gecko foot (after [44.25]) and (c) a gecko toe (after [44.9]). Each toe contains hundreds of thousands of setae and each seta contains hundreds of spatulae. Scanning electron microscope (SEM) micrographs of (d) the setae (after [44.37]) and (e) the spatula (after [44.37]) (ST – seta, SP – spatula, BR – branch)

The attachment pads on two feet of the Tokay gecko have an area of $\approx 220\,mm^2$. About three million setae on their toes that branch off into about three billion spatula on two feet can produce a clinging ability of ≈ 20 N (vertical force required to pull a lizard down a nearly vertical (85°) surface) [44.8] and allow them to climb vertical surfaces at speeds > 1 m/s with the capability to attach and detach their toes in milliseconds. In isolated setae, a 2.5 μN preload yielded adhesion of 20–40 μN, and thus the adhesion coefficient, which represents the strength of adhesion as a function of preload, ranges from 8–16 [44.26].

44.3.2 Adhesion Enhancement During Contact with Rough Surfaces

Typical rough, rigid surfaces are able to make intimate contact with a mating surface only over a very small portion of the perceived apparent area of contact. In fact, the real area of contact is typically two to six orders of magnitude less than the apparent area of contact [44.41–43]. *Autumn* et al. [44.26] proposed that divided contacts serve as a means for increasing adhesion. The surface energy approach can be used to calculate adhesion force in a dry environment in order to calculate the effect of division of contacts. If the tip of a spatula is considered as a hemisphere with radius R, the adhesion force of a single contact F_{ad}, based on the so-called Johnson–Kendall–Roberts (JKR) theory [44.44] is given by

$$F_{ad} = \frac{3}{2}\pi W_{ad} R\,, \tag{44.2}$$

where W_{ad} is the work of adhesion (units of energy per unit area). Equation (44.2) shows that the adhesion

Table 44.1 Surface characteristics of Tokay gecko feet (Young's modulus of surface material, keratin = 1–20 GPa[a,b])

Component	Size	Density	Adhesive force
Seta	30–130[c–f]/5–10[c–f] length/diameter (µm)	≈ 14 000[h,i] setae/mm²	194 µN[j] (in shear) ≈ 20 µN[j] (normal)
Branch	20–30[c]/1–2[c] length/diameter (µm)	–	–
Spatula	2–5[c]/0.1–0.2[c,g] length/diameter (µm)	100–1000[c,d] spatulae per seta	–
Tip of spatula	≈ 0.5[c,g]/0.2–0.3[c,f]/ ≈ 0.01[g] length/width/thickness (µm)	–	11 nN[k] (normal)

[a] *Russell* [44.13]
[b] *Bertram* and *Gosline* [44.45]
[c] *Ruibal* and *Ernst* [44.6]
[d] *Hiller* [44.7]
[e] *Russell* [44.12]
[f] *Williams* and *Peterson* [44.14]
[g] *Persson* and *Gorb* [44.40]
[h] *Schleich* and *Kästle* [44.15]
[i] *Autumn* and *Peattie* [44.16]
[j] *Autumn* et al. [44.25]
[k] *Huber* et al. [44.29]

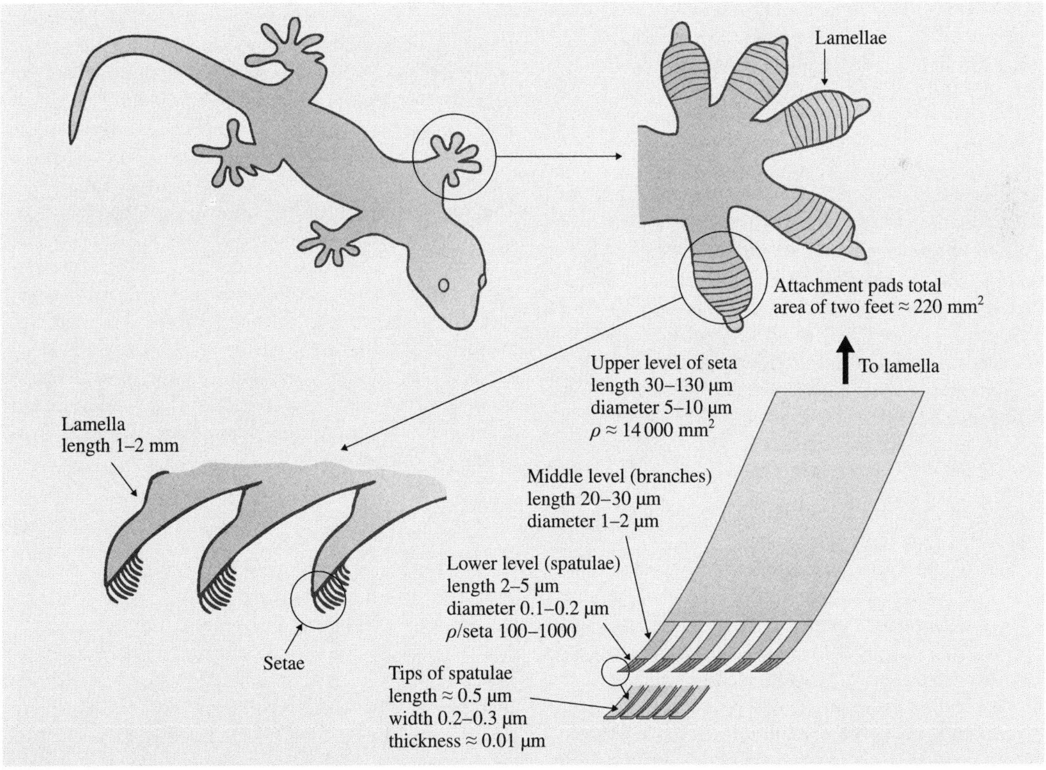

Fig. 44.4 Schematic drawings of a Tokay gecko including the overall body, one foot, a cross-sectional view of the lamellae, and an individual seta; ρ represents the number of spatulae

force of a single contact is proportional to the linear dimension of the contact. For a constant area divided into a large number of contacts or setae n the radius of a divided contact, R_1, is given by $R_1 = R/\sqrt{n}$ (self-similar scaling) [44.17]. Therefore, the adhesion force of (44.2) can be modified for multiple contacts such that

$$F'_{\mathrm{ad}} = \frac{3}{2}\pi W_{\mathrm{ad}}\left(\frac{R}{\sqrt{n}}\right)n = \sqrt{n}F_{\mathrm{ad}}\,, \qquad (44.3)$$

where F'_{ad} is the total adhesion force from the divided contacts. Thus the total adhesive force is simply the adhesion force of a single contact multiplied by the square root of the number of contacts.

For contact in a humid environment, meniscus (or capillary) forces further increase the adhesion force [44.41–43]. The attractive meniscus force (F_{m}) consists of a contribution from both Laplace pressure and surface tension (Sect. 44.4.2) [44.42, 46]. The contribution by Laplace pressure is directly proportional to the meniscus area. The other contribution is from the vertical component of the surface tension around the circumference. This force is proportional to the circumference, as is the case for the work of adhesion [44.42]. Going through the analysis presented earlier, one can show that the contribution from the component of surface tension increases on splitting into a larger number of contacts. It increases linearly with the square root of the number of contacts n (self-similar scaling) [44.2,47, 48]

$$\left(F'_{\mathrm{m}}\right)_{\text{surface tension}} = \sqrt{n}(F_{\mathrm{m}})_{\text{surface tension}}\,, \qquad (44.4)$$

where F'_{m} is the force from the divided contacts and F_{m} is the force of an individual contact. This component of meniscus force is significant if the meniscus radius is very small and the contact angles are relatively large.

During separation of two surfaces, the viscous force F_{v} of the divided contacts is given by [44.47, 48]

$$F'_{\mathrm{v}} = \frac{F_{\mathrm{v}}}{n}\,, \qquad (44.5)$$

where F'_{v} is the force from the divided contacts, and F_{v} is the force of an individual contact.

The models just presented only consider contact with a flat surface. Multiple-level hierarchical structure of the gecko provide compliance and conformability to rough surfaces in order to achieve high adhesion. The flexibility of the body provides conformability at the cm scale. Several toes on the feet provide conformability independently at the several mm scale. Lamellae on the bottom surfaces of the toes provide conformability at the mm scale. The setae on the lamellae provide conformability at the several μm scale. The tips of the setae are divided into spatulae which provide conformability at the few to several hundred nm scale. To summarize, on natural rough surfaces, the compliance and adaptability of the hierarchical structure of gecko setae allows for greater contact with a natural rough surface than a nonbranched attachment system [44.3, 30–33, 49]. Modeling of the contact between gecko setae and rough surfaces is discussed in detail in Sect. 44.6.

Table 44.2 Young's modulus of gecko skin and other materials for comparison

Material	Young's modulus
β-keratin, mostly present in gecko skin	1–20 GPa
Steel	210 GPa
Cross-linked rubber	1 MPa
Consumer adhesive tape (uncrosslinked rubber)	1 kPa

Material properties also play an important role in adhesion. A soft material is able to achieve greater contact with a mating surface than a rigid material. Although gecko skin is primarily comprised of β-keratin, a stiff material with a Young's modulus in the range 1–20 GPa [44.13, 45], the effective modulus of the setal arrays on gecko feet is ≈ 100 kPa [44.50], which is approximately four orders of magnitude lower than the bulk material. The Young's modulus of gecko skin is compared with that of various materials in Table 44.2. The surface of consumer adhesive tape has been selected to be very compliant to increase the contact area for high adhesion. Nature has selected a relatively stiff material to avoid clinging to adjacent setae. Nonorthogonal attachment angle of the seta increases the bending stiffness. Division of contacts and hierarchical structure, as discussed earlier, provide high adhesion. By combining optimal surface structure and material properties, Mother Nature has created an evolutionary superadhesive.

44.3.3 Peeling

Although geckoes are capable of producing large adhesion forces, they retain the ability to remove their feet from an attachment surface at will by peeling action. The orientation of the spatulae facilitates peeling. *Autumn* et al. [44.25] were the first to show experimentally that the adhesion force of gecko setae is dependent on the three-dimensional (3-D) orientation as well as the preload applied during attachment. Due to this fact, geckoes have developed a complex foot motion during walking. First, the toes are carefully uncurled during detachment. The maximum adhesion occurs at an attachment angle of 30° – the angle between a seta and mating surface. The gecko is then able to peel its foot off surfaces one row of setae at a time by changing the angle at which its setae contact the surface. At an attachment angle > 30°, the gecko will detach from the surface.

Shah and *Sitti* [44.51] determined the theoretical preload required for adhesion as well as the adhesion force generated for setal orientations of 30°, 40°, 50°, and 60°. We consider a solid material (elastic modulus E, Poisson's ratio ν) to make contact with the rough surface described by

$$f(x) = H\sin^2\left(\frac{\pi x}{\chi}\right) , \tag{44.6}$$

where H is the amplitude and χ is the wavelength of the roughness profile. For a solid adhesive block to achieve intimate contact with the rough surface, neglecting surface forces, it is necessary to apply a compressive stress σ_c [44.52]

$$\sigma_c = \frac{\pi E H}{2\chi\left(1-\nu^2\right)} . \tag{44.7}$$

Equation (44.7) can be modified to account for fibers oriented at an angle θ. The preload required for contact is summarized in Fig. 44.5a. As the orientation angle decreases, so does the required preload. Similarly, the adhesion strength is influenced by fiber orientation. As seen in Fig. 44.5b, the greatest adhesion force occurs at $\theta = 30°$.

Gao et al. [44.37] created a finite-element model of a single gecko seta in contact with a surface. A tensile force was applied to the seta at various angles θ, as shown in Fig. 44.5c. For forces applied at an angle < 30°, the dominant failure mode was sliding. On the contrary, the dominant failure mode for forces applied at angles > 30° was detachment. This verifies the results of *Autumn* et al. [44.25] that detachment occurs at attachment angles > 30°.

Tian et al. [44.53] have suggested that, during detachment, the angular dependence of both adhesion and friction plays a role. The pulling force of a spatula along its shaft with an angle between 0° and 90° to the substrate has a normal adhesion force produced at the

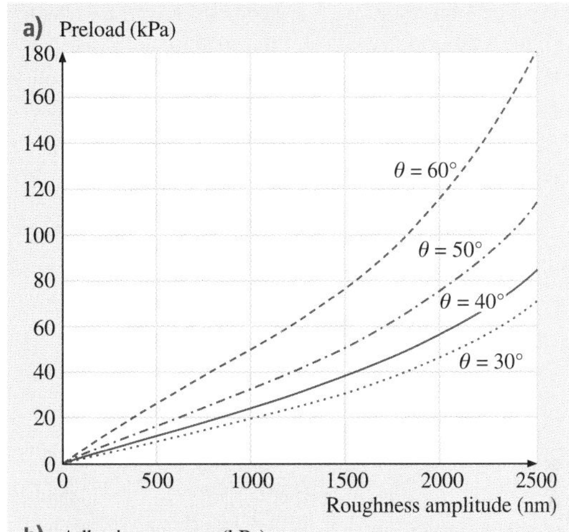

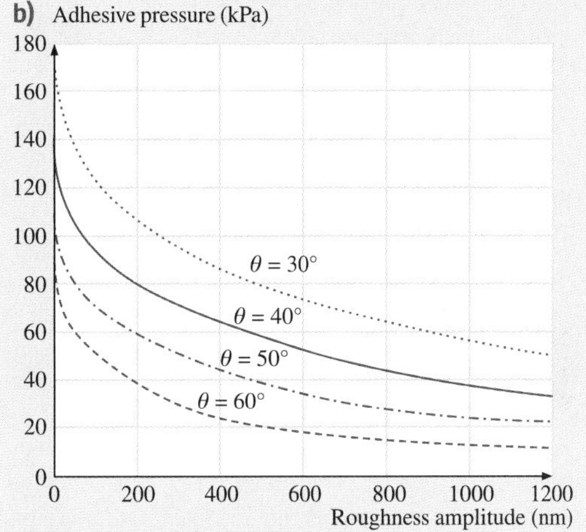

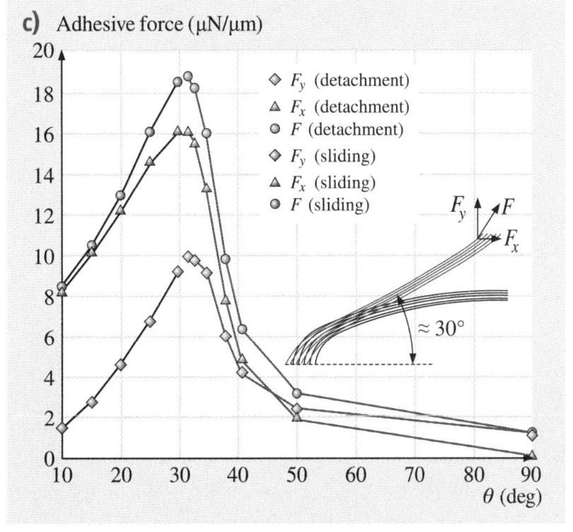

Fig. 44.5a–c Contact mechanics results for the effect of fiber orientation on (a) preload and (b) adhesive force for roughness amplitude of 0–2500 nm (after [44.51]). (c) Finite-element analysis of the adhesive force of a single seta as a function of pull direction (after [44.37]) ▶

spatula–substrate bifurcation zone and a lateral friction force contribution from the part of the spatula still in contact with the substrate. High net friction and adhesion forces on the whole gecko are obtained by rolling down and gripping the toes inward to realize small pulling angles of the large number of spatulae in contact with the substrate. To detach, the high adhesion/friction is rapidly reduced to a very low value by rolling the toes

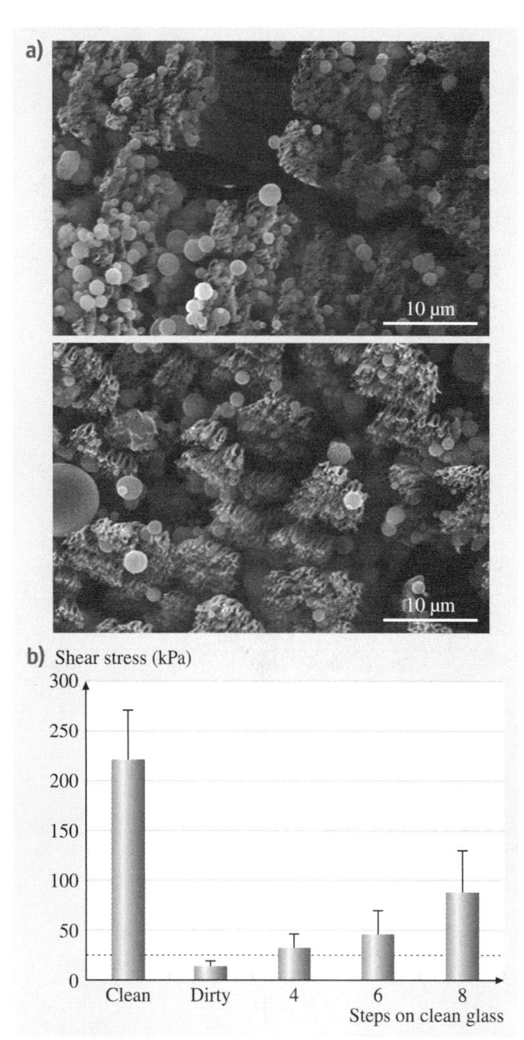

Fig. 44.6 (a) SEM image of spatulae after dirtying with microspheres (*top*) and after five simulated steps (*bottom*). (b) Mean shear stress exerted by a gecko on a surface after dirtying. The *dotted line* represents sufficient recovery to support weight by a single toe (after [44.54])

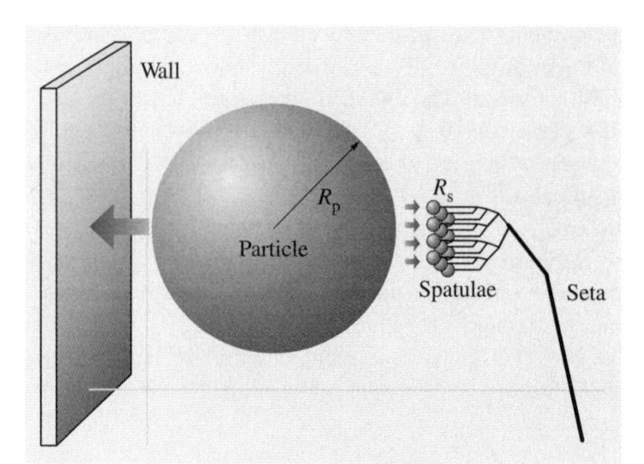

Fig. 44.7 Model of interactions between gecko spatulae of radius R_S, a spherical dirt particle of radius R_P, and a planar wall, enabling self-cleaning (after [44.54])

upward and downward, which, mediated by the lever function of the setal shaft, peels the spatula off from the substrate perpendicularly.

44.3.4 Self-Cleaning

Natural contaminants (dirt and dust) as well as man-made pollutants are unavoidable and have the potential to interfere with the clinging ability of geckoes. Particles found in the air consist of particulates that are typically $< 10\,\mu m$ in diameter, while those found on the ground can often be larger [44.55, 56]. Intuitively, it seems that the great adhesion strength of gecko feet would cause dust and other particles to become trapped in the spatulae and that they would have no way of being removed without some sort of manual cleaning action on behalf of the gecko. However, geckoes are not known to groom their feet like beetles [44.57], nor do they secrete sticky fluids to remove adhering particles like ants [44.58] and tree frogs [44.59], yet they retain adhesive properties. One potential source of cleaning is during the time when the lizards undergo molting, or the shedding of the superficial layer of epidermal cells. However, this process only occurs approximately once per month [44.60]. If molting were the sole source of cleaning, the gecko would rapidly lose its adhesive properties as it was exposed to contaminants in nature [44.54].

Hansen and *Autumn* [44.54] tested the hypothesis that gecko setae become cleaner with repeated use – a phenomenon known as self-cleaning. The cleaning ability of gecko feet was first tested experimentally by applying 2.5 μm-radius silica–alumina ceramic microspheres to clean setal arrays. Figure 44.6a shows the setal arrays immediately after dirtying and after five simulated steps. It is noted that a significant fraction of the particles have been removed after five steps. The maximum shear stress that these *dirty* arrays could withstand was measured using a sensor. After each step that the gecko took, the shear stress was once again measured. As seen in Fig. 44.6b, after four steps, the gecko foot was clean enough to withstand its own body weight.

In order to understand this cleaning process, substrate–particle interactions must be examined. The interaction energy between a dust particle and a wall and spatulae can be modeled as shown in Fig. 44.7. The interaction energy between a spherical dust particle and the wall W_{pw}, can be expressed as [44.61]

$$W_{pw} = \frac{-H_{pw} R_p}{6D_{pw}} , \tag{44.8}$$

where p and w refer to the particle and wall, respectively. H is the Hamaker constant, R_p is the radius of the particle, and D_{pw} is the separation distance between the particle and the wall. Similarly, the interaction energy between a spherical dust particle and a spatula s, assuming that the spatula tip is spherical, is [44.61]

$$W_{ps} = \frac{-H_{ps} R_p R_s}{6D_{ps} (R_p + R_s)} . \tag{44.9}$$

The ratio of the two interaction energies Z can be expressed as

$$Z = \frac{W_{pw}}{W_{ps}} = \left(1 + \frac{R_p}{R_s}\right) \frac{H_{pw} D_{ps}}{H_{ps} D_{pw}} . \tag{44.10}$$

When the energy required to separate a particle from the wall is greater than that required to separate it from a spatula ($Z > 1$), self-cleaning will occur. For small contaminants ($R_p < 0.5\,\mu m$), there are not enough spatulae available to adhere to the particle. For larger contaminants, the curvature of the particles makes it impossible for enough spatulae to adhere to it. As a result, *Hansen* and *Autumn* [44.54] concluded that self-cleaning should occur for all spherical spatulae interacting with all spherical particles.

44.4 Attachment Mechanisms

When asperities of two solid surfaces are brought into contact with each other, chemical and/or physical attraction occurs. The force developed that holds the two surfaces together is known as adhesion. In a broad sense, adhesion is considered to be either physical or chemical in nature [44.41–43, 61–66]. Chemical interactions such as electrostatic attraction charges [44.20] as well as intermolecular forces [44.7] including van der Waals and capillary forces have all been proposed as potential adhesion mechanisms in gecko feet. Others have hypothesized that geckoes adhere to surfaces through the secretion of sticky fluids [44.18, 19], suction [44.19], increased frictional force [44.21], and microinterlocking [44.22].

Through experimental testing and observations conducted over the last century and a half many potential adhesive mechanisms have been eliminated. Observation has shown that geckoes lack any glands capable of producing sticky fluids [44.18, 19], thus ruling out the secretion of sticky fluids as a potential adhesive mechanism. Furthermore, geckoes are able to create large adhesive forces normal to a surface. Since friction only acts parallel to a surface, the attachment mechanism of increased frictional force has been ruled out. *Dellit* [44.22] experimentally ruled out suction and electrostatic attraction as potential adhesive mechanisms. Experiments carried out in vacuum did not show a difference between the adhesive force at low pressures compared with under ambient conditions. Since adhesive forces generated during suction are based on pressure differentials, which are insignificant under vacuum, suction was rejected as an adhesive mechanism [44.22]. Additional testing utilized x-ray bombardment to create ionized air in which electrostatic attraction charges would be eliminated. It was determined that geckoes were still able to adhere to surfaces under these conditions, and therefore electrostatic charges could not be the sole cause of attraction [44.22]. *Autumn* et al. [44.25] demonstrated the ability of a gecko to generate large adhesive forces when in contact with a molecularly smooth SiO_2 microelectromechanical system (MEMS) semiconductor. Since surface roughness is necessary for microinterlocking to occur, it has also been ruled out as a mechanism of adhesion. Two mechanisms, van der Waals forces and capillary forces, remain as the potential sources of gecko adhesion. These attachment mechanisms are described in detail in the following sections.

44.4.1 van der Waals Forces

van der Waals bonds are secondary bonds that are weak in comparison with other physical bonds such as covalent, hydrogen, ionic, and metallic bonds. Unlike other physical bonds, van der Waals forces are always present, regardless of separation, and are effective from very large separations ($\approx 50\,\mathrm{nm}$) down to atomic separation ($\approx 0.3\,\mathrm{nm}$). The van der Waals force per unit area between two parallel surfaces f_{vdW} is given by [44.61,67,68]

$$f_{\mathrm{vdW}} = \frac{H}{6\pi D^3}\,, \quad \text{for } D < 30\,\mathrm{nm}\,, \qquad (44.11)$$

where H is the Hamaker constant and D is the separation between surfaces.

Hiller [44.7] showed experimentally that the surface energy of a substrate is responsible for gecko adhesion. One potential adhesion mechanism would then be van der Waals forces [44.24, 25]. Assuming van der Waals forces to be the dominant adhesion mechanism utilized by geckoes, the adhesion force of a gecko can be calculated. Typical values of the Hamaker constant range from 4×10^{-20} to 4×10^{-19} J [44.61]. In calculation, the Hamaker constant is assumed to be 10^{-19} J, the surface area of a spatula is taken to be $2\times10^{-14}\,\mathrm{m}^2$ [44.6, 14, 16], and the separation between the spatula and contact surface is estimated to be 0.6 nm. This equation yields the force of a single spatula to be $\approx 0.5\,\mu\mathrm{N}$. By applying the surface characteristics from Table 44.1, the maximum adhesion force of a gecko is 150–1500 N for varying spatula density of 100–1000 spatulae/seta. If an average value of 550 spatulae/seta is used, the adhesion force of a single seta is $\approx 270\,\mu\mathrm{N}$, which is in agreement with the experimental value obtained by *Autumn* et al. [44.25], which will be discussed later.

Another approach to calculate the adhesion force is to assume that the spatulae are cylinders that terminate in hemispherical tips. By using (44.2) and assuming that the radius of each spatula is $\approx 100\,\mathrm{nm}$ and that the surface energy is expected to be $50\,\mathrm{mJ/m^2}$ [44.17], the adhesive force of a single spatula is predicted to be $0.02\,\mu\mathrm{N}$. This result is an order of magnitude lower than the first approach calculated for the higher value of A. For a lower value of 10^{-20} J for the Hamaker constant, the adhesive force of a single spatula is comparable to that obtained using the surface energy approach.

Several experimental results favor van der Waals forces as the dominant adhesive mechanism, including temperature testing [44.27] and adhesion force measurements of a gecko seta with both hydrophilic and hydrophobic surfaces [44.25]. These data will be presented in Sects. 44.5.2–44.5.4.

44.4.2 Capillary Forces

It has been hypothesized that capillary forces that arise from liquid-mediated contact could be a contributing or even dominant adhesive mechanism utilized by gecko spatulae [44.7, 24]. Experimental adhesion measurements (presented later in Sects. 44.5.3 and 44.5.4) conducted on surfaces with different hydrophobicities and at various humidity values [44.28] as well as numerical simulations [44.33] support this hypothesis as a contributing mechanism. During contact, any liquid that wets or has a small contact angle on surfaces will condense from vapor in the form of an annular-shaped capillary condensate. Due to the natural humidity present in the air, water vapor will condense to liquid on the surface of bulk materials. During contact this will cause the formation of adhesive bridges (menisci) due to the proximity of the two surfaces and the affinity of the surfaces for condensing liquid [44.69–71].

In the adhesion model with capillarity by *Kim* and *Bhushan* [44.33], the tip of the spatula in a single contact was assumed to be spherical (Fig. 44.8). The total adhesion force between a spherical tip and a plane consists of the capillary force and the solid–solid interaction. The capillary force can be divided into two components: the Laplace force F_{L} and the surface ten-

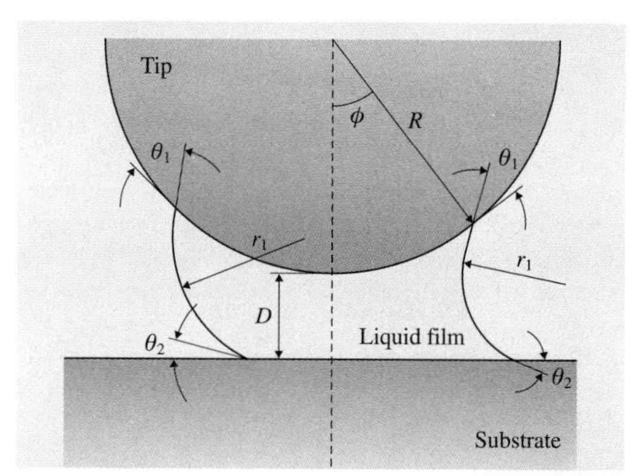

Fig. 44.8 Schematic of a sphere on a plane at a distance D with a liquid film in between, forming menisci. In this figure, R is the tip radius, ϕ is the filling angle, θ_1 and θ_2 are contact angles on the sphere and plane, respectively, and r_1 and r_2 are the two principal radii of the curved surface in two orthogonal planes (after [44.33])

sion force F_s, such that the total capillary force F_c is

$$F_c = F_L + F_s \,. \tag{44.12}$$

The Laplace force is caused by the pressure difference across the interface of a curved liquid surface (Fig. 44.8) and depends on the pressure difference multiplied by the meniscus area, which can be expressed as [44.46]

$$F_L = -\pi\kappa\gamma R^2 \sin^2\phi \,, \tag{44.13}$$

where γ is the surface tension of the liquid, R is the tip radius, ϕ is the filling angle, and κ is the mean curvature of the meniscus. From the Kelvin equation [44.61], which is the thermal equilibrium relation, the mean curvature of meniscus can be determined as

$$\kappa = \frac{\Re T}{V\gamma} \ln\left(\frac{p}{p_0}\right) \,, \tag{44.14}$$

where $\Re$ is the universal gas constant, T is the absolute temperature, V is the molecular volume, p_0 is the saturated vapor pressure of the liquid at T, and p is the ambient pressure acting outside the curved surface (p/p_0 is the relative humidity). *Orr* et al. [44.46] formulated the mean curvature of a meniscus between a sphere and a plane in terms of elliptic integrals. The filling angle ϕ can be calculated from the expression just mentioned and (44.14) using the iteration method. Then the Laplace force is calculated at a given environment using (44.13).

The surface tension of the liquid results in the formation of a curved liquid–air interface. The surface tension force acting on the sphere is [44.46]

$$F_s = 2\pi R\gamma \sin\phi \sin(\theta_1 + \phi) \,, \tag{44.15}$$

where θ_1 is the contact angle on the sphere.

Hence, the total capillary force on the sphere is

$$F_c = \pi R\gamma \{2 \sin\phi \sin(\theta_1 + \phi) - \kappa R \sin^2\phi\} \,. \tag{44.16}$$

The effect of capillarity on gecko adhesion results will be presented in Sect. 44.6.4.

44.5 Experimental Adhesion Test Techniques and Data

Experimental measurements of the adhesion force of a single gecko seta [44.25] and single gecko spatula [44.29] have been made. The effect of the environment, including temperature [44.27, 72] and humidity [44.28], has been studied. Some of the data has been used to understand the adhesion mechanism utilized by the gecko attachment system – van der Waals or capillary forces. The majority of experimental results point towards van der Waals forces as the dominant mechanism of adhesion [44.25, 27]. Recent research suggests that capillary forces can be a contributing adhesive factor [44.28, 33].

44.5.1 Adhesion Under Ambient Conditions

Two feet of a Tokay gecko are capable of producing ≈ 20 N of adhesive force with a pad area of $\approx 220\,\text{mm}^2$ [44.8]. Assuming that there are $\approx 14\,000\,\text{setae/mm}^2$, the adhesion force from a single hair should be $\approx 7\,\mu\text{N}$. It is likely that the magnitude is actually larger than this value, because it is unlikely that all setae are in contact with the mating surface [44.25]. Setal orientation greatly influences adhesive strength. This dependency was first noted by *Autumn* et al. [44.25]. It was determined that the greatest adhesion occurs at 30°. In order to determine the adhesion mechanism(s) utilized by gecko feet, it is important to know the adhesion force of a single seta. Hence, the adhesion force of gecko foot-hair has been the focus of several investigations [44.25, 29].

Adhesion Force of a Single Seta

Autumn et al. [44.25] used both a MEMS force sensor and a wire as a force gage to determine the adhesion force of a single seta. The MEMS force sensor is a dual-axis atomic force microscope (AFM) cantilever with independent piezoresistive sensors which allows simultaneous detection of vertical and lateral forces [44.73]. The wire force gage consisted of an aluminum bonding wire that displaced under a perpendicular pull. *Autumn* et al. [44.25] discovered that the setal force actually depends on the three-dimensional orientation of the seta as well as the preloading force applied during initial contact. Setae that were preloaded vertically to the surface exhibited only one-tenth of the adhesive force ($0.6 \pm 0.7\,\mu\text{N}$) compared with setae that were pushed vertically and then pulled horizontally to the surface ($13.6 \pm 2.6\,\mu\text{N}$). The dependence of the adhesion force of a single gecko spatula on perpendicular preload is illustrated in Fig. 44.9. The adhesion force increases linearly with the preload, as expected [44.41, 42, 65]. The maximum adhesion force of a single gecko foot-hair occurred when the seta was first subjected to a normal preload and then slid 5 μm along the contacting sur-

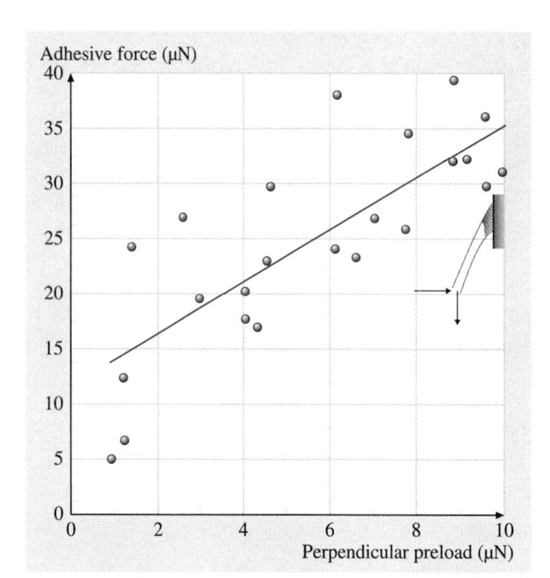

Fig. 44.9 Adhesive force of a single gecko seta as a function of applied preload. The seta was first pushed perpendicularly against the surface and then pulled parallel to the surface, as shown in the schematic (after [44.25])

face. Under these conditions, adhesion force measured 194 ± 25 μN (≈ 10 atm adhesive pressure).

Adhesive Force of a Single Spatula

Huber et al. [44.29] used atomic force microscopy to determine the adhesion force of individual gecko spatulae. A seta with four spatulae was glued to an AFM tip. The seta was then brought into contact with a surface and a compressive preload of 90 nN was applied. The force required to pull the seta off of the surface was then measured. As seen in Fig. 44.10, there are two distinct peaks on the graph – one at 10 nN and the other at 20 nN. The first peak corresponds to one of the four spatulae adhering to the contact surface, while the peak at 20 nN corresponds to two of the four spatulae adhering to the contact surface. The average adhesion force of a single spatula was found to be 10.8 ± 1 nN. The measured value is in agreement with the measured adhesive strength of an entire gecko (of the order of 10^9 spatulae on a gecko).

44.5.2 Effects of Temperature

Environmental factors are known to affect several aspects of vertebrate function, including speed of locomotion, digestion rate, and muscle contraction, and as a result several studies have been completed to investigate impact of environmental on these functions. Relationships between the environment and other properties such as adhesion are far less studied [44.27]. Only two known studies exist that examine the affect of temperature on the clinging force of the gecko [44.27, 72]. *Losos* [44.72] examined the adhesion ability of large live geckoes at temperatures up to 17 °C. *Bergmann* and *Irschick* [44.27] expanded upon this research for body temperatures ranging from 15–35 °C. The geckoes were incubated until their body temperature reached a desired level. The clinging ability of these animals was then determined by measuring the maximum force exerted by the geckoes as they were pulled off a custom-built force plate. The clinging force of a gecko for the experimental test range is plotted in Fig. 44.11. It was determined that variation in temperature does not statistically significantly affect the adhesion force of a gecko. From these results, it was concluded that the temperature independence of adhesion supports the hypothesis of clinging as a passive mechanism (i. e., van der Waals forces). Both studies only measured the overall clinging ability on the macroscale. There have not been any investigations into the effects of temperature on the clinging ability of a single seta on the microscale, and therefore testing in this area would be extremely important.

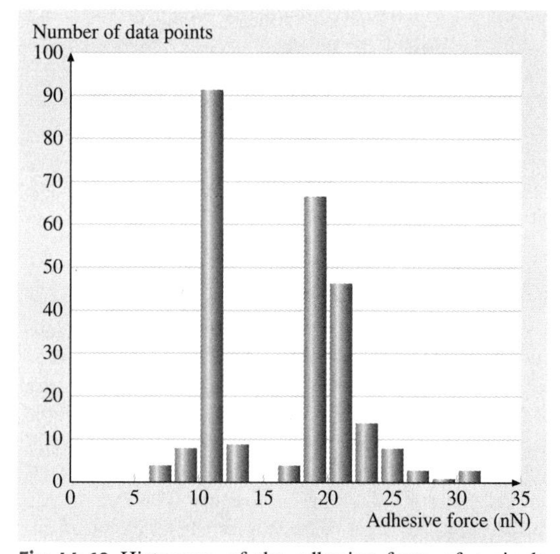

Fig. 44.10 Histogram of the adhesive force of a single gecko spatula. The *peak* at 10 nN corresponds to the adhesive force of one spatula and the *peak* at 20 nN corresponds to the adhesive force of two spatulae (after [44.29])

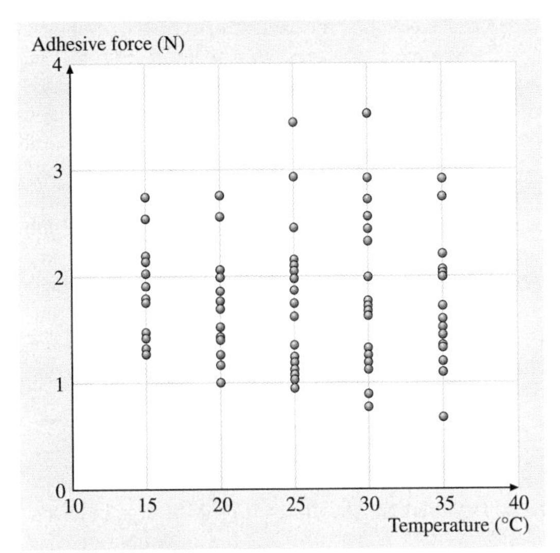

Fig. 44.11 Adhesive force of a gecko as a function of temperature (after [44.27])

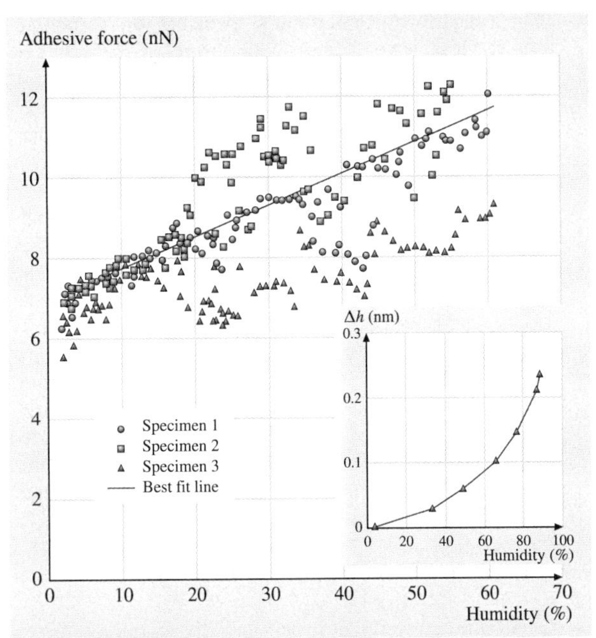

Fig. 44.12 Humidity effects on spatular pull-off force (*inset*). The increase in water film thickness on a Si wafer with increasing humidity (after [44.28])

44.5.3 Effects of Humidity

Huber et al. [44.28] employed similar methods to [44.29] (discussed previously) in order to determine

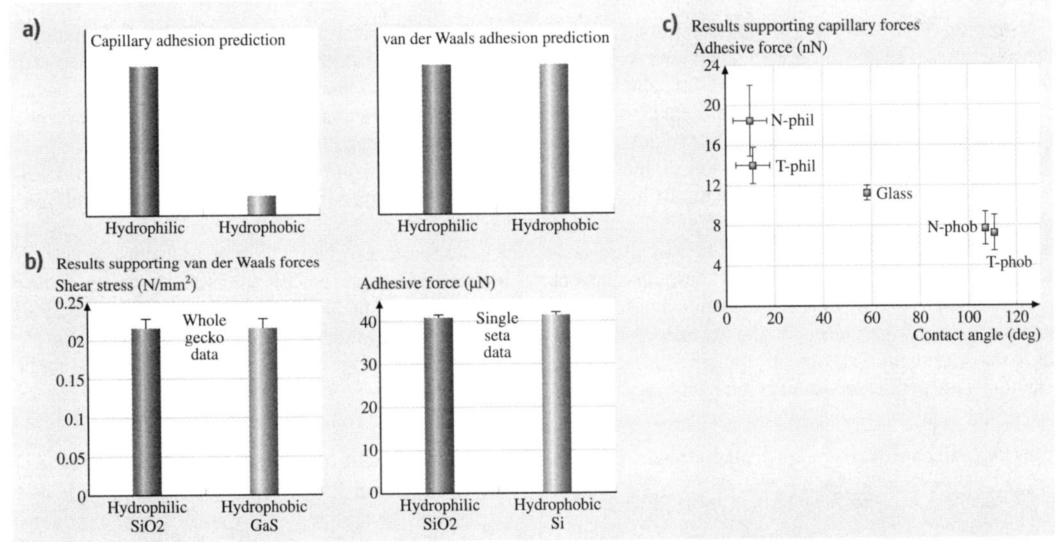

Fig. 44.13 (a) Capillary and van der Waals adhesion predictions for the relative magnitude of the adhesive force of gecko setae to hydrophilic and hydrophobic surfaces (after [44.26]). (b) Results of adhesion testing for a whole gecko and single seta with hydrophilic and hydrophobic surfaces (after [44.26]) and (c) results of adhesive force test of a single gecko spatula with surfaces with different contact angles (after [44.28])

the adhesive force of a single spatula at varying humidity. Measurements were made using an AFM placed in an airtight chamber. The humidity was adjusted by varying the flow rate of dry nitrogen into the chamber. The air was continuously monitored with a commercially available hygrometer. All tests were conducted at ambient temperature.

As seen in Fig. 44.12, even at low humidity, the adhesion force is large. An increase in humidity further increases the overall adhesion force of a gecko spatula. The pull-off force roughly doubled as the humidity was increased from 1.5% to 60%. This humidity effect can be explained in two ways: (1) by standard capillarity or (2) by a change of the effective short-range interaction due to absorbed monolayers of water – in other words, the water molecules increase the number of van der Waals bonds that are made. Based on this data, van der Waals forces are the primary adhesion mechanism, and capillary forces are a secondary adhesive mechanism.

44.5.4 Effects of Hydrophobicity

To further test the hypothesis that capillary forces play a role in gecko adhesion, the spatular pull-off force was determined for contact with both hydrophilic and hydrophobic surfaces. As seen in Fig. 44.13a, the capillary adhesion theory predicts that a gecko spatula will generate a greater adhesion force when in contact with a hydrophilic surface as compared with a hydrophobic surface, while the van der Waals adhesion theory predicts that the adhesion force between a gecko spatula and a surface will be the same regardless of the hydrophobicity of the surface [44.26]. Figure 44.13b shows the shear stress of a whole gecko and the adhesive force of a single seta on hydrophilic and hydrophobic surfaces. The data shows that the adhesion values are the same on both surfaces. This supports the van der Waals prediction of Fig. 44.13a. *Huber* et al. [44.28] found that the hydrophobicity of the attachment surface had an effect on the adhesion force of a single gecko spatula as shown in Fig. 44.13c. These results show that adhesion force has a finite value for a superhydrophobic surface and increases as the surface becomes hydrophilic. It is concluded that van der Waals forces are the primary mechanism, and capillary forces further increase the adhesion force generated.

44.6 Adhesion Modeling

With regard to the natural living conditions of the animals, the mechanics of gecko attachment can be separated into two parts: the mechanics of adhesion of a single contact to a flat surface, and the adaptation of a large number of spatulae to a natural, rough surface. Modeling of the mechanics of adhesion of spatulae to a smooth surface, in the absence of meniscus formation, was developed by *Autumn* et al. [44.26], *Jagota* and *Bennison* [44.52], and *Arzt* et al. [44.17]. As discussed in Sect. 44.3.2, the adhesion force of multiple contacts F'_{ad} can be increased by dividing the contact into a large number (n) of small contacts, while the nominal area of the contact remains the same $F'_{ad} \propto \sqrt{n} F_{ad}$. However, this model only considers contact with a flat surface. On natural, rough surfaces, the compliance and adaptability of setae are the primary sources of high adhesion. As stated earlier, the hierarchical structure of gecko setae allows for greater contact with a natural, rough surface than a nonbranched attachment system [44.49].

Bhushan et al. [44.3] and *Kim* and *Bhushan* [44.30–33] have approximated a gecko seta in contact with random rough surfaces using a hierarchical spring model. Each level of springs in their model corresponds to a level of seta hierarchy. The upper level of springs corresponds to the thicker part of gecko setae, the middle spring level corresponds to the branches, and the lower level of springs corresponds to the spatulae. The upper level is the thickest branch of the seta. It is 75 μm in

Table 44.3 Geometrical size, calculated stiffness, and typical densities of branches of seta for Tokay gecko [44.30]

Level of seta	Length (μm)	Diameter (μm)	Bending stiffness[a] (N/m)	Typical density (#/mm^2)
III upper	75	5	2.908	14×10^3
II middle	25	1	0.126	–
I lower	2.5	0.1	0.0126	$1.4–14 \times 10^6$

[a] for elastic modulus of 10 GPa with load applied at 60° to spatula long axis

length and 5 μm in diameter. The middle level, referred to as a branch, has a length of 25 μm and a diameter of 1 μm. The lower level, called a spatula, is the thinnest branch, with a length of 2.5 μm and a diameter of ≈ 0.1 μm (Table 44.3). As reported earlier, *Autumn* et al. [44.25] showed that the optimal attachment angle between the substrate and a gecko seta is 30° in the single-seta pull-off experiment. This finding is supported by the adhesion models of setae as cantilever beams [44.37, 51] (see Sect. 44.3.3 for more details). Therefore, θ was fixed at 30° in the studies by *Bhushan* et al. [44.3] and *Kim* and *Bhushan* [44.30–33] presented below.

44.6.1 Single-Spring Contact Analysis

In their analysis, *Bhushan* et al. [44.3] and *Kim* and *Bhushan* [44.30–33] assumed the tip of the spatula in a single contact to be spherical. The springs on every level of hierarchy have the same stiffness as the bending stiffness of the corresponding branches of seta. If the beam is oriented at an angle θ to the substrate and the contact load F is aligned normal to the substrate, its components along and tangential to the direction of the beam, $F\cos\theta$ and $F\sin\theta$, give rise to bending and compressive deformations, δ_b and δ_c, respectively, of [44.74]

$$\delta_b = \frac{F\cos\theta l_m^3}{3EI}\,, \quad \delta_c = \frac{F\sin\theta l_m}{A_c E}\,, \tag{44.17}$$

where $I = \pi R_m^4/4$ and $A_c = \pi R_m^2$ are the moments of inertia of the beam and the cross-sectional area, respectively, l_m and R_m are the length and radius of seta branches, respectively, and m is the level number. The net displacement, $\delta_\perp$ normal to the substrate, is given by

$$\delta_\perp = \delta_c \sin\theta + \delta_b \cos\theta\,. \tag{44.18}$$

Using (44.17) and (44.18), the stiffness of seta branches k_m is calculated as [44.75]

$$k_m = \frac{\pi R_m^2 E}{l_m \sin^2\theta\left(1 + \frac{4 l_m^2 \cot^2\theta}{3R_m^2}\right)}\,. \tag{44.19}$$

For an assumed elastic modulus E of seta material of 10 GPa with a load applied at an angle of 60° to spatulae long axis, *Kim* and *Bhushan* [44.30] calculated the stiffness of every level of seta as given in Table 44.3.

In the model, both the tips of a spatula and the asperity summits of the rough surface are assumed to be spherical with constant radius [44.3]. As a result, a single spatula adhering to a rough surface was modeled as the interaction between two spherical tips. Because β-keratin has a high elastic modulus [44.13, 45], the adhesion force between two round tips was calculated according to the Derjaguin–Muller–Toporov (DMT) theory [44.76] as

$$F_{ad} = 2\pi R_c W_{ad}\,, \tag{44.20}$$

where R_c is the reduced radius of contact, which is calculated as $R_c = (1/R_1 + 1/R_2)^{-1}$; R_1 and R_2 are the radii of the contacting surfaces: $R_1 = R_2$, $R_c = R/2$. The work of adhesion W_{ad} is then calculated using (44.21) for two flat surfaces separated by a distance D [44.61]

$$W_{ad} = -\frac{H}{12\pi D^2}\,, \tag{44.21}$$

where H is the Hamaker constant, which depends on the medium between the two surfaces. Typical values of the Hamaker constant for polymers are $H_{air} = 10^{-19}$ J in air and $H_{water} = 3.7\times10^{-20}$ J in water [44.61]. For a gecko seta, which is composed of β-keratin, the value of H is assumed to be 10^{-19} J. The work of adhesion of two surfaces in contact separated by an atomic distance $D \approx 0.2$ nm is ≈ 66 mJ/m^2 [44.61]. By assuming that the tip radius R is 50 nm, using (44.20), the adhesion force of a single contact is calculated as 10 nN [44.30]. This value is identical to the adhesion force of a single spatula measured by *Huber* et al. [44.29]. This adhesion force is used as a critical force in the model for judging whether the contact between the tip and the surface is broken or not during pull-off cycle [44.3]. If the elastic force of a single spring is less than the adhesion force, the spring is regarded as having been detached.

44.6.2 Multilevel Hierarchical Spring Analysis

In order to study the effect of the number of hierarchical levels in the attachment system on attachment ability, models with one [44.3,30,31], two [44.3,30,31], and three levels [44.30, 31] of hierarchy were simulated (Fig. 44.14). The one-level model has springs with length $l_\mathrm{I} = 2.5$ μm and stiffness $k_\mathrm{I} = 0.0126$ N/m. The length and stiffness of the springs in the two-level model are $l_\mathrm{I} = 2.5$ μm, $k_\mathrm{I} = 0.0126$ N/m and $l_\mathrm{II} = 25$ μm, $k_\mathrm{II} = 0.126$ N/m for levels I and II, respectively. The three-level model has additional upper-level springs with $l_\mathrm{III} = 75$ μm, $k_\mathrm{III} = 2.908$ N/m on the

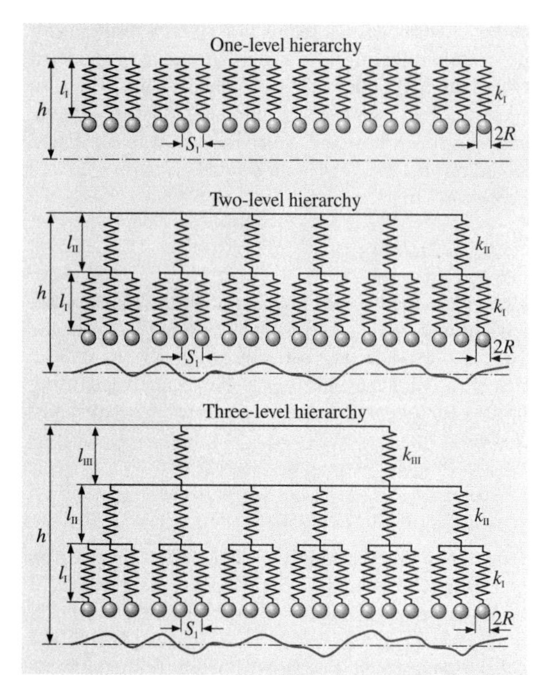

Fig. 44.14 One-, two-, and three-level hierarchical spring models for simulating the effect of hierarchical morphology on interaction of a seta with a rough surface. In this figure, $l_{\text{I,II,III}}$ are the lengths of the structures, s_{I} is the space between spatulae, $k_{\text{I,II,III}}$ are the stiffnesses of the structures, I, II, and III are level indexes, R is the tip radius, and h is the distance between the upper spring base of each model and the mean line of the rough profile (after [44.30])

springs of the two-level model, which is identical to gecko setae. The base of the springs and the connecting plate between the levels are assumed to be rigid. The distance S_{I} between the neighboring structures of level I is 0.35 μm, obtained from the average value of measured spatula density, $8\times10^6\,\text{mm}^{-2}$, calculated by multiplying 14 000 setae/mm^2 by an average of 550 spatula/seta [44.15] (Table 44.3). A 1 : 10 proportion of the number of springs in the upper level to that in the level below was assumed [44.3]. This corresponds to each spring at level III being connected to ten springs on level II, and each spring on level II being connected to ten springs on level I. The number of springs on level I considered in the model is calculated by dividing the scan length (2000 μm) by the distance S_{I} (0.35 μm), which corresponds to 5700.

The spring deflection Δl was calculated as

$$\Delta l = h - l_0 - z\,, \tag{44.22}$$

where h is the position of the spring base relative to the mean line of the surface; l_0 is the total length of a spring structure, which is $l_0 = l_{\text{I}}$ for the one-level model, $l_0 = l_{\text{I}} + l_{\text{II}}$ for the two-level model, and $l_0 = l_{\text{I}} + l_{\text{II}} + l_{\text{III}}$ for the three-level model; and z is the profile height of the rough surface. The elastic force F_{el} arising in the springs at a distance h from the surface was calculated for the one-level model as [44.3]

$$F_{\text{el}} = -k_{\text{I}} \sum_{i=1}^{p} \Delta l_i u_i\,,$$
$$u_i = \begin{cases} 1 & \text{if contact} \\ 0 & \text{if no contact} \end{cases}\,, \tag{44.23}$$

where p is the number of springs in level I of the model. For the two-level model the elastic force was calculated as [44.3]

$$F_{\text{el}} = -\sum_{j=1}^{q}\sum_{i=1}^{p} k_{ji}(\Delta l_{ji} - \Delta l_j)u_{ji}\,,$$
$$u_{ji} = \begin{cases} 1 & \text{if contact} \\ 0 & \text{if no contact} \end{cases}\,, \tag{44.24}$$

where q is the number of springs in level II of the model. For the three-level model the elastic force was calculated as [44.30]

$$F_{\text{el}} = -\sum_{k=1}^{r}\sum_{j=1}^{q}\sum_{i=1}^{p} k_{kji}(\Delta l_{kji} - \Delta l_{kj} - \Delta l_j)u_{kji}\,,$$
$$u_{kji} = \begin{cases} 1 & \text{if contact} \\ 0 & \text{if no contact} \end{cases}\,, \tag{44.25}$$

where r is the number of springs in level III of the model. The spring force when the springs approach the rough surface is calculated using either (44.23), (44.24) or (44.25) for one-, two-, and three-level models, respectively. During pull-off, the same equations are used to calculate the spring force. However, when the applied load is equal to zero, the springs do not detach due to the adhesion attraction given by (44.20). The springs are pulled apart when the pull-off force is equal to the adhesion force at the interface. The adhesion force is the lowest value of the elastic force F_{el} when the seta has detached from the contacting surface.

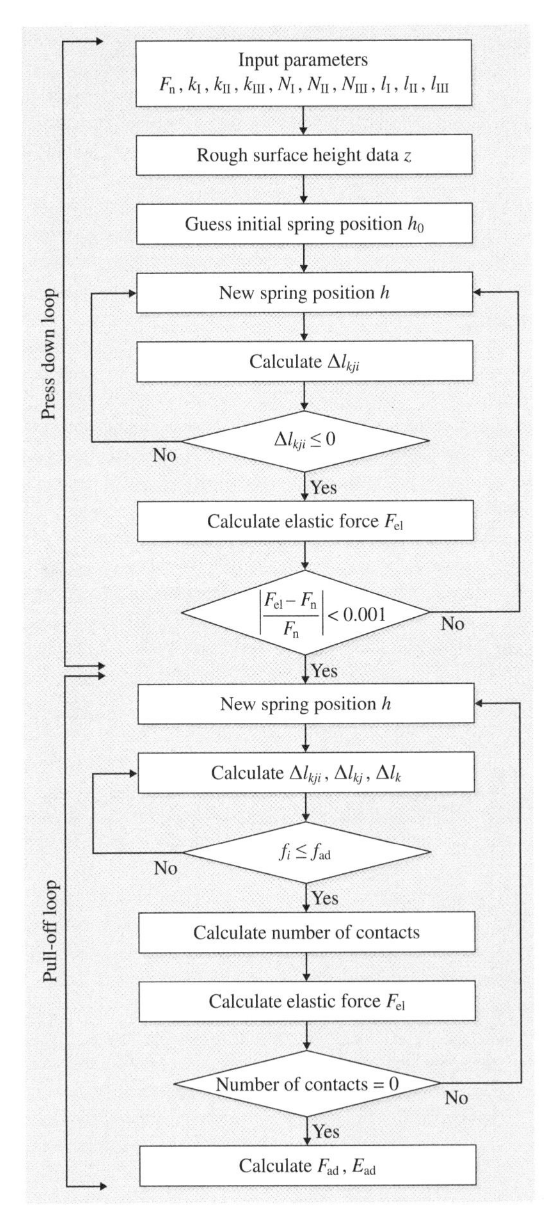

Fig. 44.15 Flow chart for the calculation of the adhesion force (F_{ad}) and the adhesion energy (E_{ad}) for the three-level hierarchical spring model. In this figure, F_n is an applied load, $k_{I,II,III}$ and $l_{I,II,III}$ are stiffnesses and lengths of structures, Δl_{kji}, Δl_{ki}, and Δl_k are the spring deformations on level I, II, and III, respectively, i, j, and k are spring indices on each level, f_i is the elastic force of a single spring and f_{ad} is the adhesion force of a single contact (after [44.30]) ◀

The adhesion energy is calculated as

$$W_{ad} = \int_{\bar{D}}^{\infty} F_{el}(D)\,dD\,, \tag{44.26}$$

where D is the distance that the spring base moves away from the contacting surface. The lower limit of the distance $\bar{D}$ is the value of D, where F_{el} is first zero when the model is pulled away from the contacting surface. Also, although the upper limit of the distance is infinity, in practice, the $F_{el}(D)$ curve is integrated to an upper limit where F_{el} increases from a negative value to zero. Figure 44.15 shows the flow chart for the calculation of the adhesion force and the adhesion energy employed by *Kim* and *Bhushan* [44.30].

The random rough surfaces used in the simulations were generated by a computer program [44.41, 42]. Two-dimensional profiles of surfaces that a gecko might encounter were obtained using a stylus profiler [44.3]. These profiles, along with the surface selection methods and surface roughness parameters [root-mean-square (RMS) amplitude σ and correlation length β^*] for scan lengths of 80, 400, and 2000 µm, are presented in Sect. 44.A. The roughness parameters are scale dependent, and therefore adhesion values also are expected to be scale dependent. As the scan length was increased, the measured values of RMS amplitude and correlation length both increased. The range of values of σ from 0.01 to 30 µm and a fixed value of $\beta^* = 200$ µm were used for modeling the contact of a seta with random rough surfaces. The chosen range covers values of roughnesses for relatively smooth, artificial surfaces to natural, rough surfaces. A typical scan length of 2000 µm was also chosen, which is comparable to a lamella length of gecko.

44.6.3 Adhesion Results of the Multilevel Hierarchical Spring Model

The multilevel hierarchical spring model was developed by *Kim* and *Bhushan* [44.30]. They obtained various useful results which will be presented next. Figure 44.16a shows the calculated spring force–distance curves for the one-, two-, and three-level hierarchical models in contact with rough surfaces of different values of RMS amplitude σ ranging from $\sigma = 0.01$ to $30\mu m$ at applied load of 1.6 µN, which was derived from the gecko's weight. When the spring model is pressed against the rough surface, contact between the spring and the rough surface occurs at point A; as the spring tip presses into the contacting surface, the force increases up to point B, B′ or B″. During pull off, the spring relaxes, and the spring force passes an equilibrium state (0 N); tips break free of adhesion forces at point C, C′ or C″ as the spring moves away from the surface. The perpendicular distance from C, C′ or C″ to zero is the adhesion force. The adhesion energy stored during contact can be obtained by calculating the area of the triangle during the unloading part of the curves (44.26).

Using the spring force–distance curves, *Kim* and *Bhushan* [44.30] calculated the adhesion coefficient, the number of contacts per unit length, and the adhesion energy per unit length of the one-, two-, and three-level models for an applied load of 1.6 μN and a wide range of RMS roughness (σ), as seen in the left graphs of Fig. 44.16b. The adhesion coefficient, defined as the ratio of the pull-off force to the applied preload, represents the strength of adhesion with respect to the preload. For the applied load of 1.6 μN, which corresponds to the weight of a gecko, the maximum adhesion coefficient is ≈ 36 when σ is smaller than 0.01 μm. This means that a gecko can generate enough adhesion force to support 36 times its body weight. However, if σ is increased to 1 μm, the adhesion coefficient for the three-level model is reduced to 4.7. It is noteworthy that the adhesion coefficient falls < 1 when the contacting surface has an RMS roughness $\sigma > 10$ μm. This implies that the attachment system is no longer capable of supporting the gecko's weight. *Autumn* et al. [44.25, 26] showed that, in isolated gecko setae contacting with the surface of a single-crystalline silicon wafer, a 2.5 μN preload yielded adhesion of 20–40 μN and thus a value of adhesion coefficient of 8–16, which supports the simulation results of *Kim* and *Bhushan* [44.30].

Figure 44.16b (top left) shows that the adhesion coefficient for the one-level model is lower than that for the three-level model, but there is only a small difference between the values of the two- and three-level models. In order to show the effect of stiffness, the results are plotted for the three-level model with springs in level III with stiffness 10 times smaller than originally. It can be seen that the three-level model with a third-level stiffness of 0.1 k_{III} has a 20–30% higher adhesion coefficient than the original three-level model. The results also show that the trends in the number of contacts are similar to that of the adhesive force. The study also investigated the effect of σ on adhesion energy. It was determined that adhesion energy decreased with in-

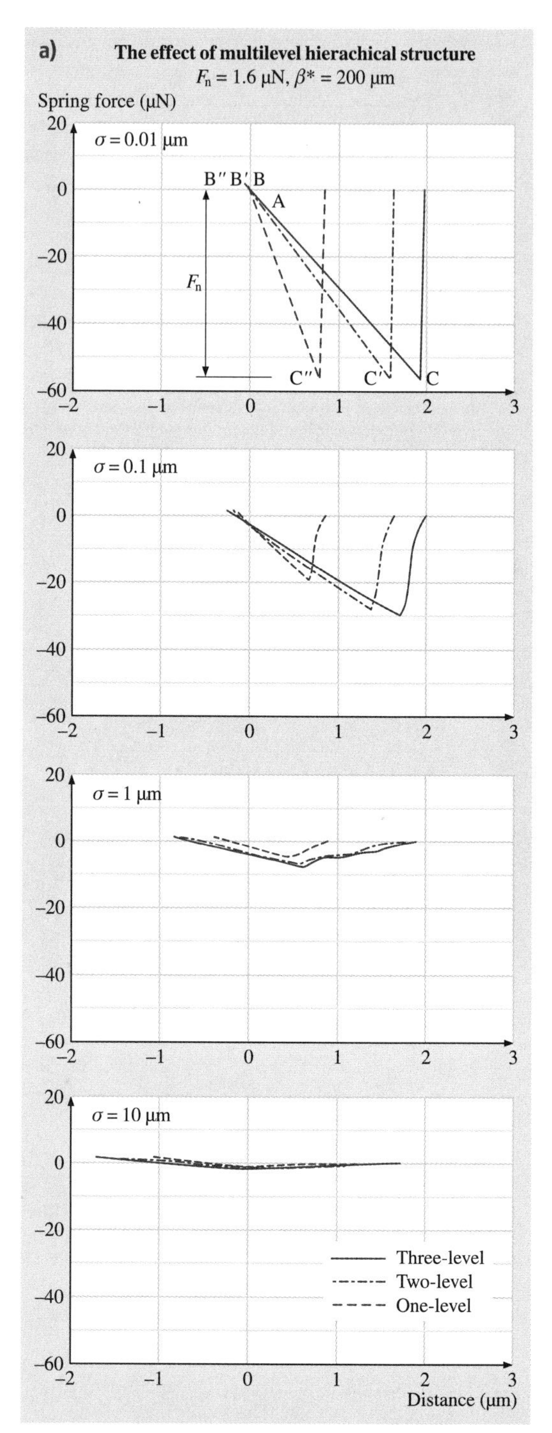

Part F | 44.6

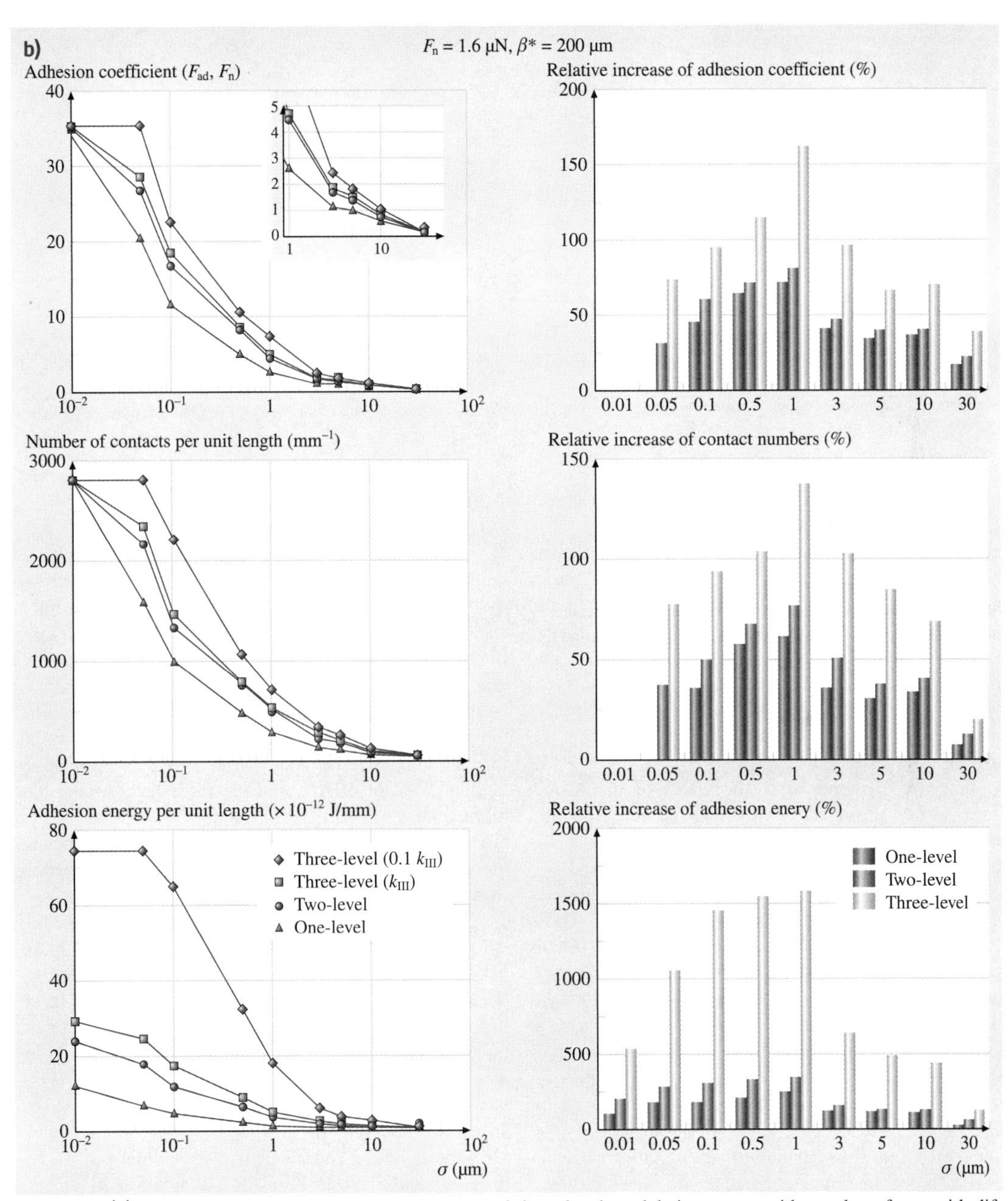

Fig. 44.16 (a) Force–distance curves of the one-, two-, and three-level models in contact with rough surfaces with different σ values for applied load of 1.6 μN. (b) Adhesion coefficient, number of contacts, and adhesion energy per unit length of profile for one- and multilevel models with increasing σ value (*left figures*), and relative increases between multi- and one-level models (*right-hand side*) for applied load of 1.6 μN. The value of k_{III} in the analysis is 2.908 N/m (after [44.30]) ◄ ▲

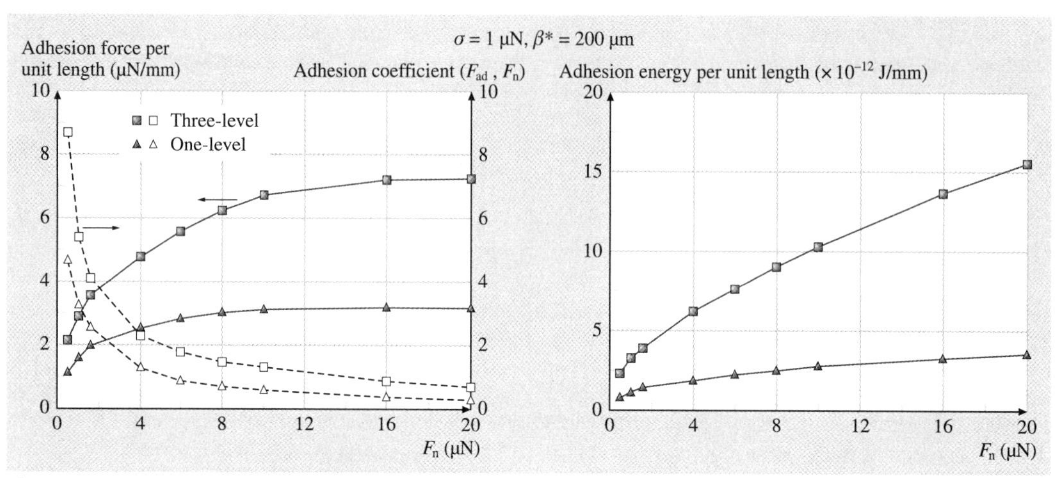

Fig. 44.17 The adhesion force, adhesion coefficient, and adhesion energy as a function of applied loads for both one- and three-level models contacting with the rough surface (after [44.30])

creasing σ. For the smooth surface with $\sigma = 0.01\,\mu\text{m}$, the adhesion energies for the two- and three-level hierarchical models are 2 and 2.4 times larger than that for the one-level model, respectively, but the adhesion energy decreases rapidly at surfaces with $\sigma > 0.05\,\mu\text{m}$; and in every model it finally decreases to zero at surfaces with $\sigma > 10\,\mu\text{m}$. The adhesion energy for the three-level model with $0.1\,k_{\text{III}}$ is 2–3 times higher than that for the original three-level model.

In order to demonstrate the effect of the hierarchical structure on adhesion enhancement, *Kim* and *Bhushan* [44.30] calculated the increases in the adhesion coefficient, the number of contacts, and the adhesion energy of the two-, three- and three-level (with $0.1\,k_{\text{III}}$) models relative to the one-level model. These results are shown on the right side of Fig. 44.16b. It was found that, for the two- and three-level models, the adhesion coefficient increases slowly with increasing σ and has maximum values of ≈ 70 and 80% at $\sigma = 1\,\mu\text{m}$, respectively, and then decreases for surfaces with $\sigma > 3\,\mu\text{m}$. The condition at which a significant enhancement occurs is related to the maximum spring deformation, which is the applied load divided by the spring stiffness. If the maximum spring deformation is greater than 2–3 times larger than the σ value of the surface roughness, significant adhesion enhancement occurs. The three-level model with $0.1\,k_{\text{III}}$ shows significant adhesion enhancement. The relative increase of the adhesion coefficient and adhesion energy for the three-level model with $0.1\,k_{\text{III}}$ has maximum values at $\sigma = 1\,\mu\text{m}$.

Figure 44.17 shows the variation of adhesion force and adhesion energy as a function of applied load for both one- and three-level models contacting a surface with $\sigma = 1\,\mu\text{m}$. It is shown that, as the applied load increases, the adhesion force increases up to a certain applied load and then has a constant value, whereas adhesion energy continues to increase with increasing applied load. The one-level model has a maximum value of adhesion force per unit length of $\approx 3\,\mu\text{N/mm}$ at an applied load of $10\,\mu\text{N}$, and the three-level model has a maximum value of $\approx 7\,\mu\text{N/mm}$ at an applied load of $16\,\mu\text{N}$. However, the adhesion coefficient continues to decrease at higher applied loads because the adhesion force is constant even if the applied load increases.

The simulation results for the three-level model, which is close to gecko setae, presented in Fig. 44.16 show that roughness reduces the adhesion force. At surfaces with $\sigma > 10\,\mu\text{m}$, the ratio of the adhesion force to the gecko weight indicates that it cannot support itself. However, in practice, a gecko can cling to or crawl on the surface of ceilings with higher roughness. *Kim* and *Bhushan* [44.30] did not consider the effect of lamellae in their study. The authors state that the lamellae can adapt to the waviness of a surface while the setae allow for adaptation to micro- or nanoroughness and expect that adding the lamellae of gecko skin to the model would lead to higher adhesion over a wider range of roughness. In addition, their hierarchical model only considers deformation normal to the surface and the motion of setae. It should be noted that the measurements of the adhesion force of a single gecko seta made by *Autumn* et al. [44.25] demonstrated that a load applied normal to the surface was insufficient for effective attachment of seta.

Finally the effects of spring stiffness and the number of springs on the adhesion enhancement of the multi-level hierarchical model (one- and three-level models with four different spring stiffnesses and three different numbers of springs) were analyzed by *Kim* and *Bhushan* [44.31]. The stiffness k_{I} was taken equal to 0.0126 N/m, as before. Other stiffnesses, k_{II} and k_{III} were normalized with respect to k_{I}. The three-level model with $k_{\mathrm{III}}/k_{\mathrm{I}} = 100$ and $k_{\mathrm{II}}/k_{\mathrm{I}} = 10$ had similar stiffness values to gecko seta, as presented in Table 44.3 and used in the previous example (Figs. 44.16 and 44.17). The left part in Fig. 44.18 shows the adhesion coefficient, number of contacts, and adhesion energy per unit length for the one- and three-level models with four different spring stiffnesses as a function of σ value for an applied load of 1.6 μN. Trends as a function of σ are the same as observed previously in Fig. 44.16b. For the case of $k_{\mathrm{III}} = k_{\mathrm{II}} = k_{\mathrm{I}}$, one gets the highest value (36) of the adhesion coefficient on a rough surface, and it remains high up to σ value of $\approx 1\,\mu$m, and then starts to decrease. As the stiffness values k_{II} and k_{III} increase, the adhesion coefficient starts to decrease at lower values of σ, and it decreases rapidly with increasing σ. The number of contacts and adhesion energy per unit length as a function of σ have trends similar to that of adhesion coefficient. The right part in Fig. 44.18 shows the relative increase between the one- and three-level models. The trends are the same as discussed earlier.

To study the effect of the number of springs, three different cases of the number of springs in the upper level compared with in the lower level were considered. The three-level model with $N_{\mathrm{I}}/N_{\mathrm{II}} = 10$ and $N_{\mathrm{II}}/N_{\mathrm{III}} = 10$ is closest to the case of the gecko's setae (discussed earlier). Figure 44.19 shows the adhesion force, adhesion coefficient, and adhesion energy as a function of applied loads for one- and three-level models with different numbers of springs contacting with the rough surface. The variation of the number of springs on each level affects the equivalent stiffness of the model. As the number of springs on the lower level increases, the equivalent stiffness decreases. The figure shows that the three-level model with $N_{\mathrm{I}}/N_{\mathrm{II}} = 100$ and $N_{\mathrm{II}}/N_{\mathrm{III}} = 10$ gives the largest adhesion force and adhesion energy among the models, because the equivalent stiffness is lowest.

44.6.4 Capillary Effects

Kim and *Bhushan* [44.33] investigated the effects of capillarity on gecko adhesion by considering the capillary force as well as the solid–solid interaction. The Laplace and surface tension components of the capillary force are treated according to Sect. 44.4.2. The solid–solid adhesive force was calculated by DMT theory according to (44.20) and will be denoted by F_{DMT}.

The work of adhesion was then calculated by (44.21). *Kim* and *Bhushan* [44.33] assumed typical values of the Hamaker constant to be $H_{\mathrm{air}} = 10^{-19}$ J in air and $H_{\mathrm{water}} = 6.7\times10^{-19}$ J in water [44.61]. The work of adhesion of two surfaces in contact separated by an atomic distance $D \approx 0.2$ nm [44.61] is $\approx 66\,\mathrm{mJ/m^2}$ in air and $44\,\mathrm{mJ/m^2}$ in water. Assuming the tip radius R to be 50 nm, the DMT adhesion force F_{DMT} of a single contact in air and in water is $F_{\mathrm{DMT}}^{\mathrm{air}} = 11$ nN and $F_{\mathrm{DMT}}^{\mathrm{water}} = 7.3$ nN, respectively. As the humidity increases from 0% to 100%, the DMT adhesion force will take a value between $F_{\mathrm{DMT}}^{\mathrm{air}}$ and $F_{\mathrm{DMT}}^{\mathrm{water}}$. To calculate the DMT adhesion force for intermediate humidity, an approximation method by *Wan* et al. [44.77] was used. The work of adhesion W_{ad} for the intermediate humidity can be expressed as

$$W_{\mathrm{ad}} = \int_D^\infty \frac{H}{6\pi h^3}\,\mathrm{d}h = \int_D^{h_{\mathrm{f}}} \frac{H_{\mathrm{water}}}{6\pi h^3}\,\mathrm{d}h + \int_{h_{\mathrm{f}}}^\infty \frac{H_{\mathrm{air}}}{6\pi h^3}\,\mathrm{d}h\,, \tag{44.27}$$

where h is the separation along the plane. h_{f} is the water film thickness at a filling angle of ϕ, which can be calculated as

$$h_{\mathrm{f}} = D + R(1-\cos\phi)\,. \tag{44.28}$$

Therefore, using (44.20), (44.27), and (44.28), the DMT adhesion force for intermediate humidity is given by

$$F_{\mathrm{DMT}} = F_{\mathrm{DMT}}^{\mathrm{air}}\left\{1 - \frac{1}{\left[1+R(1-\cos\phi)/D\right]^2}\right\} + F_{\mathrm{DMT}}^{\mathrm{water}}\left\{\frac{1}{\left[1+R(1-\cos\phi)/D\right]^2}\right\}. \tag{44.29}$$

Finally, *Kim* and *Bhushan* [44.33] calculated the total adhesion force F_{ad} as the sum of (44.16) and (44.29)

$$F_{\mathrm{ad}} = F_{\mathrm{c}} + F_{\mathrm{DMT}}\,. \tag{44.30}$$

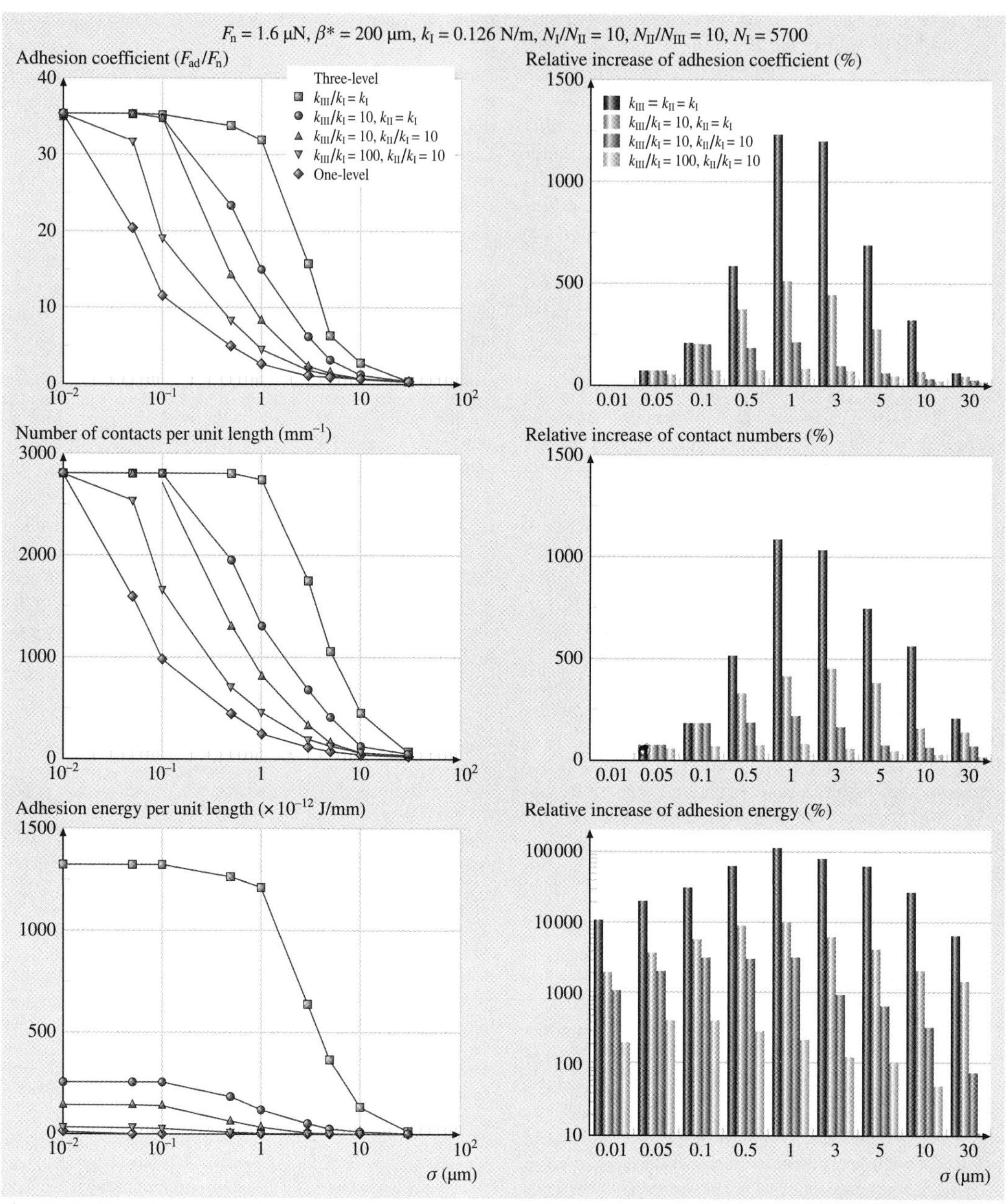

Fig. 44.18 The adhesion coefficient, number of contacts, and adhesion energy per unit length of profile for one- and three-level models with different spring stiffnesses as a function of σ value (*left column*), and relative increases between one- and three-level models (*right column*) for applied load of 1.6 μN (after [44.31])

Kim and *Bhushan* [44.33] then used the total adhesion force as a critical force in the three-level hierarchical spring model discussed previously. In the spring model for gecko seta, if the force applied upon

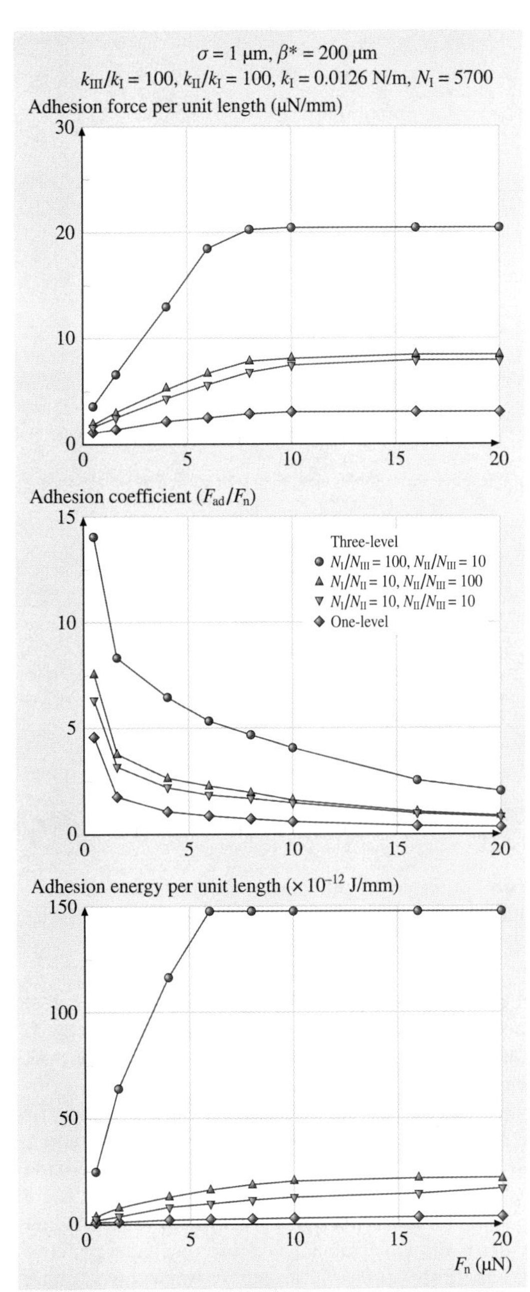

Fig. 44.19 The adhesive force, adhesion coefficient, and adhesion energy as a function of applied loads for one- and three-level models with different number of springs contacting with the rough surface (after [44.31])

spring deformation is greater than the adhesion force, the spring is regarded as having been detached.

To simulate the capillary contribution to the adhesion force for a gecko spatula, *Kim* and *Bhushan* [44.33] set the contact angle on a gecko spatula tip to be $\theta_1 = 128°$ [44.28]. It was assumed that the spatula tip radius $R = 50$ nm, the ambient temperature $T = 25\,°C$,

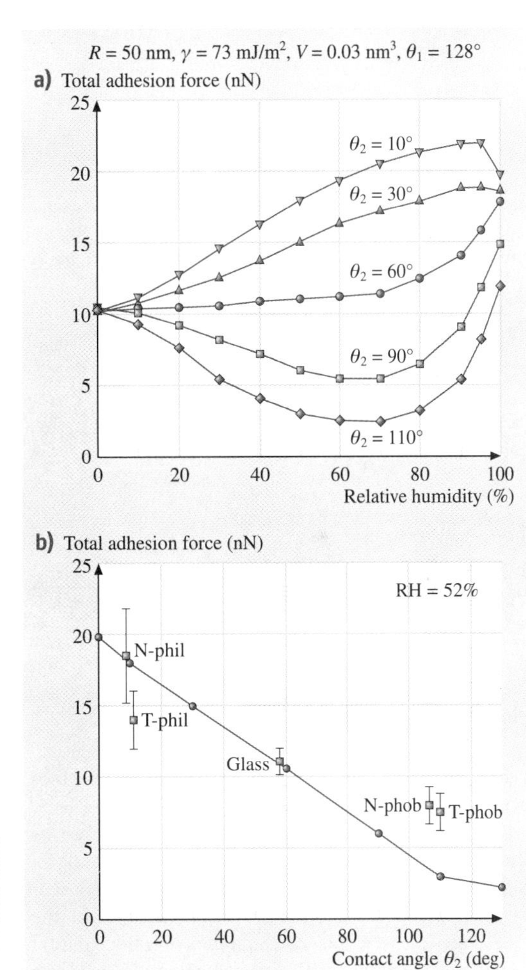

Fig. 44.20 (a) Total adhesion force as a function of relative humidity for a single spatula in contact with surfaces with different contact angles. (b) Comparison of the simulation results of *Kim* and *Bhushan* [44.33] with the measured data obtained by *Huber* et al. [44.28] for a single spatula in contact with hydrophilic and hydrophobic surfaces (after [44.33])

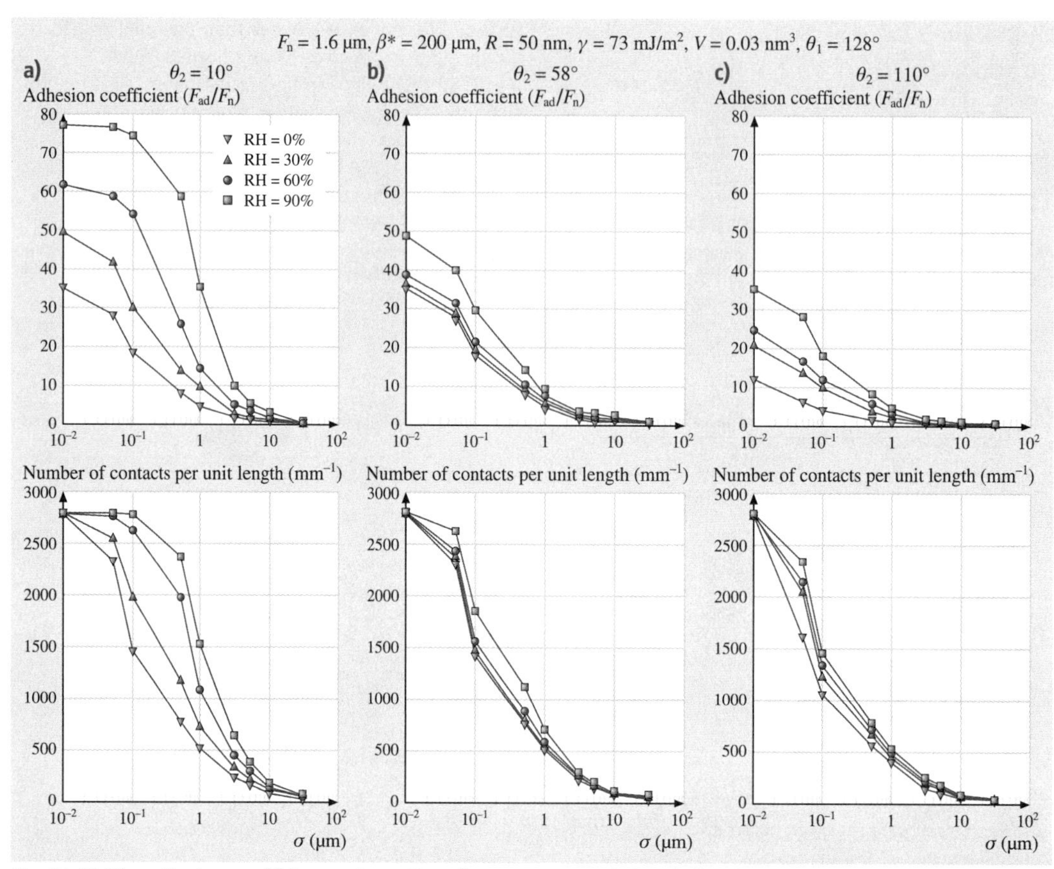

Fig. 44.21 The adhesion coefficient and number of contacts per unit length for the three-level hierarchical model in contact with rough surfaces with different values of RMS amplitudes σ and contact angles for different relative humidities (after [44.33])

the surface tension of water $\gamma = 73\,\mathrm{mJ/m^2}$, and the molecular volume of water $V = 0.03\,\mathrm{nm^3}$ [44.61].

Figure 44.20a shows the total adhesion force as a function of relative humidity for a single spatula in contact with surfaces with different contact angles. Total adhesion force decreases with an increase in the contact angle on the substrate, and the difference of total adhesion force among different contact angles is larger in the intermediate-humidity regime. As the relative humidity increases, the total adhesion force for surfaces with contact angles $< 60°$ has a higher value than the DMT adhesion force not considering wet contact, whereas for contact angles $> 60°$, the total adhesion force has lower values at most relative humidity.

The simulation results of *Kim* and *Bhushan* [44.33] are compared with the experimental data by *Huber* et al. [44.28] in Fig. 44.20b. *Huber* et al. [44.28] measured the pull-off force of a single spatula in contact with four different types of Si wafer and glass at ambient temperature of 25 °C and relative humidity of 52%. According to their description, wafer families "N" and "T" in Fig. 44.20b differ by the thickness of the top amorphous Si oxide layer. The "phil" type is cleaned Si oxide surface, which is hydrophilic with a water contact angle of $\approx 10°$, whereas the "phob" type is a Si wafer covered with a hydrophobic monolayer, resulting in a water contact angle of $> 100°$. The glass has a water contact angle of 58°. *Huber* et al. [44.28] showed that the adhesion force of a gecko spatula rises significantly for substrates with increasing hydrophilicity (adhesive force increases by a factor of two as the mating surfaces go from hydrophobic to hydrophilic). As shown in Fig. 44.20b, the simulation results of *Kim* and *Bhushan* [44.33] closely match the experimental data of *Huber* et al. [44.28].

Kim and *Bhushan* [44.33] carried out adhesion analysis for a three-level hierarchical model for gecko seta. Figure 44.21 shows the adhesion coefficient and number of contacts per unit length for the three-level hierarchical model in contact with rough surfaces with different values of the RMS amplitude σ ranging from $\sigma = 0.01$ to $30\,\mu m$ for different relative humidity values and contact angles of the surface. It can be seen that, for a surface with contact angle $\theta_2 = 10°$, the adhesion coefficient is greatly influenced by relative humidity. At 0% relative humidity the maximum adhesion coefficient is ≈ 36 at a value of $\sigma < 0.01\,\mu m$ compared with 78 for 90% relative humidity for the same surface roughness. As expected the effect of relative humidity on increasing the adhesion coefficient decreases as the contact angle becomes larger. For hydrophobic surfaces, relative humidity decreases the adhesion coefficient. Similar trends can be noticed in terms of the number of contacts. Thus, the conclusion can be drawn that hydrophilic surfaces are beneficial to gecko adhesion enhancement.

44.7 Modeling of Biomimetic Fibrillar Structures

The mechanics of adhesion between a fibrillar structure and a rough surface as it relates to the design of biomimetic structures has been a topic of investigation by many researchers [44.31, 32, 37, 49, 52, 75, 78–80]. *Kim* and *Bhushan* [44.32] developed a convenient, general, and useful guideline for understanding biological systems and for improving biomimetic attachment. This adhesion database was constructed by modeling fibers as oriented cylindrical cantilever beams with spherical tips. The authors then carried out numerical simulation of the attachment system in contact with random rough surfaces considering three constraint conditions: buckling, fracture, and sticking of the fiber structure. For a given applied load and roughnesses of contacting surface and fiber material, a procedure to find an optimal fiber radius and aspect ratio for the desired adhesion coefficient was developed.

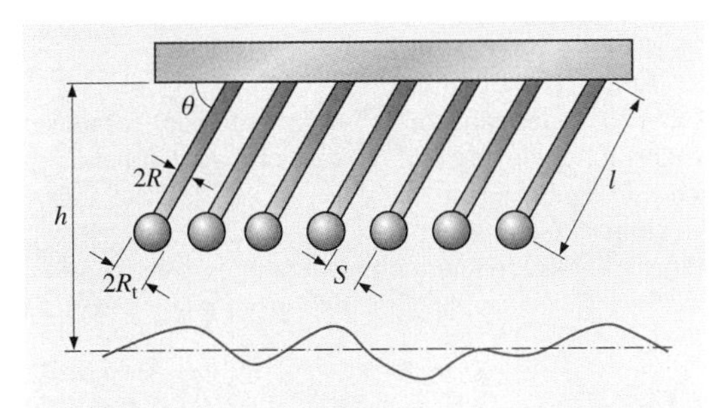

Fig. 44.22 Single-level attachment system with oriented cylindrical cantilever beams with spherical tip. In this figure, l is the length of fibers, θ is the fiber orientation, R is the fiber radius, R_t is the tip radius, S is the spacing between fibers, and h is the distance between the base of the model and the mean line of the rough profile (after [44.32])

The model of *Kim* and *Bhushan* [44.32] is used to find the design parameters for fibers of a single-level attachment system capable of achieving desired properties, i.e., high adhesion coefficient and durability. The design variables for an attachment system are as follows: fiber geometry (radius and aspect ratio of fibers, tip radius), fiber material, fiber density, and fiber orientation. The optimal values for the design variables to achieve the desired properties should be selected for the fabrication of a biomimetic attachment system.

44.7.1 Fiber Model

The fiber model of *Kim* and *Bhushan* [44.32] consists of a simple idealized fibrillar structure consisting of a single-level array of micro/nanobeams protruding from a backing, as shown in Fig. 44.22. The fibers are modeled as oriented cylindrical cantilever beams with spherical tips. In Fig. 44.22, l is the length of fibers, θ is the fiber orientation, R is the fiber radius, R_t is the tip radius, S is the spacing between fibers, and h is the distance between the upper spring base of each model and the mean line of the rough profile. The end terminal of the fibers is assumed to be a spherical tip with constant radius and constant adhesion force.

44.7.2 Single-Fiber Contact Analysis

Kim and *Bhushan* [44.32] modeled an individual fiber as a beam oriented at an angle θ to the substrate, and the contact load F aligned normal to the substrate. The net displacement normal to the substrate can be calculated according to (44.17) and (44.18). The fiber stiffness ($k = F/\delta_\perp$) is given by [44.75]

$$k = \frac{\pi R^2 E}{l \sin^2\theta \left(1 + \frac{4l^2 \cot^2\theta}{3R^2}\right)} = \frac{\pi R E}{2\lambda \sin^2\theta \left(1 + \frac{16\lambda^2 \cot^2\theta}{3}\right)}, \quad (44.31)$$

where $\lambda = l/2R$ is the aspect ratio of the fiber and θ is fixed at 30°.

Two alternative models dominate the world of contact mechanics: the Johnson–Kendall–Roberts (JKR) theory [44.44] for compliant solids, and the Derjaguin–Muller–Toporov (DMT) theory [44.76] for stiff solids. Although gecko setae are composed of β-keratin with a high elastic modulus [44.13, 45], which is close to the DMT model, in general the JKR theory prevails for biological or artificial attachment systems. Therefore the JKR theory was applied in the subsequent analysis of *Kim* and *Bhushan* [44.32] to compare materials with wide ranges of elastic modulus. The adhesion force between a spherical tip and a rigid flat surface is thus calculated using the JKR theory as [44.44]

$$F_{ad} = \frac{3}{2}\pi R_t W_{ad}, \quad (44.32)$$

where R_t is the radius of spherical tip and W_{ad} is the work of adhesion, calculated according to (44.25). *Kim* and *Bhushan* [44.32] used this adhesion force as a critical force. If the elastic force of a single spring is less than the adhesion force, they regarded the spring as having been detached.

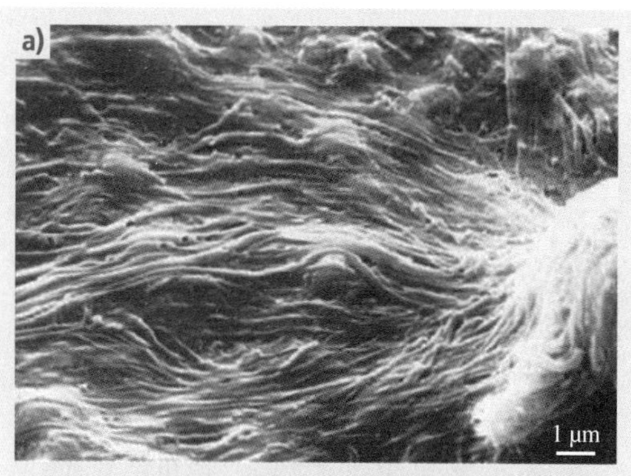

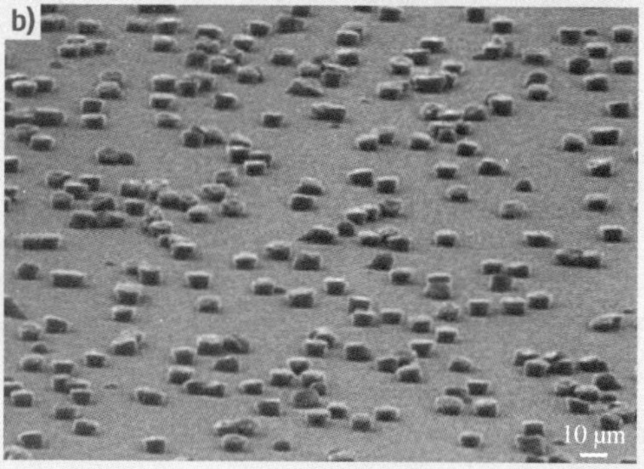

Fig. 44.23a,b SEM micrographs of **(a)** high-aspect-ratio polymer fibrils that have collapsed under their own weight and **(b)** low-aspect-ratio polymer fibrils that are incapable of adapting to rough surfaces (after [44.49])

44.7.3 Constraints

In the design of fibrillar structures, a trade-off exists between the aspect ratio of the fibers and their adaptability to a rough surface. If the aspect ratio of the fibers is too large, they can adhere to each other or even collapse under their own weight, as shown in Fig. 44.23a. If the aspect ratio is too small (Fig. 44.23b), the structures will lack the compliance necessary to conform to a rough surface. The spacing between the individual fibers is also important. If the spacing is too small, adjacent fibers can attract each other through intermolecular forces, which will lead to bunching. Therefore, *Kim* and *Bhushan* [44.32] considered three necessary conditions in their analysis: buckling, fracture, and sticking of fiber structure, which constrain the allowed geometry.

Nonbuckling Condition

A fibrillar interface can deliver a compliant response while still employing stiff materials because of bending and microbuckling of fibers. Based on classical Euler buckling, *Glassmaker* et al. [44.75] established a stress–strain relationship and a critical compressive strain for buckling ε_{cr} for a fiber oriented at an angle θ to the substrate

$$\varepsilon_{cr} = -\frac{b_c \pi^2}{3\left(Al^2/3I\right)} \left(1 + \frac{A_c l^2}{3I} \cot^2\theta\right), \quad (44.33)$$

where A_c is the cross-sectional area of the fibril, and b_c is a factor that depends on boundary conditions. The factor b_c has a value of 2 for pinned–clamped microbeams. For fibers having a circular cross-section, ε_{cr}

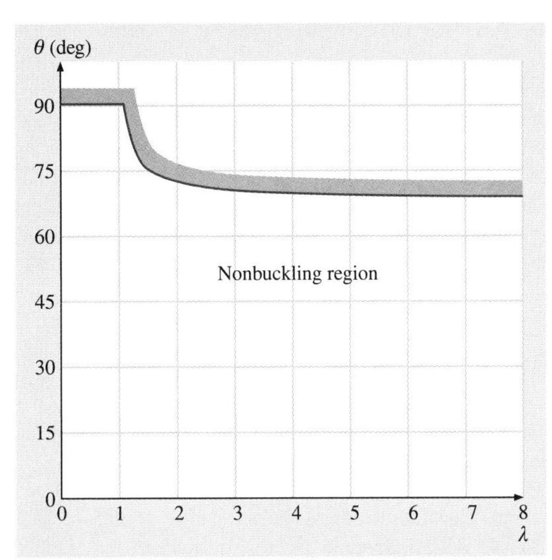

Fig. 44.24 Critical fiber orientation as a function of aspect ratio λ for the nonbuckling condition for pinned–clamped microbeams ($b_c = 2$) (after [44.32])

is calculated as

$$\varepsilon_{cr} = -\frac{b_c\pi^2}{3\left(4l^2/3R^2\right)}\left(1+\frac{4l^2}{3R^2}\cot^2\theta\right) = -b_c\pi^2\left(\frac{1}{16\lambda^2}+\frac{\cot^2\theta}{3}\right). \quad (44.34)$$

In (44.34), ε_{cr} depends on both the aspect ratio λ and the orientation θ of fibers. If $\varepsilon_{cr} = 1$, which means the fiber deforms up to the backing, buckling does not occur. Figure 44.24 plots the critical orientation θ as a function of aspect ratio for the case of $\varepsilon_{cr} = 1$. The critical fiber orientation for buckling is 90° at $\lambda < 1.1$. This means that buckling does not occur regardless of the orientation of the fiber at $\lambda < 1.1$. For $\lambda > 1.1$, the critical fiber orientation for buckling decreases with an increase in λ, and has a constant value of 69° at $\lambda > 3$. *Kim* and *Bhushan* [44.32] used a fixed value at 30° for θ, because as stated earlier, the maximum adhesive force is achieved at this orientation, and buckling is not expected to occur.

No-Fiber-Fracture Condition

For small contacts, the strength of the system will eventually be determined by fracture of the fibers. *Spolenak* et al. [44.81] suggested the limit of fiber fracture as a function of the adhesion force. The axial stress σ_f in a fiber is limited by its theoretical fracture strength σ_{th}^f as

$$\sigma_f = \frac{F_{ad}}{R^2\pi} \le \sigma_{th}^f . \quad (44.35)$$

Using (44.32), a lower limit for the useful fiber radius, R, is calculated as

$$R \ge \sqrt{\frac{3R_t W_{ad}}{2\sigma_{th}^f}} \approx \sqrt{\frac{15R_t W_{ad}}{E}} , \quad (44.36)$$

where the theoretical fracture strength is approximated by $E/10$ [44.82]. The lower limit of fiber radius for fiber fracture by the adhesion force depends on elastic modulus. By assuming $W_{ad} = 66\,\mathrm{mJ/m^2}$ as stated earlier, *Kim* and *Bhushan* [44.32] calculated the lower limits of fiber radius for $E = 1\,\mathrm{MPa}$, 0.1 GPa, and 10 GPa to be 0.32, 0.032, and 0.0032 μm, respectively.

The contact stress cannot exceed the ideal contact strength transmitted through the actual contact area at the instant of tensile instability [44.81]. *Kim* and *Bhushan* [44.32] used this condition (44.35) to extract the limit of tip radius R_t,

$$\sigma_c = \frac{F_{ad}}{a_c^2\pi} \le \sigma_{th} , \quad (44.37)$$

where σ_c is the contact stress, σ_{th} is the ideal strength of van der Waals bonds, which equals $\approx W_{ad}/b$, b is the characteristic length of surface interaction, and a_c is the contact radius. Based on the JKR theory, for the rigid contacting surface, a_c at the instant of pull-off is calculated as

$$a_c = \left(\frac{9\pi W_{ad} R_t^2 \left(1-\upsilon^2\right)}{8E}\right)^{1/3} , \quad (44.38)$$

where ν is Poisson's ratio. The tip radius can then be calculated by combining (44.37) and (44.38) as

$$R_t \ge \frac{8b^3E^2}{3\pi^2(1-\nu^2)^2 W_{ad}^2} . \quad (44.39)$$

The lower limit of tip radius also depends on the elastic modulus. Assuming $W_{ad} = 66\,\mathrm{mJ/m^2}$ and $b = 2\times10^{-10}\,\mathrm{m}$ [44.82], the lower limit of tip radius for $E = 1\,\mathrm{MPa}$, 0.1 GPa, and 10 GPa is calculated as 6×10^{-7}, 6×10^{-3}, and 60 nm, respectively. In this study, *Kim* and *Bhushan* [44.32] fixed the tip radius at 100 nm, which satisfies the tip radius condition throughout a wide range of elastic modulus up to 10 GPa.

Nonsticking Condition

A high density of fibers is also important for high adhesion. However, if the space S between neighboring fibers is too small, the adhesion forces between them become stronger than the forces required to bend the fibers. Then, fibers might stick to each other and get entangled. Therefore, to prevent fibers from sticking to each other, they must be spaced apart and be stiff enough to prevent sticking or bunching. Several authors (e.g., [44.49]) have formulated a nonsticking criterion. *Kim* and *Bhushan* [44.32] adopted the approach of *Sitti* and *Fearing* [44.49]. Both adhesion and elastic forces will act on bent structures. The adhesion force between two neighboring round tips is calculated as

$$F_{ad} = \frac{3}{2}\pi R_t' W_{ad} , \tag{44.40}$$

where R_t' is the reduced radius of contact, which is calculated as $R_t' = (1/R_{t1} + 1/R_{t2})^{-1}$; R_{t1}, R_{t2} – radii of contacting tips; for the case of similar tips, $R_{t1} = R_{t2}$, $R_t' = 2/R_t$.

The elastic force of a bent structure can be calculated by multiplying the bending stiffness ($k_b = 3\pi R^4 E/4l^3$) by a given bending displacement δ as

$$F_{el} = \frac{3}{4}\frac{\pi R^4 E\delta}{l^3} . \tag{44.41}$$

The condition for the prevention of sticking is $F_{el} > F_{ad}$. By combining (44.40) and (44.41), a requirement for the minimum distance S between structures which will prevent sticking of the structures is given as [44.32]

$$S > 2\delta = 2\left(\frac{4}{3}\frac{W_{ad} l^3}{ER^3}\right) = 2\left(\frac{32}{3}\frac{W_{ad}\lambda^3}{E}\right) . \tag{44.42}$$

The constant 2 takes into account the two nearest structures. Using the distance S, the fiber density ρ is calculated as

$$\rho = \frac{1}{(S+2R)^2} . \tag{44.43}$$

Equation (44.43) was then used to calculate the allowed minimum density of fibers without sticking or bunching. In (44.42), it is shown that the minimum distance S depends on both the aspect ratio λ and the elastic modulus E. A smaller aspect ratio and higher elastic modulus allow for greater packing density. However, fibers with a low aspect ratio and high modulus are not desirable for adhering to rough surfaces due to lack of compliance.

44.7.4 Numerical Simulation

The simulation of adhesion of an attachment system in contact with random rough surfaces was carried out numerically. In order to conduct two-dimensional (2-D) simulations it is necessary to calculate the applied load F_n as a function of the applied pressure P_n as an input condition. Using ρ calculated by the nonsticking condition, *Kim* and *Bhushan* [44.32] calculated F_n as

$$F_n = \frac{P_n p}{\rho} , \tag{44.44}$$

where p is the number of springs in the scan length L, which equals $L/(S+2R)$.

Fibers of the attachment system are modeled as a one-level hierarchy of elastic springs (Fig. 44.14) [44.32]. The deflection of each spring and the elastic force arising in the springs are calculated according to (44.22) and (44.23), respectively. The adhesion force is the lowest value of the elastic force F_{el} when the fiber has detached from the contacting surface. *Kim* and *Bhushan* [44.32] used an iterative process to obtain the optimal fiber geometry in terms of fiber radius and aspect ratio. If the applied load, roughness of the contacting surface, and fiber material are given, the procedure for calculating the adhesion force is iterated until the desired adhesion force is satisfied. In order to simplify the design problem, the fiber material is regarded as a known variable. The next step is constructing the design database. Figure 44.25a shows the flow chart for the construction of the adhesion design database, and Fig. 44.25b shows the calculation of the adhesion force, which is part of the procedure to construct the adhesion design database.

44.7.5 Results and Discussion

Figure 44.26 shows an example of the adhesion design database for biomimetic attachment systems consisting of single-level cylindrical fibers with an orientation angle of 30° and spherical tips of $R_t = 100$ nm constructed by *Kim* and *Bhushan* [44.32]. The minimum fiber radius calculated by using the no-fiber-fracture condition, which plays a role of the lower limit of optimized fiber radius, is also added to the plot. The plots in Fig. 44.26 cover all applicable fiber materials from a soft elastomer material such as poly(dimethylsiloxane) (PDMS) to stiffer polymers such as polyimide and β-keratin. The dashed lines in

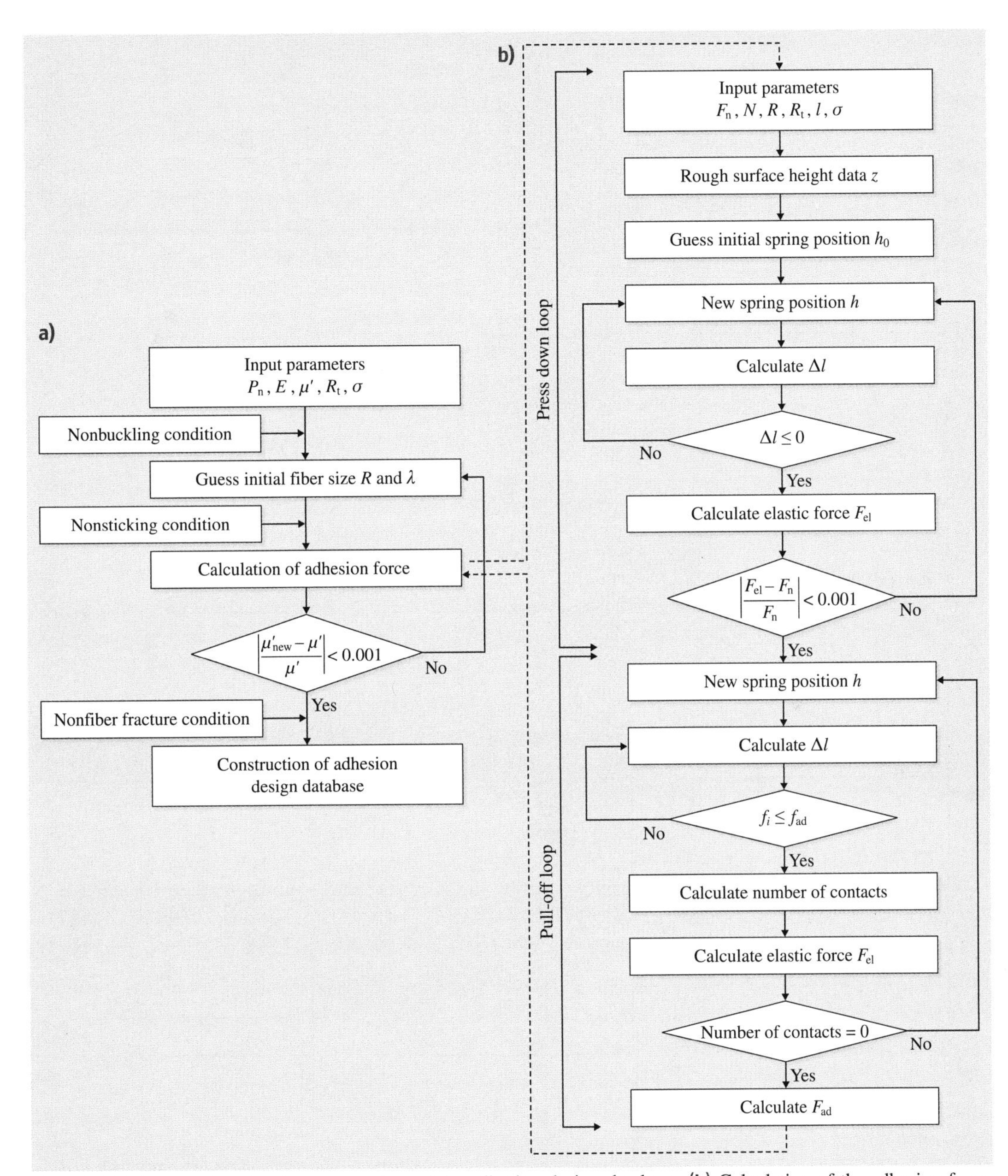

Fig. 44.25 (a) Flow chart for the construction of the adhesion design database. (b) Calculation of the adhesion force. P_n is the applied pressure, E is the elastic modulus, μ' is the adhesion coefficient, R_t is the tip radius, σ is the RMS amplitude, R is the fiber radius, λ is the fiber aspect ratio, F_n is the applied load, N is the number of springs, k and l are the stiffness and length of structures, Δl is the spring deformation, f_i is the elastic force of a single spring, and f_{ad} is the adhesion force of a single contact (after [44.32])

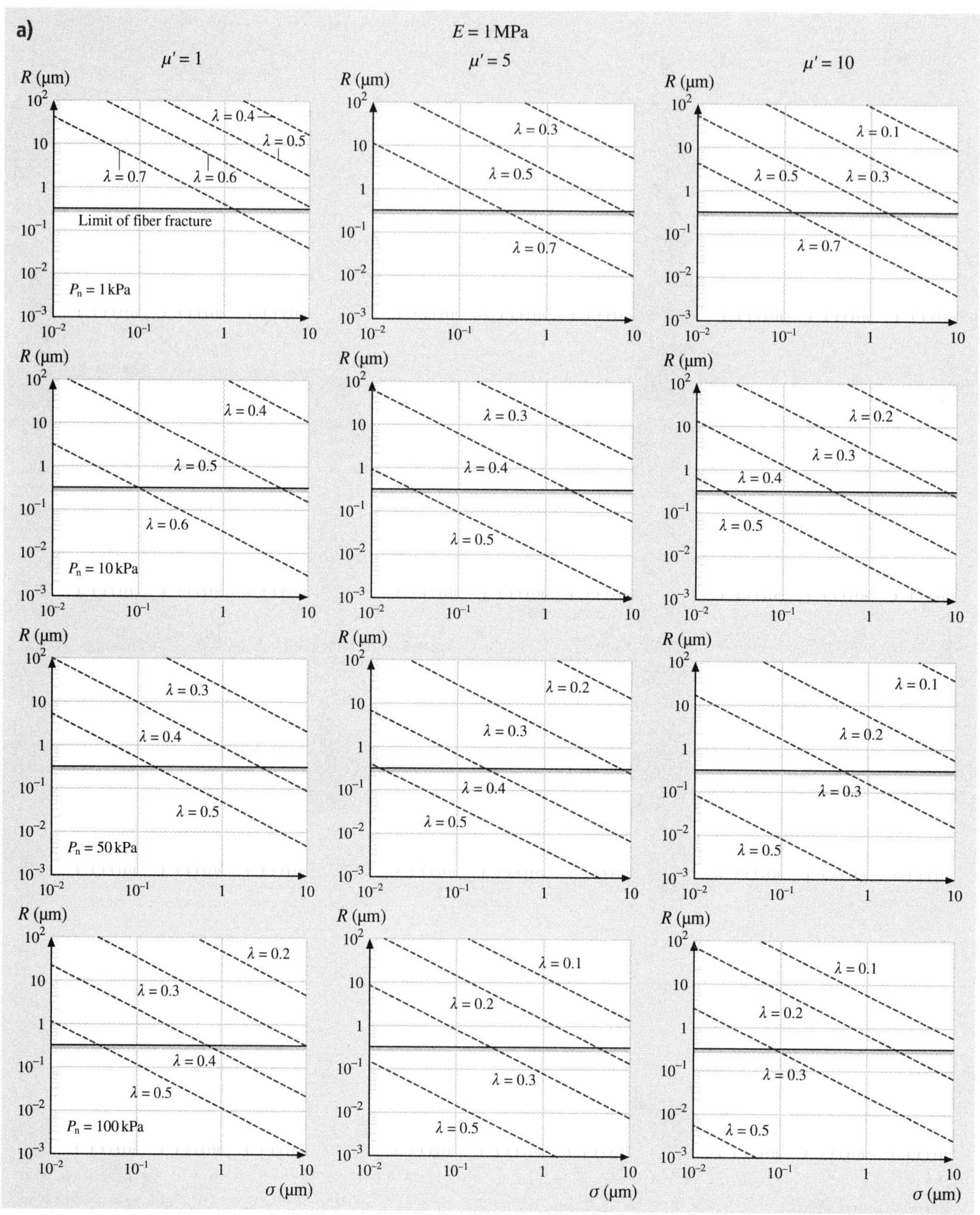

Fig. 44.26a–c Adhesion design database for biomimetic attachment systems consisting of single-level cylindrical fibers with orientation angle of 30° and spherical tips of 100 nm for elastic modulus of (**a**) 1 MPa, (**b**) 100 MPa, and (**c**) 10 GPa (after [44.32]). The *solid lines* shown in (**b**) and (**c**) correspond to the case studies I and II in the text, respectively, which satisfy the specified requirements (after [44.32]) ▲ ▶

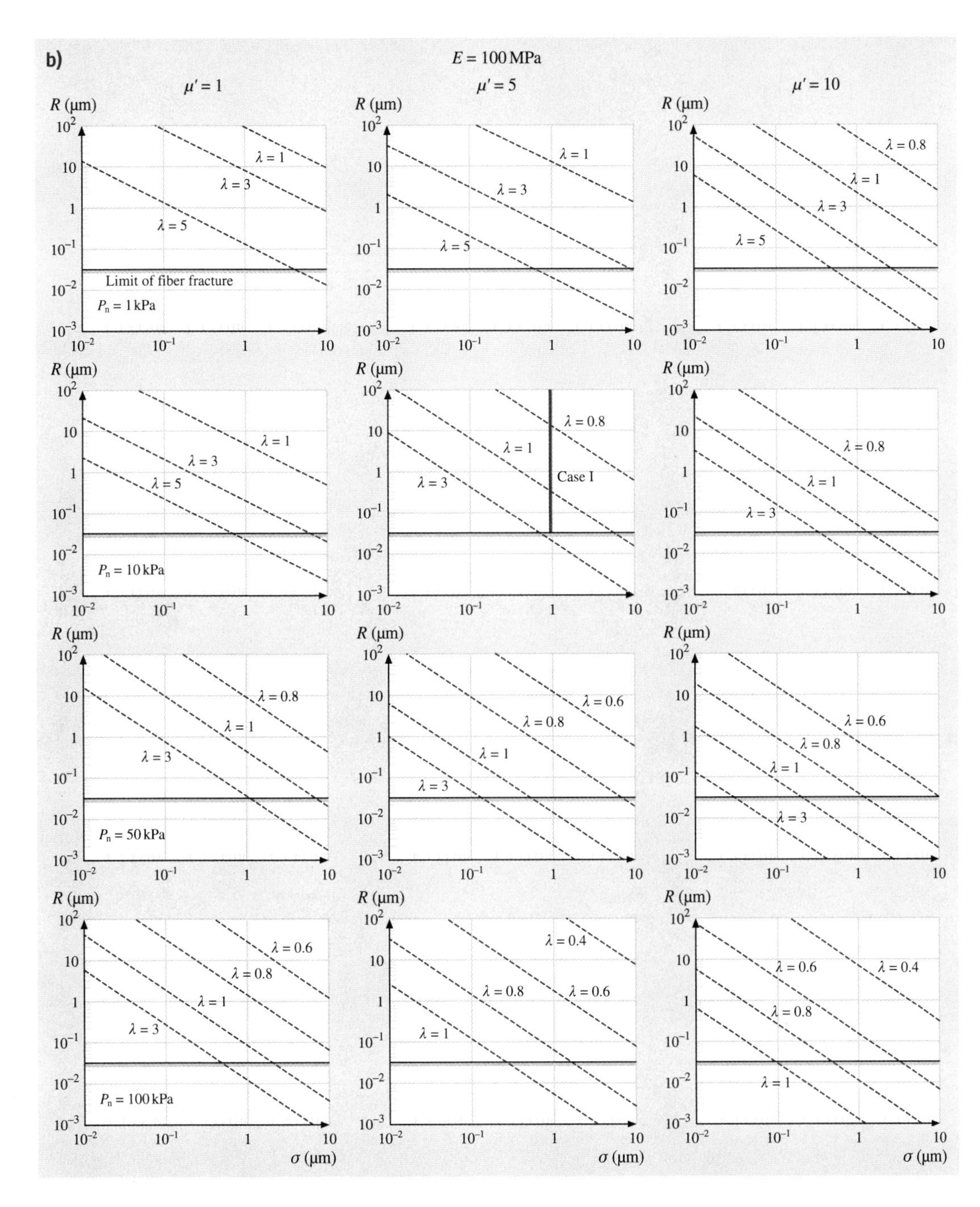

each plot represent the limits of fiber fracture due to the adhesion force. For a soft material with $E = 1$ MPa in Fig. 44.26a, the range of the desirable fiber radius is > 0.3 μm and that of the aspect ratio is $\approx< 1$. As elastic modulus increases, the feasible range of both fiber radius and aspect ratio also increase, as shown in

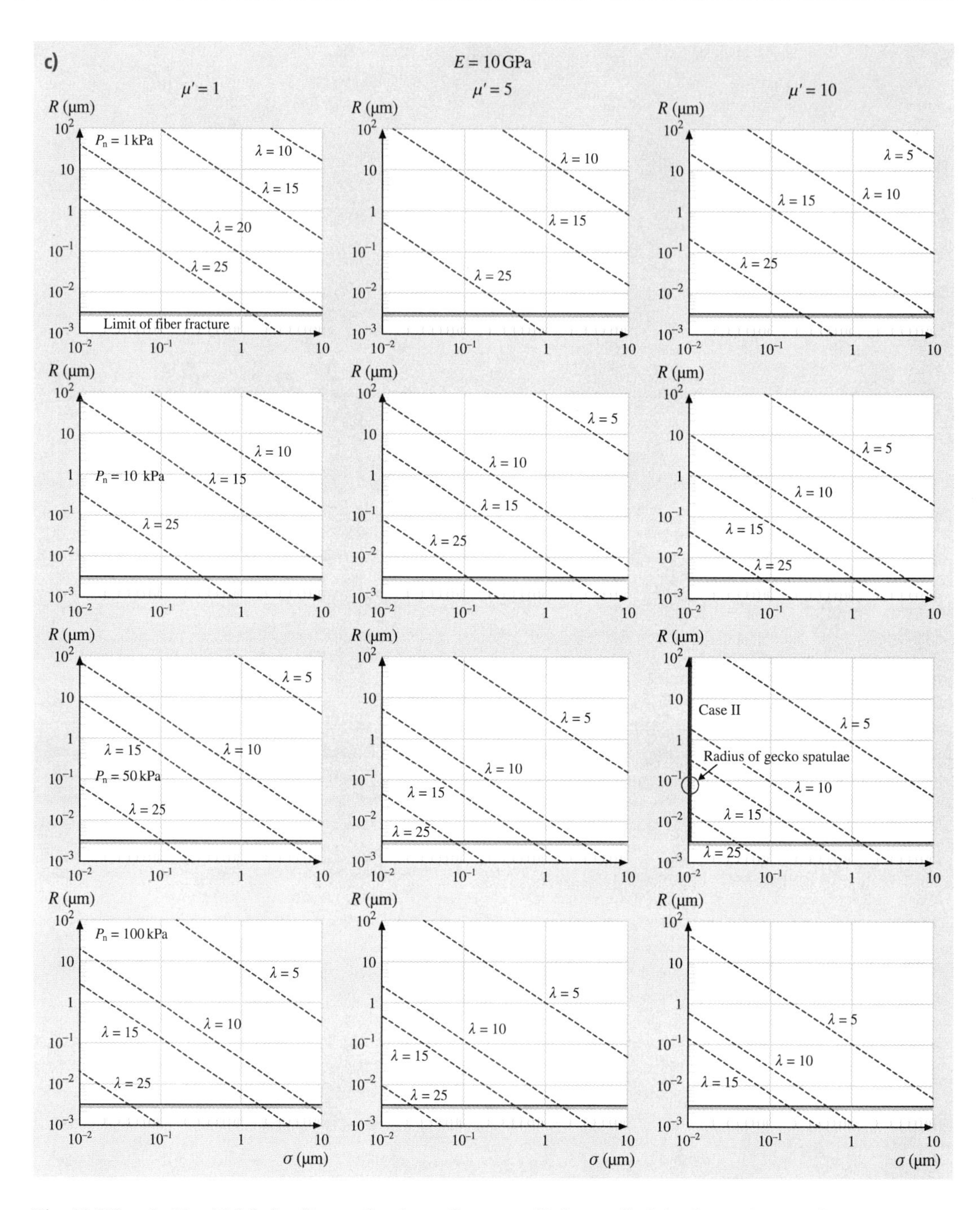

Fig. 44.26b,c. In Fig. 44.26, the fiber radius has a linear relation with the surface roughness on a logarithm scale.

If the applied load, roughness of the contacting surface, and elastic modulus of the fiber material are specified, the optimal fiber radius and aspect ratio for the desired adhesion coefficient can be selected from this design database. The adhesion databases are useful for understanding biological systems and for guiding the fabrication of biomimetic attachment systems. Two case studies [44.32] are discussed below.

Case study I: Select the optimal size of fibrillar adhesive for a wall-climbing robot with the requirements:

- Material: polymer with $E \approx 100\,\text{MPa}$
- Applied pressure by weight $< 10\,\text{kPa}$
- Adhesion coefficient ≈ 5
- Surface roughness $\sigma < 1\,\mu\text{m}$.

The subplot of the adhesion database that satisfies these requirements is found in the second column and second row in Fig. 44.26b. From this subplot, any values on the marked line can be selected to meet the requirements. For example, fiber radius of $0.4\,\mu\text{m}$ with an aspect ratio of 1 or fiber radius of $10\,\mu\text{m}$ with an aspect ratio of 0.8 would satisfy the specified requirements.

Case study II: Comparison with the adhesion test for a single gecko seta [44.25, 26]:

- Material: β-keratin with $E \approx 10\,\text{GPa}$
- Applied pressure $= 57\,\text{kPa}$ ($2.5\,\mu\text{N}$ on an area of $43.6\,\mu\text{m}^2$)
- Adhesion coefficient $= 8{-}16$
- Surface roughness $\sigma < 0.01\,\mu\text{m}$.

Autumn et al. [44.25, 26] showed that, in isolated gecko setae contacting the surface of a single-crystalline silicon wafer, a $2.5\,\mu\text{N}$ preload yielded adhesion of $20{-}40\,\mu\text{N}$ and thus a value of adhesion coefficient of 8–16. The region that satisfies the above requirements is marked in Fig. 44.26c. The spatulae of gecko setae have an approximate radius of $0.05\,\mu\text{m}$ with an aspect ratio of 25. However, the radius corresponding to $\lambda = 25$ for the marked line is $\approx 0.015\,\mu\text{m}$. This discrepancy is due to the difference between the simulated fiber model and the real gecko setae model. Gecko setae are composed of a three-level hierarchical structure in practice, so higher adhesion can be generated than in a single-level model [44.3, 30, 31]. Given the simplification in the fiber model, this simulation result is very close to the experimental result.

44.8 Fabrication of Biomimetic Gecko Skin

Based on the studies reported in the literature, the dominant adhesion mechanism utilized by gecko and spider attachment systems appears to be van der Waals forces. The hierarchical structure involving complex divisions of the gecko skin (lamellae–setae–branches–spatulae) enable a large number of contacts between the gecko skin and mating surface. As shown in previous calculations, the van der Waals adhesive force for two parallel surfaces is inversely proportional to the cube of the distance between two surfaces. These hierarchical fibrillar microstructured surfaces would be capable of reusable dry adhesion and would have uses in a wide range of applications from everyday objects such as adhesive tapes, fasteners, toys, microelectronic, and space applications, and treads of wall-climbing robots. The development of nanofabricated surfaces capable of replicating this adhesion force developed in nature is limited by current fabrication methods. Many different techniques have been used in an attempt to create and characterize bioinspired adhesive tapes. Attempts are being made to develop climbing robots using gecko-inspired structures [44.84–87].

44.8.1 Single-Level Roughness Structures

One of the simplest approaches is to create a pattern by various micro/nanofabrication techniques and use it as a master template and mold with a liquid polymer to create micro/nanostructured replicas. *Sitti* and *Fearing* [44.49] employed an AFM tip to create a set of dimples on a wax surface. These dimples served as a mold for creating polymer micropillars, shown in Fig. 44.27.

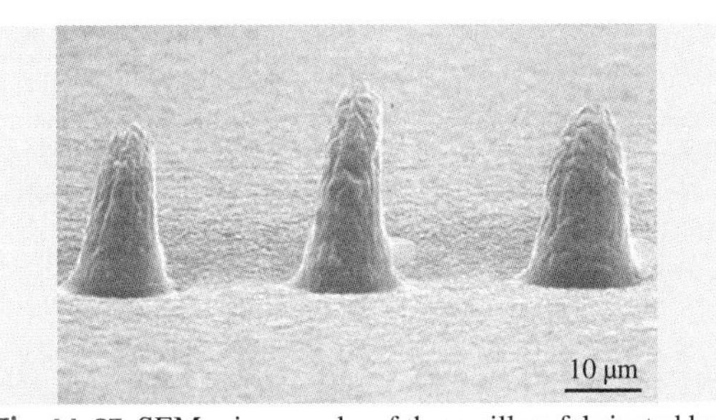

Fig. 44.27 SEM micrographs of three pillars fabricated by molding from dimples created by AFM-tip indentation (after [44.83])

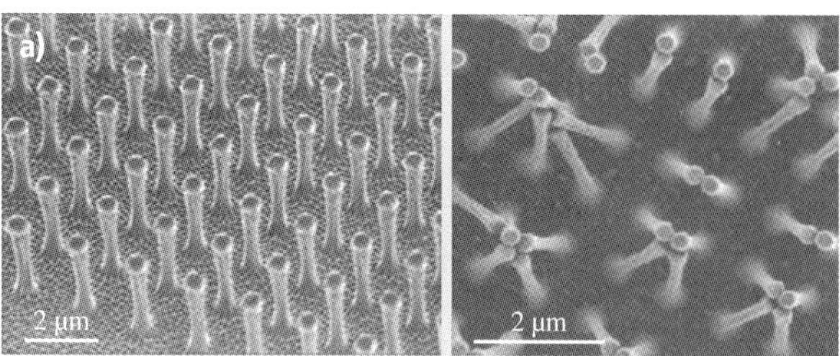

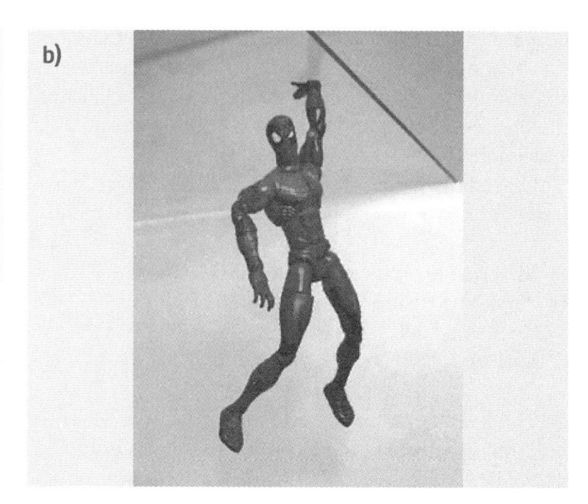

Fig. 44.28 (a) (*left*) An array of polyimide nanohairs and (*right*) bunching of the nanohairs, which leads to a reduction in adhesive force. (b) A spiderman toy (about 0.4 N) with a hand covered with the molded polymer nanohairs, clinging to a glass plate (after [44.89])

The adhesive force to an individual pyramidal pillar was measured using another AFM cantilever. Although each pillar of the material was found to be capable of producing large adhesion forces, the surface failed to replicate gecko adhesion on a macroscale. This was due to the lack of flexibility in the pillars. In order to ensure that the largest possible area of contact occurs between the tape and the mating surface, a soft, compliant fibrillar structure would be desired. Compliant fibrillar structures enable more fibrils to be in close proximity to a mating surface to increase the van der Waals forces. *Sitti* and *Fearing* [44.49] and *Cho* and *Choi* [44.88] used nanoporous anodic alumina and polycarbonate membranes as a template to create polymeric nanofibers.

Geim et al. [44.89] created arrays of polyimide nanofibers using electron-beam lithography and dry etching in oxygen plasma (Fig. 44.28a, left). By using electron-beam lithography, thermal evaporation of an aluminum film, and lift-off, an array of nanoscale aluminum disks was prepared. These patterns were then transferred to the polyimide film by dry etching in oxygen plasma. A 1 cm^2 sample was able to create 3 N of adhesive force under the new arrangement. This is approximately one-third the adhesive strength of a gecko. They fabricated a Spiderman toy (≈ 0.4 N) with a hand covered with molded polymer nanohairs (Fig. 44.28b). They demonstrated that it could cling to a glass plate. Bunching of the nanohairs (as described earlier) if they are closely spaced was determined to greatly reduce the both the adhesive strength and durability of the polymer tape. The bunching can be clearly seen in Fig. 44.28a (right). Therefore, an optimal geometry is required.

Davies et al. [44.90] fabricated mushroom-headed microfibers made of PDMS. In one of the fabrication strategies, a silicon wafer with a thickness which defined the stalk length was obtained, and the masks with mushroom-head features were first used to pattern one side of the silicon wafer with resist. Features were etched to a depth equal to that of the thickness of the mushroom head. Next, the smaller-diameter mask was used to pattern the other side of the wafer, which was then etched to produce holes through the entire thickness of the wafer, meeting the mushroom-headed cavities. This mold was first coated in a fluorocarbon release agent. A PDMS solution was then spun onto this mold and cured to produce mushroom-headed microfibers. The resulting casting comprising stalks and mushroom heads was then pulled through the mold in a single peeling process. To create angled microfiber arrays found in biological attachments using photolithography, *Aksak* et al. [44.91] simply varied the ultraviolet (UV) exposure angle by tilting the wafer during exposure. The fibers were formed at an angle not perpendicular to the substrate surface (Fig. 44.29a). This master template of angled SU-8 fibers was then used to form many copies of the fiber arrays from curable polyurethanes by molding. They reported that angled fibers exhibited reduced adhesion compared with similar vertical fibers due to a peeling moment. However, angled fibers are favored in biological attachment systems. *Murphy* et al. [44.92] modified angled fiber arrays by adding soft spherical and spatula-shaped tips via dipping in a liquid polymer of interest (Fig. 44.29b). To add tips to the fibers, the fiber array sample attached to a micropositioning stage was dipped into a liquid polyurethane layer and retracted, retaining some of the liquid polymer on the tips of the fibers. To form spherical tips, the sample was placed with the fibers facing up and allowed to cure. To form spatula tips, the fiber sample was placed onto a smooth low-

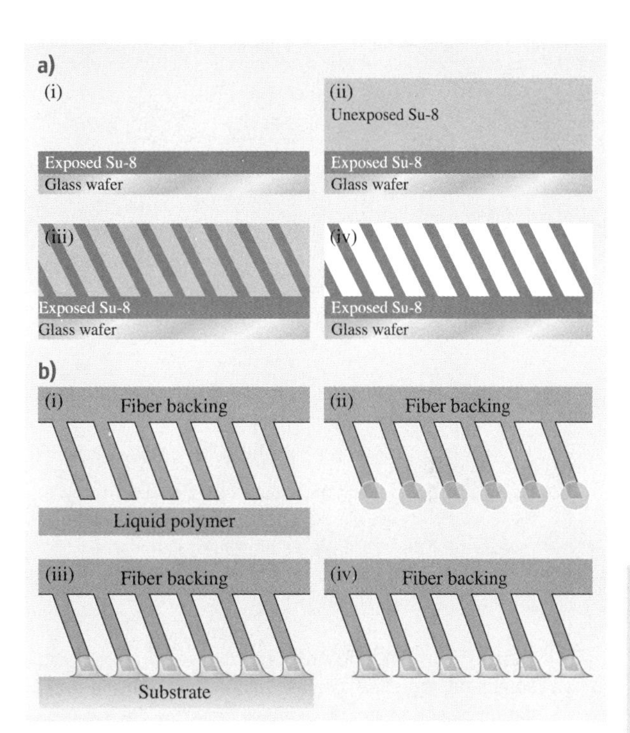

Fig. 44.29 (**a**) The process steps of the polymer fiber orientation: (i) a thin layer of SU-8 is spun on a glass substrate, then exposed and cured; (ii) a thicker layer of SU-8 is spun, which will become the fibers; (iii) the thick layer is patterned with UV exposure by tilting the wafer; (iv) the SU-8 photoresist is developed, leaving the desired angled fiber array (after [44.91]). (**b**) Fiber tip fabrication process: (i) bare fibers are aligned with a layer of liquid polymer; (ii) the fibers are tipped into the liquid and retracted; (iii) the fibers are brought into contact with a substrate; (iv) the fibers are peeled away from the substrate after curing (after [44.92]) ▶

energy surface and then peeled away after curing. They reported very high adhesion of these fibers with soft tips because of increased contact area.

Del Campo et al. [44.93, 94] fabricated pillar arrays with controlled 3-D tip geometries resembling those found in biological attachments. The fabrication strategy was based on complete or partial soft molding on 2-D masters made by lithography with elastomeric precursors followed in some cases by inking and microprinting steps. The patterned master with high-aspect-ratio cylindrical holes was produced by photolithography using SU-8 photoresist films. The SU-8 masters were filled with elastomeric precursors (PDMS supplied as Sylgard 184 by Dow Corning) to produce arrays of cylindrical pillars (Fig. 44.30a). Arrays of pillars with spherical and spatular tips were obtained by inking the Sylgard 184-structured substrates in a thin film of Sylgard 184 precursor. Curing of arrays in upside-down orientation yielded hemispher-

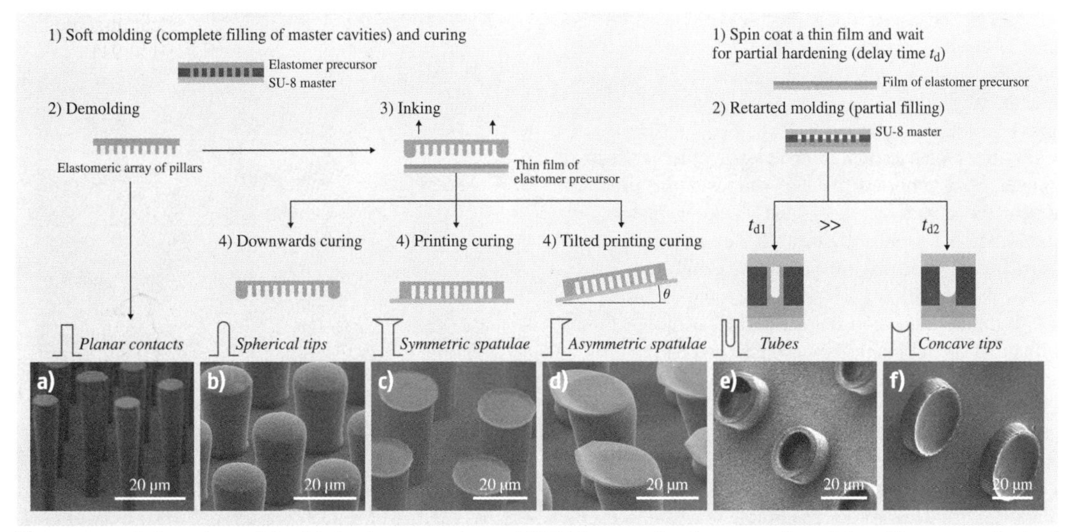

Fig. 44.30a–f Overview of the fabrication strategies and SEM images showing examples of the pillar arrays obtained with controlled 3-D tip geometries (after [44.93])

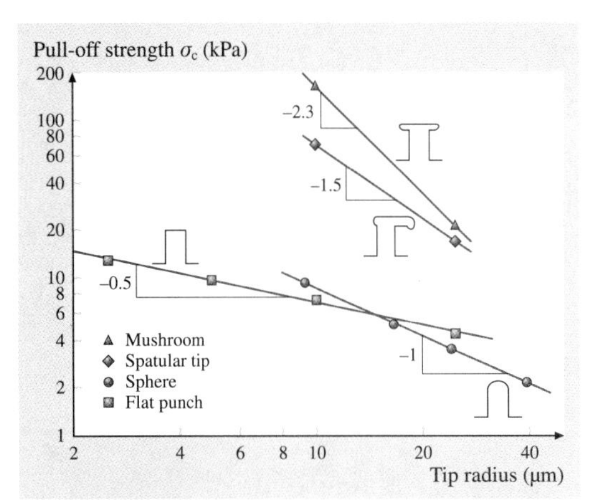

Fig. 44.31 Tip radius dependence of the pull-off force for flat, spherical, spatular, and mushroom-like contacts at preload of 1 mN. In the case of spherical tips, the radius corresponds to the tip radius. For all other geometries, the pillar radius is used (after [44.94])

ical tips as a consequence of gravity and surface tension acting on the fluid drop (Fig. 44.30b). Alternatively, the inked stamp can be pressed against a flat substrate and then cured. This leads to pillars with a flat top (Fig. 44.30c). The top can be symmetric or asymmetric depending on the tilt of the substrate during curing (Fig. 44.30c,d). They also used silicones used for dental impressions. These materials possess higher initial viscosities and faster cross-linking kinetics than Sylgard 184, which results in incomplete cavity filling. By soft-molding these materials after selected delay times after mixing, arrays of tubes and pillars with concave tips (Fig. 44.30e,f) were obtained. They performed adhesion tests on various geometries against a sapphire sphere. They reported that the shape of the pillar tip affects the contact area and adhesion behavior. Figure 44.31 shows pull-off strength data as a function of tip radius for various tip geometries. For a given tip radius, pillars with the flat punch geometry have significantly higher adhesion than spherical contacts. Pillars with mushroom tips have the highest adhesion.

Gorb et al. [44.96] and *Bhushan* and *Sayer* [44.95] characterized two polyvinylsiloxane (PVS) samples from Gottlieb Binder Inc., Holzgerlingen, Germany, one consisting of mushroom-shaped pillars (Fig. 44.32a) and the other an unstructured control surface (Fig. 44.32b). The structured sample is inspired by the micropatterns found in the attachment systems of male beetles from the family Chrysomelidae and is easier to fabricate. Both sexes possess adhesive hairs on their tarsi; however, males bear hair extremely specialized for adhesion to the smooth surface of female's covering wings during mating. The hairs have broad flattened tips with grooves under the tip to provide flexibility. The mushroom shape provides a larger contact area. The structured samples were produced at room temperature by pouring two-compound polymerizing PVS into the holed template lying on a smooth glass support. The fabricated sample is comprised of pillars that are arranged in a hexagonal order to allow maximum packing density. They are $\approx 100\,\mu m$ in height, $60\,\mu m$ in base diameter, $35\,\mu m$ in middle diameter, and $25\,\mu m$ in diameter at the narrowed region just below the terminal contact plates. These plates were $\approx 40\,\mu m$ in diameter and $2\,\mu m$ in thickness at the lip edges. The adhesion force of the two samples in contact with a smooth flat glass substrate was measured by *Gorb* et al. [44.96] using a microtribometer. Results revealed that the structured specimens featured an adhesion force more than twice that of the unstructured specimens. The adhesion force was also found to be independent of the

a)
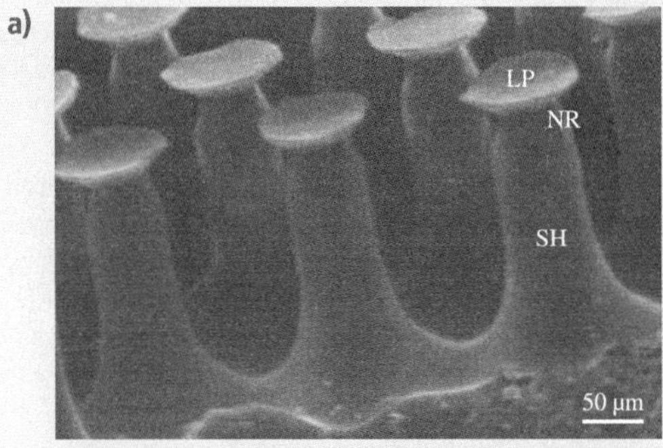

b)

Fig. 44.32a,b SEM micrographs of (**a**) structured and (**b**) unstructured PVS samples (SH – shaft, NR – neck region, LP – lip) (after [44.95])

preload. Moreover, it was found that the adhesive force of the structured sample was more tolerant to contamination compared with the control, and it could be easily cleaned with a soap solution.

Bhushan and *Sayer* [44.95] characterized the surface roughness, friction force, and contact angle of the structured sample and compared the results with an unstructured control. As shown in Fig. 44.33a, the macroscale coefficient of kinetic friction of the structured sample was found to be almost four times greater than that of the unstructured sample. This increase was determined to be a result of the structured roughness of the sample and not the random nanoroughness. It is also noteworthy that the static and kinetic coefficients of friction are approximately equal for the structured sample. It is believed that the divided contacts allow the broken contacts of the structured sample to constantly recreate contact. As seen in Fig. 44.33b, the pillars also increased the hydrophobicity of the structured sample in comparison with the unstructured sample, as expected due to the increased surface roughness [44.98–100]. A large contact angle is important for self-cleaning [44.101], which agrees with the findings of *Gorb* et al. [44.96] that the structured sample is more tolerant of contamination than the unstructured sample.

Directed self-assembly has been proposed as a method to produce regularly spaced fibers [44.83, 102]. In this technique, a thin liquid polymer film is coated on a flat conductive substrate. As demonstrated in Fig. 44.34, a closely spaced metal plate is used to apply a direct-current (DC) electric field to the polymer film. Due to instabilities in the film, pillars will begin to grow until they touch the upper metal plate. Self-assembly is desirable because the components spontaneously assemble, typically by bouncing around in a solution or gas phase until a stable structure of minimum energy is reached.

Vertically aligned multiwalled carbon nanotubes (MWCNT) have been used to create nanostructures on polymer surfaces. *Yurdumakan* et al. [44.97] used chemical vapor deposition (CVD) to grow vertically aligned MWCNT that are 50–100 μm in length on quartz or silicon substrates. A catalyst was deposited on the silicon oxide surface as patches using photolithography. The MWCNT grew selectively on the patches with controlled thickness and length and were vertically aligned. The sample with MWCNT sites facing up was then dipped in methyl methacrylate solution. After polymerization, poly(methyl methacrylate) (PMMA)-MWCNT sheets are peeled off from the silicon substrate. The MWCNTs are exposed from the silicon-facing side of the PMMA matrix by etching the top 25 μm with a solvent. SEM images of the MWCNT grown on a silicon substrate as well as transferred into a PMMA ma-

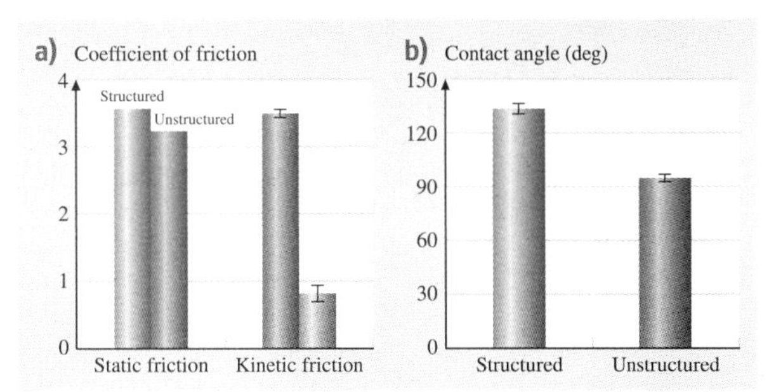

Fig. 44.33 (**a**) Coefficients of static and kinetic friction for structured and unstructured samples sliding against magnetic tape with normal load of 130 mN. (**b**) Water contact angle for the structured and unstructured samples (after [44.95])

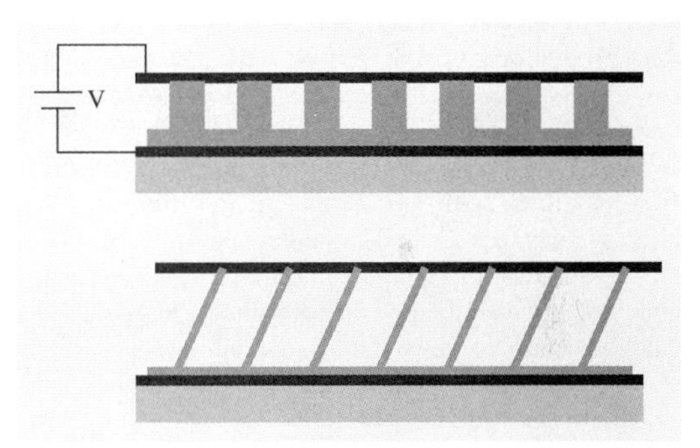

Fig. 44.34 Directed self-assembly-based method of producing high-aspect-ratio micro/nanofibers (after [44.83])

Fig. 44.35a,b Multiwalled carbon nanotube structures: (**a**) grown on silicon by chemical vapor deposition, (**b**) transferred into a PMMA matrix and then exposed on the surface after solvent etching (after [44.97])

trix and then exposed on the surface can be seen in Fig. 44.35. On the nanoscale, the MWCNT surface was able to achieve adhesive forces two orders of magnitude greater than those of gecko foot-hairs. These structures provided high adhesion on the nanometer level and were not capable of producing high adhesion forces on the macroscale. *Ge* et al. [44.103] and others have fabricated nanostructures by transferring micropatterned, vertically aligned MWCNT arrays onto flexible polymer tape. They reported high adhesion on the macroscale. They also performed peeling experiments. Durability of the adhesive tape is an issue, as some of the nanotubes can detach from the substrate with repeated use. *Qu* et al. [44.104] measured adhesion on vertically aligned MWCNT arrays on Si substrate and reported high adhesion on the nanoscale.

44.8.2 Multilevel Hierarchical Structures

The aforementioned fabricated surfaces only have one level of roughness. Although these surfaces are capable of producing high adhesion on the micro/nanoscale, they are not expected to produce large-scale adhesion due to a lack of compliance and bunching.

Sitti [44.83] proposed a molding technique for creating structures with two levels. In this method two different molds are created – one with pores of the order of magnitude of micrometers in diameter and a second with pores of nanometer-scale diameter. One potential mold material is porous anodic alumina (PAA), which has been demonstrated to produce ordered pores on the nanometer scale of equal size. Pore-widening techniques could be used to create micrometer-scale pores. As seen in Fig. 44.36, the two molds would be bonded to each other and then filled with a liquid polymer. *Del Campo* and *Greiner* [44.105] fabricated a hierarchical structure by multilevel photolithography. Figure 44.37 shows a schematic of the process and an example of the two-level SU-8 patterns obtained.

Northen and *Turner* [44.106, 107] created a multilevel compliant structure by employing a microelectromechanical-based approach. The multiscale structures consist of arrays of organic-looking photoresist nanorods (organorods), $\approx 2\,\mu m$ tall and 50–200 nm in diameter (comparable in size to gecko spatulae) (Fig. 44.38a), atop photolithographically defined $2\,\mu m$-thick SiO_2 platforms 100–150 μm on a side (Fig. 44.38b). The platforms of various geometries are supported by single high-aspect-ratio pillars down to 1 μm in diameter and with heights up to $\approx 50\,\mu m$ (Fig. 44.38c). The structures are fabricated out of 100 mm single-crystal wafers using standard bulk micromachining techniques. An array of four-fingered platform structures is shown in Fig. 44.38d. Adhesion testing was performed using a nanorod surface on a solid substrate and on the multilevel structures by *Northen* and *Turner* [44.107]. They reported that the adhesive pressure of the multilevel structures was about four times higher than that of surfaces with only one level of hierarchy. The durability of the multilevel structure was also much greater than the single-level structure. The adhesion of the multilevel structure did not change between iterations one and five. During the same number of iterations, the adhesive pressure of the single-level structure decreased to zero.

In summary, literature clearly indicates that, in order to create a dry superadhesive, a fibrillar surface construction is necessary to maximize the van der Waals forces by using so-called division of contacts. Hierarchical structure provides compliance for adaptability to a variety of rough surfaces. A material must be soft enough to conform to rough surfaces yet hard enough to

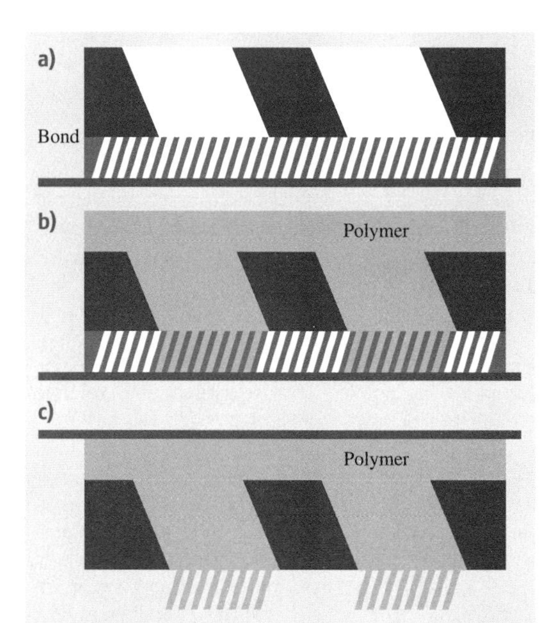

Fig. 44.36a–c Proposed process for creating multilevel structures using molding. Micro- and nanometer-sized pore membranes are bonded together (**a**) and filled with liquid polymer through the micropore membrane site (**b**), followed by curing of the polymer and etching the array of both membranes in order to leave (**c**) the polymer surface (after [44.83])

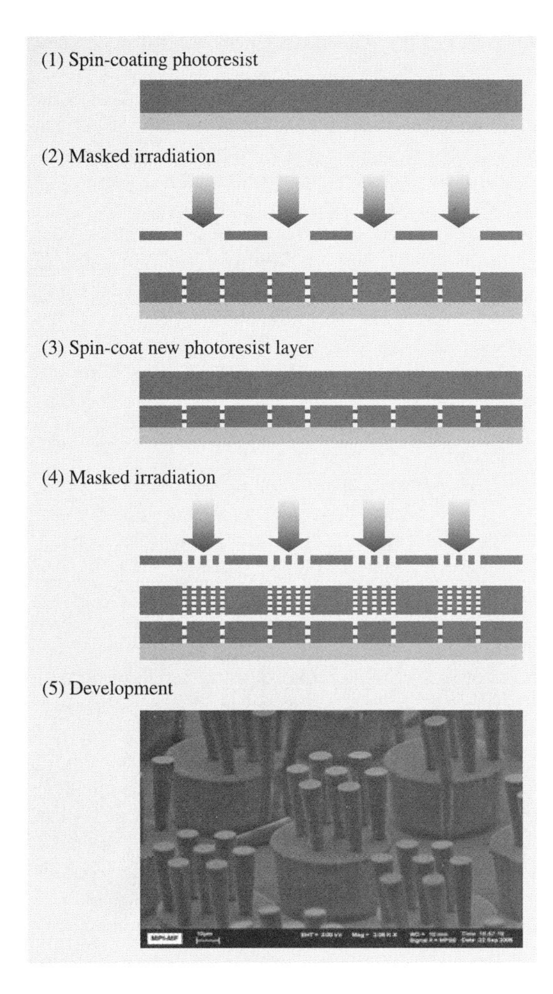

Fig. 44.37 Layer-by-layer structuring method and example of fabricated hierarchical structure in SU-8. Base pillars have 50 μm diameter and 40 μm height and the top pillars have 9 μm diameter and 35 μm height (after [44.105]) ◂

Fig. 44.38a–d Multilevel fabricated adhesive structure composed of (a) organorods atop (b) silicon dioxide platforms. The platforms are supported by (c) support pillars. (d). This structure was repeated multiple times over a silicon wafer (after [44.106])

avoid bunching, which will decrease the adhesive force. It is also desirable to have a superhydrophobic surface in order to utilize self-cleaning. Inspired by previous work on adding tips to the fibrillar structures, the end of the fibers could be modified to enhance adhesion. For example, a soft adhesive could be used to coat fiber ends to provide added adhesion using conventional adhesives.

44.9 Conclusion

The adhesive properties of geckoes and other creatures such as flies, beetles, and spiders are due to the hierarchical structures present on each creature's hairy attachment pads. Geckoes have developed the most intricate adhesive structures of any of the aforementioned creatures. The attachment system consists of ridges called lamellae that are covered in microscale setae that branch off into nanoscale spatulae, of which there are about three billion on two feet. The so-called division of contacts provide high dry adhesion. Multiple-level hierarchically structured surface construction plays an important role in adapting to surface roughness, bringing the spatulae in close proximity to the mating surface. These structures, as well as material properties, allow the gecko to obtain a much larger real area of contact between its feet and a mating surface than is possible with a nonfibrillar material. Two feet of a Tokay gecko have $\approx 220\,mm^2$ of attachment pad area, on which the gecko is able to generate $\approx 20\,N$ of adhesion force. Although capable of generating high adhesion forces, a gecko is able to detach from a surface at will – an ability known as smart adhesion. Detachment is achieved by a peeling motion of the gecko's feet from a surface.

Experimental results have supported the adhesion theories of intermolecular forces (van der Waals) as a primary adhesion mechanism and capillary forces as a secondary mechanism, and have been used to rule out several other mechanisms of adhesion including the secretion of sticky fluids, suction, and increased frictional forces. Atomic force microscopy has been employed by several investigators to determine the adhesion strength of gecko foot hairs. The measured values of the lateral force required to pull parallel to the surface for a single seta (194 μN) and the adhesive force (normal to the surface) of a single spatula (11 nN) are comparable to the van der Waals prediction of 270 μN and 11 nN for a seta and spatula, respectively. The adhesion force generated by seta increases with preload and reaches a maximum when both perpendicular and parallel preloads are applied. Although gecko feet are strong adhesives, they remain free of contaminant particles through self-cleaning. Spatular size along with material properties enables geckoes to easily expel any dust particles that come into contact with their feet.

The recent creation of a three-level hierarchical model for a gecko lamella consisting of setae, branches, and spatulae has brought more insight into the adhesion of biological attachment systems. One-, two-, and three-level hierarchically structured spring models for the simulation of a seta contacting with random rough surfaces were considered. The simulation results show that the multilevel hierarchical structure has a higher adhesion force as well as higher adhesion energy than the one-level structure for a given applied load, due to better adaptation and attachment ability. It is concluded that the multilevel hierarchical structure produces adhesion enhancement, and this enhancement increases with increasing applied load and decreasing stiffness of the springs. The condition at which significant adhesion enhancement occurs appears to be related to the maximum spring deformation. The result shows that significant adhesion enhancement occurs when the maximum spring deformation is greater than two to three times larger than the σ value of the surface roughness. As the applied load increases, the adhesion force increases up to a certain applied load and then has a constant value, whereas adhesion energy continues to increase with increasing applied load. For the effect of spring stiffness, the adhesion coefficient increases with a decrease in the stiffness of springs. A hierarchical model with softer springs can generate greater adhesion enhancement for lower applied load. As the number of springs in the lower level increases, the equivalent stiffness decreases. Therefore, the three-level model with a larger number of springs in the lowest level gives a larger adhesion force and energy. Inclusion of capillary forces in the spring model shows that the total adhesion force decreases with increasing contact angle of water on the substrate, and the difference of total adhesion force among different contact angles is larger in the intermediate-humidity regime. In addition, the simulation results match the measured data for a single spatula in contact with both hydrophilic and hydrophobic surfaces, which further supports van der Waals forces as the dominant mechanism of adhesion and capillary forces as a secondary mechanism.

There is great interest among the scientific community in creating surfaces that replicate the adhesion strength of gecko feet. These hierarchical fibrillar microstructured surfaces would be capable of reusable dry adhesion and would have uses in a wide range of applications from everyday objects such as adhesive tapes, fasteners, toys, microelectronic, space applications, and treads of wall-climbing robots. In the design of fibrillar structures, it is necessary to ensure that the fibrils are compliant enough to deform easily to the mating surface's roughness profile, yet rigid enough not to collapse under their own weight. Spacing between the individual fibrils is also important. If the spacing is too small, adjacent fibrils can attract each other through intermolecular forces, which will lead to bunching. The adhesion design database developed by *Kim* and *Bhushan* [44.32] serves as a reference for choosing design parameters.

Nanoindentation, lithography, self-assembly, and carbon nanotube arrays are some of the methods that have been used to create fibrillar structures. The limitations of current machining methods on the micro/nanoscale have resulted in the majority of fabricated surfaces consisting of only one level of hierarchy. Bunching, lack of compliance, and lack of durability are some of the problems that may arise with the aforementioned structures. A multilayered compliant system has been created using a microelectromechanical-based approach in combination with nanorods. Multilevel photolithography has also been used to fabricate hierarchical fibrillar structures. Fibrillar structures show great promise for the creation of adhesive structures. Some of the structures have been incorporated into the design of treads of climbing robots.

44.A Typical Rough Surfaces

Several natural (sycamore tree bark and siltstone) and artificial surfaces (dry wall, wood laminate, steel, aluminum, and glass) were chosen to determine the surface parameters of typical rough surfaces that a gecko might encounter. An Alpha-step 200 (Tencor Instruments, Mountain View) was used to obtain surface profiles for three different scan lengths: 80 μm, which is approximately the size of a single gecko seta; 2000 μm, which is close to the size of a gecko lamella; and an intermediate scan length of 400 μm. The radius of the stylus tip was 1.5–2.5 μm, and the applied normal load was 3 mg. The surface profiles were then analyzed using a specialized computer program to determine the root-mean-square amplitude σ, correlation length β^*, peak to valley distance P–V, skewness Sk, and kurtosis K.

Sample surface profiles and their corresponding parameters at a scan length of 2000 μm can be seen in Fig. 44.39a. The roughness amplitude σ varies from as low as 0.01 μm in glass to as high as 30 μm in tree bark. Similarly, the correlation length varies from 2 to 300 μm. The scan length dependence of the surface parameters is illustrated in Fig. 44.39b. As the scan length of the profile increases, so do the roughness amplitude and correlation length. Table 44.4 summarizes the scan-length-dependent parameters σ and β^* for all seven

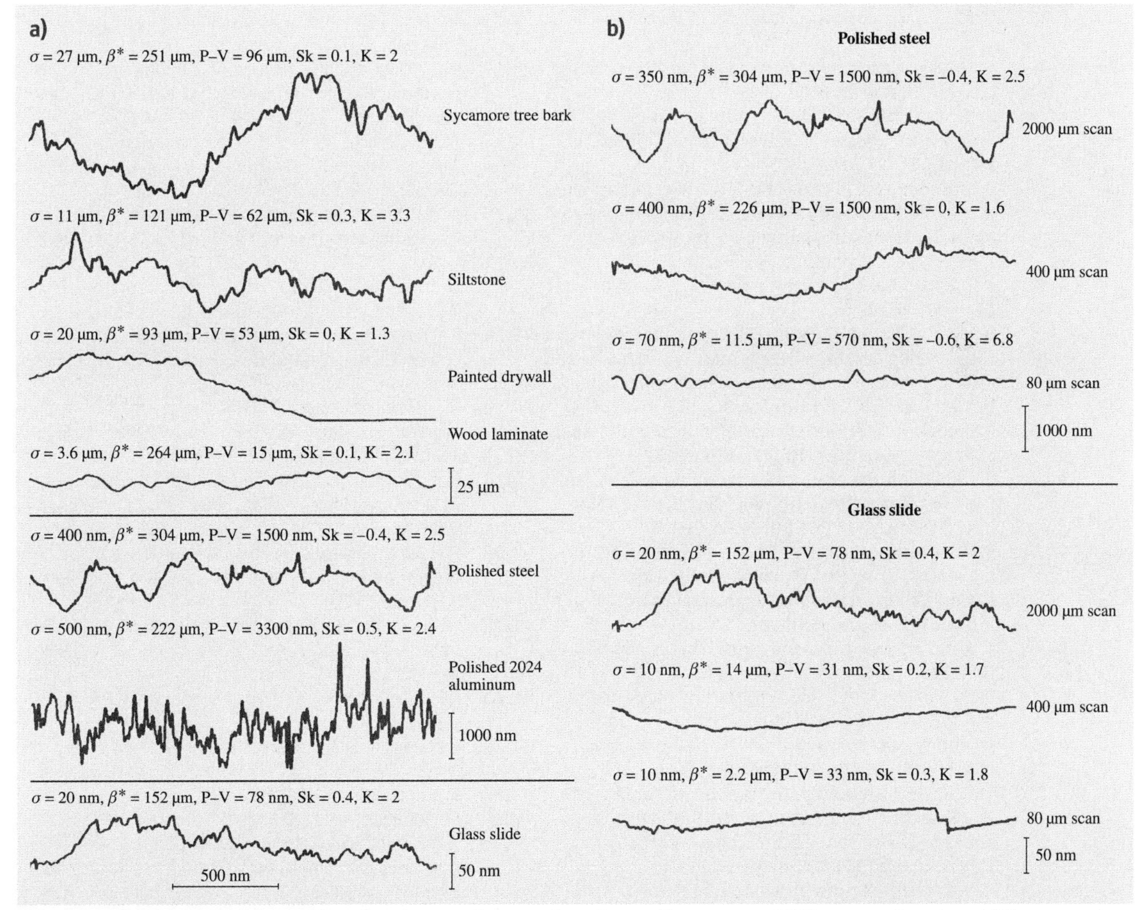

Fig. 44.39 **(a)** Surface height profiles of various random rough surfaces of interest at a 2000 μm scan length and **(b)** a comparison of the profiles of two surfaces at 80, 400, and 2000 μm scan lengths (after [44.3])

Scan length Surface	80 μm		2000 μm	
	σ (μm)	β^* (μm)	σ (μm)	β^* (μm)
Sycamore tree bark	4.4	17	27	251
Siltstone	1.1	4.8	11	268
Painted drywall	1	11	20	93
Wood laminate	0.11	18	3.6	264
Polished steel	0.07	12	0.40	304
Polished 2024 aluminum	0.40	6.5	0.50	222
Glass	0.01	2.2	0.02	152

Table 44.4 Scale dependence of surface parameters σ and β^* for rough surfaces at scan lengths of 80 and 2000 μm 44.3 ◄

sampled surfaces. At a scale length of 80 μm (size of seta), the roughness amplitude does not exceed 5 μm, while at a scale length of 2000 μm (size of lamella), the roughness amplitude is as high as 30 μm. This suggests that setae should adapt to surfaces with roughness on the order of several micrometers, while lamellae should adapt to roughness on the order of tens of micrometers. Larger roughness values would be adapted to by the skin of the gecko. The spring model of *Bhushan* et al. [44.3] verifies that setae are only capable of adapting to roughnesses of a few micrometers and suggests that lamellae are responsible for adaptation to rougher surfaces.

References

44.1 S. Gorb: *Attachment Devices of Insect Cuticles* (Kluwer, Dordrecht 2001)

44.2 B. Bhushan: Adhesion of multilevel hierarchical attachment systems in gecko feet, J. Adhes. Sci. Technol. **21**, 1213–1258 (2007)

44.3 B. Bhushan, A.G. Peressadko, T.W. Kim: Adhesion analysis of two-level hierarchical morphology in natural attachment systems for 'smart adhesion', J. Adhes. Sci. Technol. **20**, 1475–1491 (2006)

44.4 A.G. Kluge: Gekkotan lizard taxonomy, Hamadryad **26**, 1–209 (2001)

44.5 D. Han, K. Zhou, A.M. Bauer: Phylogenetic relationships among gekkotan lizards inferred from C-mos nuclear DNA sequences and a new classification of the Gekkota, Biol. J. Linn. Soc. **83**, 353–368 (2004)

44.6 R. Ruibal, V. Ernst: The structure of the digital setae of lizards, J. Morphol. **117**, 271–294 (1965)

44.7 U. Hiller: Untersuchungen zum Feinbau und zur Funktion der Haftborsten von Reptilien, Z. Morphol. Tiere **62**, 307–362 (1968), in German

44.8 D.J. Irschick, C.C. Austin, K. Petren, R.N. Fisher, J.B. Losos, O. Ellers: A comparative analysis of clinging ability among pad-bearing lizards, Biol. J. Linn. Soc. **59**, 21–35 (1996)

44.9 K. Autumn: How gecko toes stick, Am. Sci. **94**, 124–132 (2006)

44.10 D.W. Tinkle: Gecko. In: *Encyclopedia Americana*, Vol. 12 (Grolier, Norwich 1992) p. 359

44.11 Aristotle: *Historia Animalium* (*The History of Animals*) (1918), transl. by D.A.W. Thompson, http://classics.mit.edu/Aristotle/history_anim.html

44.12 A.P. Russell: A contribution to the functional morphology of the foot of the Tokay, *Gekko gecko*, J. Zool. London **176**, 437–476 (1975)

44.13 A.P. Russell: The morphological basis of weight-bearing in the scansors of the Tokay gecko, Can. J. Zool. **64**, 948–955 (1986)

44.14 E.E. Williams, J.A. Peterson: Convergent and alternative designs in the digital adhesive pads of scincid lizards, Science **215**, 1509–1511 (1982)

44.15 H.H. Schleich, W. Kästle: Ultrastrukturen an Gecko-Zehen, Amphib. Reptil. **7**, 141–166 (1986), in German

44.16 K. Autumn, A.M. Peattie: Mechanisms of adhesion in geckos, Integr. Comp. Biol. **42**, 1081–1090 (2002)

44.17 E. Arzt, S. Gorb, R. Spolenak: From micro to nano contacts in biological attachment devices, Proc. Natl. Acad. Sci. USA **100**, 10603–10606 (2003)

44.18 J. Wagler: *Natürliches System der Amphibien* (Cotta'sche Buchhandlung, Munich 1830), in German

44.19 G. Simmermacher: Untersuchungen über Haftapparate an Tarsalgliedern von Insekten, Z. Wiss. Zool. **40**, 481–556 (1884), in German

44.20 H.R. Schmidt: Zur Anatomie und Physiologie der Geckopfote, Jena. Z. Naturwiss. **39**, 551 (1904), in German

44.21 S.L. Hora: The adhesive apparatus on the toes of certain geckos and tree frogs, J. Asiat. Soc. Beng. **9**, 137–145 (1923)

44.22 W.D. Dellit: Zur Anatomie und Physiologie der Geckozehe, Jena. Z. Naturwiss. **68**, 613–658 (1934), in German

44.23 J.G.J. Gennaro: The gecko grip, Nat. Hist. **78**, 36–43 (1969)

44.24 N.E. Stork: Experimental analysis of adhesion of *Chrysolina polita* on a variety of surfaces, J. Exp. Biol. **88**, 91–107 (1980)

44.25 K. Autumn, Y.A. Liang, S.T. Hsieh, W. Zesch, W.P. Chan, T.W. Kenny, R. Fearing, R.J. Full: Adhesive force of a single gecko foot-hair, Nature **405**, 681–685 (2000)

44.26 K. Autumn, M. Sitti, Y.A. Liang, A.M. Peattie, W.R. Hansen, S. Sponberg, T.W. Kenny, R. Fearing,

J.N. Israelachvili, R.J. Full: Evidence for van der Waals adhesion in gecko setae, Proc. Natl. Acad. Sci. USA **99**, 12252–12256 (2002)
44.27 P.J. Bergmann, D.J. Irschick: Effects of temperature on maximum clinging ability in a diurnal gecko: evidence for a passive clinging mechanism?, J. Exp. Zool. A **303**, 785–791 (2005)
44.28 G. Huber, H. Mantz, R. Spolenak, K. Mecke, K. Jacobs, S.N. Gorb, E. Arzt: Evidence for capillarity contributions to gecko adhesion from single spatula and nanomechanical measurements, Proc. Natl. Acad. Sci. USA **102**, 16293–16296 (2005)
44.29 G. Huber, S.N. Gorb, R. Spolenak, E. Arzt: Resolving the nanoscale adhesion of individual gecko spatulae by atomic force microscopy, Biol. Lett. **1**, 2–4 (2005)
44.30 T.W. Kim, B. Bhushan: The adhesion analysis of multilevel hierarchical attachment system contacting with a rough surface, J. Adhes. Sci. Technol. **21**, 1–20 (2007)
44.31 T.W. Kim, B. Bhushan: Effect of stiffness of multilevel hierarchical attachment system on adhesion enhancement, Ultramicroscopy **107**, 902–912 (2007)
44.32 T.W. Kim, B. Bhushan: Optimization of biomimetic attachment system contacting with a rough surface, J. Vac. Sci. Technol. A **25**, 1003–1012 (2007)
44.33 T.W. Kim, B. Bhushan: The adhesion model considering capillarity for gecko attachment system, J. R. Soc. Interface **5**, 319–327 (2008)
44.34 W. Federle: Why are so many adhesive pads hairy?, J. Exp. Biol. **209**, 2611–2621 (2006)
44.35 A.B. Kesel, A. Martin, T. Seidl: Adhesion measurements on the attachment devices of the jumping spider *Evarcha arcuata*, J. Exp. Biol. **206**, 2733–2738 (2003)
44.36 A.M. Peattie, R.J. Full: Phylogenetic analysis of the scaling of wet and dry biological fibrillar adhesives, Proc. Natl. Acad. Sci. USA **104**, 18595–18600 (2007)
44.37 H. Gao, X. Wang, H. Yao, S. Gorb, E. Arzt: Mechanics of hierarchical adhesion structures of geckos, Mech. Mater. **37**, 275–285 (2005)
44.38 P.F.A. Maderson: Keratinized epidermal derivatives as an aid to climbing in gekkonid lizards, Nature **2003**, 780–781 (1964)
44.39 N. Rizzo, K. Gardner, D. Walls, N. Keiper-Hrynko, D. Hallahan: Characterization of the structure and composition of gecko adhesive setae, J. R. Soc. Interface **3**, 441–451 (2006)
44.40 B.N.J. Persson, S. Gorb: The effect of surface roughness on the adhesion of elastic plates with application to biological systems, J. Chem. Phys. **119**, 11437–11444 (2003)
44.41 B. Bhushan: *Principles and Applications of Tribology* (Wiley, New York 1999)
44.42 B. Bhushan: *Introduction to Tribology* (Wiley, New York 2002)
44.43 B. Bhushan (Ed.): *Nanotribology and Nanomechanics – An Introduction*, 2nd edn. (Springer, Berlin, Heidelberg 2008)
44.44 K.L. Johnson, K. Kendall, A.D. Roberts: Surface energy and the contact of elastic solids, Proc. R. Soc. A **324**, 301–313 (1971)
44.45 J.E.A. Bertram, J.M. Gosline: Functional design of horse hoof keratin: the modulation of mechanical properties through hydration effects, J. Exp. Biol. **130**, 121–136 (1987)
44.46 F.M. Orr, L.E. Scriven, A.P. Rivas: Pendular rings between solids: meniscus properties and capillary forces, J. Fluid. Mech. **67**, 723–742 (1975)
44.47 S. Cai, B. Bhushan: Effects of symmetric and asymmetric contact angles and division of menisci on meniscus and viscous forces during separation, Philos. Mag. **87**, 5505–5522 (2007)
44.48 S. Cai, B. Bhushan: Meniscus and viscous forces during separation of hydrophilic and hydrophobic surfaces with liquid mediated contacts, Mater. Sci. Eng. R **61**, 78–106 (2008), (invited)
44.49 M. Sitti, R.S. Fearing: Synthetic gecko foot-hair for micro/nano structures as dry adhesives, J. Adhes. Sci. Technol. **17**, 1055–1073 (2003)
44.50 K. Autumn, C. Majidi, R.E. Groff, A. Dittmore, R. Fearing: Effective elastic modulus of isolated gecko setal arrays, J. Exp. Biol. **209**, 3558–3568 (2006)
44.51 G.J. Shah, M. Sitti: Modeling and design of biomimetic adhesives inspired by gecko foot-hairs, IEEE Int. Conf. Robot. Biomim. (2004) pp. 873–878
44.52 A. Jagota, S.J. Bennison: Mechanics of adhesion through a fibrillar microstructure, Integr. Comp. Biol. **42**, 1140–1145 (2002)
44.53 Y. Tian, N. Pesika, H. Zeng, K. Rosenberg, B. Zhao, P. McGuiggan, K. Autumn, J. Israelachvili: Adhesion and friction in gecko toe attachment and detachment, Proc. Natl. Acad. Sci. USA **103**, 19320–19325 (2006)
44.54 W.R. Hansen, K. Autumn: Evidence for self-cleaning in gecko setae, Proc. Natl. Acad. Sci. USA **102**, 385–389 (2005)
44.55 W.C. Hinds: *Aerosol Technology* (Wiley, New York 1982)
44.56 R. Jaenicke: Atmospheric aerosol size distribution. In: *Atmospheric Particles*, ed. by R.M. Harrison, R. van Grieken (Wiley, New York 1998) pp. 1–29
44.57 N.E. Stork: A comparison of the adhesive setae on the feet of lizards and arthropods, J. Nat. Hist. **17**, 829–835 (1983)
44.58 W. Federle, M. Riehle, A.S.G. Curtis, R.J. Full: An integrative study of insect adhesion: Mechanics of wet adhesion of pretarsal pads in ants, Integr. Comp. Biol. **42**, 1100–1106 (2002)
44.59 G. Hanna, W.J.P. Barnes: Adhesion and detachment of the toe pads of tree frogs, J. Exp. Biol. **155**, 103–125 (1991)

44.60 W.G. Van der Kloot: Molting. In: *Encyclopedia Americana*, Vol. 19 (Grolier, Norwich 1992) pp. 336–337
44.61 J.N. Israelachvili: *Intermolecular and Surface Forces*, 2nd edn. (Academic, San Diego 1992)
44.62 J.J. Bikerman: *The Science of Adhesive Joints* (Academic, New York 1961)
44.63 W.A. Zisman: Influence of constitution on adhesion, Ind. Eng. Chem. **55**(10), 18–38 (1963)
44.64 R. Houwink, G. Salomon: Effect of contamination on the adhesion of metallic couples in ultra high vacuum, J. Appl. Phys. **38**, 1896–1904 (1967)
44.65 B. Bhushan: *Tribology and Mechanics of Magnetic Storage Devices*, 2nd edn. (Springer, New York 1996)
44.66 B. Bhushan (Ed.): *Springer Handbook of Nanotechnology*, 2nd edn. (Springer, Berlin, Heidelberg 2007)
44.67 H.C. Hamaker: London van der Waals attraction between spherical bodies, Physica **4**, 1058–1072 (1937)
44.68 J.N. Israelachvili, D. Tabor: The measurement of Van der Waals dispersion forces in the range of 1.5 to 130 nm, Proc. R. Soc. A **331**, 19–38 (1972)
44.69 A.D. Zimon: *Adhesion of Dust and Powder* (Plenum, New York 1969), transl. from Russian by M. Corn
44.70 P.L. Fan, M.J. O'Brien: Adhesion in deformable isolated capillaries, Adhes. Sci. Technol. **9A**, 635 (1975)
44.71 P.B.P. Phipps, D.W. Rice: Role of water in atmospheric corrosion. In: *Corrosion Chemistry*, ACS Symp. Ser., Vol. 89, ed. by G.R. Brubaker, P.B.P. Phipps (Am. Chem. Soc., Washington DC 1979) pp. 235–261
44.72 J.B. Losos: Thermal sensitivity of sprinting and clinging performance in the Tokay gecko (*Gekko gecko*), Asiat. Herpetol. Res. **3**, 54–59 (1990)
44.73 B.W. Chui, T.W. Kenny, H.J. Mamin, B.D. Terris, D. Rugar: Independent detection of vertical and lateral forces with a sidewall-implanted dual-axis piezoresistive cantilever, Appl. Phys. Lett. **72**, 1388–1390 (1998)
44.74 W.C. Young, R. Budynas: *Roark's Formulas for Stress and Strain*, 7th edn. (McGraw-Hill, New York 2001)
44.75 N.J. Glassmaker, A. Jagota, C.Y. Hui, J. Kim: Design of biomimetic fibrillar interfaces: 1. Making contact, J. R. Soc. Interface **1**, 23–33 (2004)
44.76 B.V. Derjaguin, V.M. Muller, Y.P. Toporov: Effect of contact deformation on the adhesion of particles, J. Colloid Interface Sci. **53**, 314–326 (1975)
44.77 K.T. Wan, D.T. Smith, B.R. Lawn: Fracture and contact adhesion energies of mica-mica, silica-silica, and mica-silica interfaces in dry and moist atmospheres, J. Am. Ceram. Soc. **75**, 667–676 (1992)
44.78 B.N.J. Persson: On the mechanism of adhesion in biological systems, J. Chem. Phys. **118**, 7614–7621 (2003)
44.79 N.J. Glassmaker, A. Jagota, C.Y. Hui: Adhesion enhancement in a biomimetic fibrillar interface, Acta Biomater. **1**, 367–375 (2005)
44.80 H. Yao, H. Gao: Mechanics of robust and releasable adhesion in biology: bottom-up designed hierarchical structures of gecko, J. Mech. Phys. Solids **54**, 1120–1146 (2006)
44.81 R. Spolenak, S. Gorb, E. Arzt: Adhesion design maps for bio-inspired attachment systems, Acta Biomater. **1**, 5–13 (2005)
44.82 G.E. Dieter: *Mechanical Metallurgy* (McGraw-Hill, London 1988)
44.83 M. Sitti: High aspect ratio polymer micro/nano-structure manufacturing using nanoembossing, nanomolding and directed self-assembly, Proc. IEEE/ASME Adv. Mechatron. Conf., Vol. 2 (2003) pp. 886–890
44.84 K. Autumn, A. Dittmore, D. Santos, M. Spenko, M. Cutkosky: Frictional adhesion, a new angle on gecko attachment, J. Exp. Biol. **209**, 3569–3579 (2006)
44.85 K.A. Daltorio, S. Gorb, A. Peressadko, A.D. Horchler, R.E. Ritzmann, R.D. Quinn: A robot that climbs walls using micro-structured polymer adhesive, Proc. 30th Annu. Meet. Adhes. Soc. (2007) pp. 329–331
44.86 B. Aksak, M.P. Murphy, M. Sitti: Gecko inspired micro-fibrillar adhesives for wall climbing robots on micro/nanoscale rough surfaces, Proc. ICRA 2008, Pasadena (2008) pp. 3058–3063
44.87 M.R. Cutkosky, S. Kim: Design and fabrication of multi-materials structures for bio-inspired robots, Philos. Trans. R. Soc. A **367**, 1799–1813 (2009)
44.88 W.K. Cho, I.S. Choi: Fabrication of hairy polymeric films inspired by geckos: wetting and high adhesion properties, Adv. Func. Mater. **18**, 1089–1096 (2007)
44.89 A.K. Geim, S.V. Dubonos, I.V. Grigorieva, K.S. Novoselov, A.A. Zhukov, S.Y. Shapoval: Microfabricated adhesive mimicking gecko foot-hair, Nat. Mater. **2**, 461–463 (2003)
44.90 J. Davies, S. Haq, T. Hawke, J.P. Sargent: A practical approach to the development of a synthetic gecko tape, Int. J. Adhes. Adhes. **29**, 380–390 (2008)
44.91 B. Aksak, M.P. Murphy, M. Sitti: Adhesion of biologically inspired vertical and angled polymer microfiber arrays, Langmuir **23**, 3322–3332 (2007)
44.92 M.P. Murphy, B. Aksak, M. Sitti: Adhesion and anisotropic friction enhancement of angeled heterogeneous micro-fiber arrays with spherical and spatula tips, J. Adhes. Sci. Technol. **21**, 1281–1296 (2007)
44.93 A. del Campo, C. Greiner, I. Alvares, E. Arzt: Patterned surfaces with pillars with controlled 3-D tip geometry mimicking bioattachment devices, Adv. Mater. **19**, 1973–1977 (2007)

44.94 A. del Campo, C. Greiner, E. Arzt: Contact shape controls adhesion of bioinspired fibrillar surfaces, Langmuir **23**, 10235–10243 (2007)

44.95 B. Bhushan, R.A. Sayer: Surface characterization and friction of a bio-inspired reversible adhesive tape, Microsyst. Technol. **13**, 71–78 (2007)

44.96 S. Gorb, M. Varenberg, A. Peressadko, J. Tuma: Biomimetic mushroom-shaped fibrillar adhesive microstructures, J. R. Soc. Interface **4**, 271–275 (2007)

44.97 B. Yurdumakan, N.R. Raravikar, P.M. Ajayan, A. Dhinojwala: Synthetic gecko foot-hairs from multiwalled carbon nanotubes, Chem. Commun. **30**, 3799–3801 (2005)

44.98 R.N. Wenzel: Resistance of solid surfaces to wetting by water, Ind. Eng. Chem. **28**, 988–994 (1936)

44.99 Z. Burton, B. Bhushan: Hydrophobicity, adhesion, and friction properties of nanopatterned polymers and scale dependence for micro- and nanoelectromechanical systems, Nano Lett. **5**, 1607–1613 (2005)

44.100 B. Bhushan, Y.C. Jung: Wetting, adhesion, and friction of superhydrophobic and hydrophilic leaves and fabricated micro/nanopatterned surfaces, J. Phys. Condens. Matter **20**, 22510 (2008)

44.101 M. Nosonovsky, B. Bhushan: *Multiscale Dissipative Mechanisms and Hierarchical Surfaces, NanoScience and Technology* (Springer, Berlin, Heidelberg 2008)

44.102 E. Schäffer, T. Thurn-Albrecht, T.P. Russell, U. Steiner: Electrically induced structure formation and pattern transfer, Nature **403**, 874–877 (2000)

44.103 L. Ge, S. Sethi, L. Ci, M. Ajayan, A. Dhinojwale: Carbon nanotube-based synthetic gecko tape, Proc. Natl. Acad. Sci. USA **104**, 10792–10795 (2007)

44.104 L. Qu, L. Dai, M. Stone, Z. Xia, Z.L. Wang: Carbon nanotube arrays with strong shear binding-on and easy normal lifting-off, Science **322**, 238–242 (2008)

44.105 A. del Campo, C. Greiner: SU-8: A photoresist for high-aspect-ratio and 3-D submicron lithography, J. Micromech. Microeng. **17**, R81–R95 (2007)

44.106 M.T. Northen, K.L. Turner: A batch fabricated biomimetic dry adhesive, Nanotechnology **16**, 1159–1166 (2005)

44.107 M.T. Northen, K.L. Turner: Meso-scale adhesion testing of integrated micro- and nanoscale structures, Sens. Actuators A **130/131**, 583–587 (2006)